NEW TRENDS IN ALLOY DEVELOPMENT, CHARACTERIZATION AND APPLICATION

Edited by **Zaki Ahmad**

New Trends in Alloy Development, Characterization and Application
http://dx.doi.org/10.5772/59221
Edited by Zaki Ahmad

Contributors

Rajat Roy, Kazu-Masa Yamada, Nobuaki Matsuhashi, Sebastian Feliu Batlle, Amir A. El hadad, Violeta Barranco, Irene Llorente, Federico R García-Galván, Antonia Jiménez-Morales, Juan Carlos Galván, Harutaka Mekaru, Amir Farzaneh, Salih Okur, Zaki Ahmad, Mustafa Can, Omer Mermer, Ali Kemal Havare, Farid Umarov, K.M. Mukashev, Adilhan Shokanov, Ildiko Peter, mario rosso, Naila Riaz Mastoi, Asadullah Khan, Robina Farooq

Published by InTech

Janeza Trdine 9, 51000 Rijeka, Croatia

Publishing Process Manager Ivona Lovric

Technical Editor InTech DTP team

Cover InTech Design team

First published September, 2015

Printed in Croatia

Additional hard copies can be obtained from orders@intechopen.com

New Trends in Alloy Development, Characterization and Application, Edited by Zaki Ahmad

p. cm.

ISBN 978-953-51-2171-8

For EU product safety concerns:
IN TECH d.o.o., Prolaz Marije Krucifikse Kozulić 3, 51000 Rijeka, Croatia,
info@intechopen.com or visit our website at intechopen.com.

Meet the editor

Dr. Zaki Ahmad obtained his PhD in metallurgy from the University of Leeds in 1971. He is a chartered metallurgical engineer; a fellow of the Institute of Materials, Minerals and Mining; and a fellow of the Institute of Nanotechnology. He is the editor of three books on aluminium alloys and author of books on corrosion engineering published by Elsevier. He was awarded the best researcher award for the Middle East in 2010 by Energy Exchange. He is professor emeritus of King Fahad University of Petroleum and Minerals, KSA, and adjunct professor in the Chemical Engineering Department, Comsats Lahore. He has to his credit over 120 papers, and he is on the editorial board of InTech and two international journals. His main areas of interest are aluminium alloys, and he has been working on the development of these alloys for the last several years.

Contents

Preface

Despite the revolution in industry and dominance of emerging technologies, such as communication, transportation, construction, automotive, and green technologies, and the appearance of millions of products in the world market, aluminium and aluminium alloys have not been overshadowed by new developments and new products. On the contrary, the new research in aluminium alloys is going hand in hand with the new technologies that would become crystal clear to the reader of this book. One great virtue of aluminium alloys has been the extent of capabilities and the range of applications it offers. The breadth of individual alloys is very broad and would need a book to highlight it. The book *New Trends in Alloy Development, Characterization and Application* describes the new research ideas that have been put in practice and to elevate the structural, mechanical, physical, and chemical properties in alloy development. The philosophy of this book is that the new characteristics, new methodologies, and analytical procedures have to be a basic consideration before developing an alloy. The chapter on rapid solidification technology by Dr. Roy Rajat demonstrates the new technologies and the importance of rapid solidification technologies. The new RSP technology can have a dramatic effect on grain size, tensile strength, microstructural refinement, and overcoming interdendritic shrinkage. Among the light alloys, AZ31 and AZ61 are making a dramatic progress because of their favorable chemical composition, structure, and thickness of the native oxide film. These oxide films protect the alloys from deterioration. This has been described well by Dr. Feliu Batlle Sebastian in his chapter *Native Oxide Films on AZ31 and AZ61 Commercial Magnesium*. The technology of Ni-P in molding materials contains the latest information on the methodology and application of molding materials by Dr. Mekaru Harutaka in his chapter on the application of Ni-P alloys to a mold material. The chapter on Mössbauer spectroscopy appears to be a bit out of place from a long distance because it is based on quantum mechanics. Mössbauer effects provide a link between nuclear and solid-state physics. This technology has been applied to metals, alloys, organometallics, glass, and minerals. The chapter by Prof. Umarov Farid presents an interesting interpretation of data and explanation related to various materials. According to Mössbauer's concluding remarks, the finding would extend the existing knowledge and take it into existing work of unknown phenomenon. Prof. Umarov Farid has powered that point. The last two chapters are devoted to the work done on superhydrophobicity in recent years. The new findings in hydrophobicity, including dust and water-repellent surfaces, antigermicidal coatings, biofouling substrates, fuel consumption phenomenon of drag reduction, strong and conversion of energy, and the methods of fabricating hydrophobic surfaces and their theories, are presented with plenty of illustration. Aluminium has been the focus of attention in Chapter 8. The work on hydrophobicity has opened a new era in aluminium alloy by mitigating corrosion and bio fouling.

I hope the variety of new topics would be of interest to the readers and fulfill our objectives of writing the book.

I am highly grateful to all my colleagues: Dr. M. A. Bodla (Director COMSATS, Lahore, Pakistan), Dr. Asad Ullah Khan (H.O.D, Chemical Engineering Department, COMSATS, Lahore, Pakistan), Dr. Robina Farooq (COMSATS, Lahore, Pakistan), Mr. Tahir Saif (COMSATS, Lahore, Pakistan), Ms. Naila Riaz (COMSATS, Lahore, Pakistan), and my colleagues at KFUPM, Abdul Aleem, Faheemuddin, and Habib Abualhamayel. I thank Dr. M. Shoaib and Dr. Nasir Aqeeli who always supported me at KFUPM. Last but not the least, it is the spirit of my most beloved son Intekhab Ahmad (deceased) who inspired me throughout my life to write. I wholeheartedly thank Dr. Intesar Ahmad, Head of the Department of Electrical Engineering, LCWU, Lahore, and Mr. Manzar Ahmad, Associate Professor, South Asia University. I appreciate the support given by Ms. Ana Pantar of InTech, and I acknowledge the support of Ms. Ivona Lovric.

Ignorance is the curse of God; knowledge is the wing wherewith we fly to heaven. Knowledge only comes from the books.

—with apologies to Shakespeare

Prof. Dr. Zaki Ahmad
King Fahd University of Petroleum and Minerals,
Dharan, Saudi Arabia

Section 1

Development of Light Alloys

Chapter 1

Light Alloys — From Traditional to Innovative Technologies

Ildiko Peter and Mario Rosso

Additional information is available at the end of the chapter

http://dx.doi.org/10.5772/60769

Abstract

Selection of materials with the expected, application-dependent characteristics constitutes a very important point in any industrial application. In the automotive and aeronautical industries, the current tendency is to use light metals and their alloys for production of various components. For example, some of the problems related to fuel consumption and weight reduction could be partially solved by using such alloys as an alternative to traditional iron-based alloy components. Due to their very attractive properties, the most commonly employed light materials for producing high-stressed components are aluminium, magnesium and their alloys. Al-based alloys have a high strength/weight ratio, good formability, excellent combination of castability and mechanical properties which together with an excellent corrosion resistance make them very appropriate for a large variety of applications. There are two important families of aluminium alloys: (i) wrought alloys, firstly cast as ingots and/or billets and then mechanically hot- and/or cold-worked into the preferred shape, and (ii) cast alloys, directly cast into their final form through different traditional or innovative processes.

At the same time, there is continuing interest in Mg alloys for engineering proposals because of their lowest density directly connected to a weight saving of about 40% compared to steel and cast iron and 20% compared to aluminium for the same component performance. Their high specific strength, good castability and machinability, high thermal conductivity, high dimensional stability, good electromagnetic shielding property, high damping characteristics and full recyclability place them in a particular position for production of different types of components. High-pressure die casting is the most widely used technique and represents about 50% of all light alloy casting production. The tendency to use low-pressure die casting is increasing (it accounts for about 20% of all production). Gravity die casting, a small but growing contribution derived from a vacuum die casting and squeeze casting process, represents the remaining share of manufacturing processes. Generally, during solidification in a traditional manufacturing process, volume contraction is observed. This is due to the wrong feeding system and/or gas development, which in turn generate some voids or cavities within a casting, which are in turn responsible for the presence of defects in the casting components. For these reasons, the research community and manufacturing industries are giving a high level of attention to the development of innovative production procedures. In this context, semi-solid metal processing is able to attain at least the same level of properties and performances as those obtained by conventional techniques. Progresses in materials development represent a valid support for enhancing the life of an engineering component and its reliability.

In this chapter, a general overview of the actual scenario concerning the production and use of light alloys will be presented, including a short history and description of state-of-the-art techniques integrated with some results of the current research in this field carried out by the authors.

Keywords: Al alloys, Mg alloys, High-pressure die casting, Low-pressure die casting, modified squeeze casting, semi-solid metal processing, innovative rheocasting process, self-hardening Al alloy, microstructure, mechanical performance

1. Introduction

1.1. Overview on light metal alloys

Currently, a great variety of materials are available with their own features, values, applications and obviously their restrictions. Ferrous and non-ferrous metals and their alloys, ceramics, composites, plastics and various other materials with unique properties are employed today in manufacturing, either separately or in combination. Selection of the materials with the expected application-dependent characteristics constitutes a very important aspect in any industrial application. After material selection, the manufacturing processes to be used based on the targeted application also become very important. As concerns materials, the general tendency is to better control the alloys' compositions and the presence of defects due to impurities, inclusions and faults. Progresses in materials development represent a valid support for enhancing the life of an engineering component and its reliability. As regards manufacturing procedures, the efficient management of the productivity is very important, minimizing the cost and maintaining the high quality of the product.

In this chapter, a general overview of the present scenario concerning the production and use of light alloys will be presented. It is based on the data available in the literature integrated with some results currently obtained during the research carried out by the authors. Although

http://dx.doi.org/10.5772/60769

the widespread production of light alloys considers traditionally developed processes and commonly employed alloys, there is a general tendency to provide and develop innovative production procedures and new alloy compositions able to guarantee excellent performances, in a more cost-effective and/or timely manner. These aspects are directly correlated to mass reduction and to fuel economy and they constitute a central aspect for many industrial applications. In automotive and aeronautical fields, the current tendency is to use light metals and their alloys to produce different components: the most frequently employed Al alloys belong to the AlSi7Mg, AlSi7Cu3Mg and AlSi7Cu3Mg systems. As regards Mg alloys in the automotive industry, the preferred selection is related to Zr-free alloys. The commercially available alloys are Mg-Al, Mg-Al-Zn, Mg-Al-Si, while the alloys belonging to the Mg-Al-Rare Earths, Mg-Al-(Alkaline Earths) systems are semi-commercial or being developed [1].

Al alloys reveal a high strength /weight ratio, good formability, excellent combination of castability and mechanical properties, which associated to an excellent corrosion resistance, make them very suitable for a large variety of applications. Al alloys are basically classified in two ways: on one hand, as (i) heat-treatable and (ii) non-heat-treatable alloys and, on the other hand, according to the process by which they get their properties: (a) wrought alloys, firstly cast as ingots and/or billets and then mechanically hot- and/or cold-worked into the preferred form, and (b) cast alloys, directly cast into their final form through a variety of traditional or innovative processes. Al alloys have a particular position concerning its use as a structural material in automotive applications for producing cylinder heads, brake rotors, engine blocks, for manufacturing aircraft components or marine engines [2-10]. To produce components of complex shapes, casting or forging are the most convenient processes [2, 3, 6, 10]: forged parts have better quality and mechanical properties and castings are cheaper and the foundry route is generally preferred.

Currently, the interest in Mg alloys for engineering proposals continues because of their lower density directly connected to a weight saving of about 40% compared to steel and cast iron and 20% compared to aluminium for the same component performance. Their high specific strength, good castability and machinability, high thermal conductivity, high dimensional stability, good electromagnetic shielding property, high damping characteristics and full recyclability place them in a particular position for different types of component production. Some drawbacks, i.e. their high tendency to galvanic corrosion when they are in contact with dissimilar metals or electrolyte, the difficulty to deform them by cold-working and last but not least their high cost seems to be limiting their spread. Most Mg alloy components are produced by a high-pressure die casting process, which is one of the most efficient and growing methods for Mg production of complex shape components, but defects, i.e. shrinkage and gas porosity, are often observed in the cast parts. These defects deteriorate the mechanical properties of the casting components, limiting the application of Mg alloys [1, 11-17].

1.1.1. Heat-treatable and non-heat-treatable Al alloys

Generally, the mechanical properties of Al alloys with a specific composition can vary drastically depending on their thermo-mechanical processing, applied either during production or as a post-production step. The non-heat-treatable alloys have their strength through different levels of cold-working or strain hardening applied to them that can be performed by rolling, drawing through dies, stretching or similar operations where area reduction is obtained. Regulating the amount of total reduction in area of the material controls its final properties. In the case of heat-treatable or precipitation hardening alloys, achievement of their strength and other properties occurs after applying different heat treatments.

For the non-heat-treatable alloys, the highest mechanical properties are achieved by hot- or cold-working mechanisms during their production. The initial strength of these alloys depends on the hardening effect of some elements, i.e. manganese, silicon, iron and magnesium, individually or in various combinations. Strength can further be improved by a cold-working process, as the aluminium has got a high ductility, and consequently can be easily deformed. For example, alloys containing appreciable amount of magnesium are usually given a final elevated-temperature treatment called stabilizing to ensure the stability of their properties.

Heat-treatable alloys are those whose characteristics depend on heat treatments and age hardening. In this case, significant strengthening can be achieved by heating and cooling. The initial strength of these alloys is also influenced by the addition of alloying elements to the pure aluminium. The series that belong to this group are copper, magnesium, silicon alloys and zinc alloys.

The presence of these elements determines higher solid solubility as the temperature increases, producing further important reinforcement by solution heat treatment (quenching, age hardening, artificial or natural aging) [18-20].

1.1.2. Casting and wrought Al alloys

Casting aluminium alloys were already very popular, and they are still finding new applications in many industrial fields. About 80% of all aluminium casting products are derived from aluminium scrap, a fraction that is significantly higher than for wrought products. In the past 10 years, casting technologies have been considerably developed and at present a high product quality is achieved. There are several requirements for casting aluminium alloys:

a. good corrosion resistance;

b. high level of mechanical properties, i.e. yield strength, tensile strength and elongation;

c. good castability, which is correlated to excellent fluidity in liquid state, and low shrinkage porosity.

The last condition is particularly important, and porosity is still the most critical problem. There are two principal ways to approach it, using on the manufacturing route (1) or alloy composition (2):

1. enhancement of the available casting processes and generation of innovative production technologies which provide a guarantee for excellent castings with low castability;

2. elaboration of a new alloy composition employing conventional production processes.

http://dx.doi.org/10.5772/60769

Due to the high temperature used in the casting process, the solubility of these alloys is higher than that of wrought alloys.

Wrought aluminium alloys represent about 85% of aluminium alloy applications. They are initially cast as ingots or billets and then hot- and/or cold-worked mechanically into the desired shape. Their crystal structure gives them good cold formability. The addition of alloying elements improves most of their mechanical properties. Although they contain relatively small amount of alloying elements, the structure of wrought alloys offers better mechanical properties than cast alloys. Adopting plastic deformation techniques, a high level of grain refinement and homogenized microstructure has been obtained. There are four main processes applied to wrought alloys to obtain different final products:

1. rolled plates, fat sheets, coiled sheets, and foils;
2. extruded rods, solid and hollow shapes, profiles, or tubes;
3. forming products: rolled or extruded products are formed to achieve complex shapes;
4. forged products: they have complex shapes with superior mechanical properties.

1.1.3. Casting and wrought Mg alloys

The development of Mg alloys started shortly after the first industrial production of this light metal, but unlike Al wrought alloys their development was not constant. Similar to Al alloys, it can be differentiated between cast or wrought Mg alloys. From a technological point of view, the capability of plastic deformation is of vital importance with wrought alloys, while casting alloys should exhibit good mould-filling features. On one hand, Mg has excellent casting properties, but its deformation at room temperature is reduced due to the hexagonal crystal lattice structure. Mg alloys can be deformed by classic deformation processes (rolling, extruding, forging), depending on the alloy composition, above 200°C. More casting alloys than wrought alloys are available. On the other hand, wrought Mg alloys have specific advantages because they are not porous and excellent mechanical properties can be achieved by special thermos-mechanical treatment. This leads to some differences in alloy content, but they are not as significant as in the case of Al alloys [10, 21-28].

1.2. Traditional manufacturing processes for light metal alloys

There are several types of die casting processes. Various processes are now in use to achieve both economically and technologically viable Al castings production. The variety of methods results from the different ways in which gas can be eliminated from the cavity, how the injection system works or how much heat is lost during the process. Die casting is a metal casting process characterized by forcing the molten metal under pressure into a mould. The mould cavity is created using two hardened tool steel dies. A die casting die is a complex tool usually made up of two halves called a cover and an ejector die: the cover die is fixed to the motionless part of the injection machine, while the ejector die is secured to a moveable ejector plate. The die cavity in which the castings are formed is made of cavity inserts fixed to each of the two die halves, fixed or moveable cores, as well as other moveable parts.

At the end of the casting process, the die opens and the ejector pins extract the part from the die; furthermore, a die can be constructed as a simple die in order to produce a single casting for each shot or as a multiple-cavity die to produce multiple castings of identical shape per each shot [24-26, 29].

Non-ferrous metals, like Al, Zn, Cu, Mg, are the most processed materials using this technique, because of their relatively low melting temperature that steel can withstand. Depending on the casted metal, there are two kinds of available processes:

1. hot-chamber pressure die casting (Figure 1a): in this case, the pressure chamber is joined to the die cavity and is permanently dipped in the liquid metal. The inlet port of the pressurizing cylinder is uncovered as the plunger moves to the open, unpressurized, position, consenting the filling of the die cavity with fresh melt. This process is used for metals with a low melting point and high fluidity such as Sn, Zn and Pb.
2. cold-chamber pressure die casting (Figure 1b) : the liquid metal is placed into the cold chamber before each shot and is suitable for Al and its alloys, Cu and its alloys where combination with Fe at high temperature occurs with no difficulty [30-34].

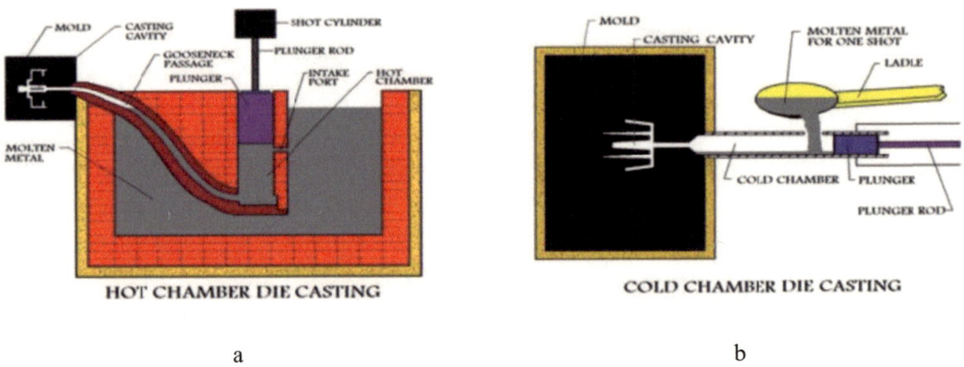

Figure 1. Scheme of the hot-chamber and cold-chamber die casting machine

1.2.1. Cold and hot die casting: Differences and similarities

Compared to the hot die casting process, the cold die casting process requires the application of higher pressure; in fact, the pressure at which the molten metal is forced into the die cavity and fills it properly is an order of magnitude higher in cold die casting than in hot die casting. The cold-chamber die casting production process involves a pressure (20-350 MPa) which leads to a product with thin walls and good mechanical properties. From an economical point of view, in this case a substantial initial investment is required. The key dissimilarities of the two methods belong to the mode in which the liquid metal is introduced into the tool.

The selection of material for die casting is principally centered on material characteristics, i.e. the density, the melting point, the strength and the corrosion resistance of the alloy. The hot-chamber process is used for metals of low melting point and high fluidity (i.e. Sn, Zn or Pb), but not for Al, Mg or steel. Cold-chamber die casting makes it possible to increase the temperature of liquid metal and a wide range of metals with different sizes can be manufactured by this process.

Preheating the die to 100-200°C is fundamental in order to reduce the thermal shock when liquid metal comes in contact with the tool steel. Two operative parts in particular are in motion during the casting cycle: a metal shot chamber (cold chamber) located at the entrance of the mold, and a piston connected to a power cylinder injects a fixed quantity of liquid metal.

The application of pressure, which is released just at the beginning of the solidification process, causes the liquid metal to fill all the sections of the die that evenly guarantees great surface detail. The mold is then opened and the casting can be removed by the action of the ejector pins. The mold is closed after spraying its internal walls with lubricant and the piston is retreated in the shot chamber for the next cycle of production [29]. The type of tool depends on the casted metal properties.

In hot-chamber die casting, the supply of molten metal is close to the die casting machine and it is part of it. The shot cylinder provides the power for the injection of the stroke, situated above the liquid metal. The plunger rod moves down to the plunger, which interacts with the liquid material. Initially, the plunger is situated at the upper part of the hot chamber and die filling takes place when the port opens. When the phase initiates, the power cylinder forces the plunger down, cutting off the flow of liquid metal to the hot chamber.

At this moment the correct amount of molten material should be in the chamber for the 'shot' that will be used to fill the mold and produce the casting: the plunger travels further downward forcing the molten metal into the die. The pressure exerted on the liquid metal to fill the die in hot-chamber die casting manufacture usually varies from about 5-35 MPa. The pressure is held long enough for the casting to solidify while the plunger travels upward to the hot chamber exposing the intake ports again and allowing the chamber to refill with molten material. Hot-chamber die casting has the advantage of being a very productive manufacturing process although it has some drawbacks, because the setup requires critical parts of the mechanical apparatus to be continuously submersed in molten material. Continuous submersion at a high enough temperature will cause thermal-related damage to these components with a risk of breaks.

However, at industrial level, the disadvantages of the hot-chamber process usually make it the most suitable choice for lower melting point alloys. In the cold-chamber die casting process, the material has to be carried in for every shot or cycle of production and this slows down the production rate.

Aluminium casting alloys were first produced using processes that were historically used to cast other metals. It is generally believed that the art of metal casting was first practiced more than 5500 years ago. Pressure die casting began in the 1820s to deal with the high demand for casting parts. Initially, the high pressure of the injection was achieved manually, but it was later replaced by pneumatic and hydraulic systems. The improvements aimed not only to reduce production time but also to improve products to make them last longer. In 1870, the process was considerably automated [31, 35, 36].

Magnesium alloy components are usually produced by various casting processes. The most appropriate methods are high-pressure die casting and gravity casting, particularly sand and permanent mold casting. Other important production technologies are squeeze casting, thixocasting and thixomolding [32, 11, 15]. The use of die-casting Mg alloys in automotive components continues to grow and in terms of long-term potential growth high-pressure die casting continues to remain the best choice [16, 17, 34].

Die casting process are in continuous development to improve production efficiency and to find new applications. Engineers are able to optimize current designs and use aluminium in place of other heavier materials.

1.2.2. Low-pressure die casting process

Among the most interesting techniques, low-pressure die casting constitutes an excellent compromise between quality, costs, productivity and geometrical feasibility. The first patent, concerning casting of Pb alloys, was deposited in England at the beginning of the 19th century, and its industrial application started 30 years ago [36]. Nowadays, it is also adopted for casting Al and Mg alloys. Low-pressure die casting is especially suitable for the production of components characterized by a symmetry with respect to an axis of rotation (body of revolution). In the automotive field and in many other fields there are several applications; however, this process is frequently wrongly associated only with the development of wheels. The initial economic investment is generally situated between the gravity casting and the high-pressure die casting processes. The tendency to use low-pressure die casting is increasing: it accounts for about 20% of the whole production. This technique is considered as a competitive casting method in cases of relatively small production mass and/or when heat treatment is needed after casting to increase the mechanical properties. The tolerances and the surface finishing properties are about the same as those obtained by gravity die casting. Thanks to the absence of feeders, the reduction of machining costs is significant. However, the initial machine cost is slightly higher than that requested for sand casting. The principle of low-pressure die casting is relatively simple, but the design of the die, control of the cooling circuits and of the solidification step need to be accurately designed and controlled. A schematic outline of the instrument is reported in Figure 3. In this process, the permanent die and the filling system are placed over the furnace containing the liquid alloy and the cavity is filled by forcing the liquid metal into a ceramic tube (stalk), which connects the die to the furnace. The forcing is done using a pressurized gas, typically ranging from 0.3 to 1.5 bars. Once the die cavity is filled, the overpressure in the furnace is taken out and the casting is finally removed. The low injection rate and the relatively high cycle time lead to good control of the fluid-dynamics of the process, avoiding the defects caused by turbulence phenomena [36-40].

Low-pressure die casting has been well established for aluminium casting with some commercially available casting machines. Producing Mg alloy components using low-pressure die casting has the potential benefits of low porosity and the possibility of semi-automatic production, offering superior casting quality and high productivity. Even if many casting processes can potentially be used for structural cast Mg component production with many

http://dx.doi.org/10.5772/60769

advantages, the low-pressure casting process is not widely adopted as yet and its use for mass production has still to be consolidated [41-44].

1.2.3. High-pressure die casting process

High-pressure die casting is the most widely used technique and represents about 50% of all light alloy casting production. Due to its flexibility, it is the widest used method for casting metals: almost all possible sizes and shapes can be produced, from the smallest electronic components, which require precise engineering and weigh less than one gram, to the largest automotive components requiring durability and high mechanical strength. This technique is characterized by the introduction of the alloy at high pressure into a metal mold. A schematic outline of the instrument is reported in Figure 4.

Production is faster with this compared to other methods, making total costs per casting lower. In addition, the high integrity of the components and their external surfaces help to reduce finishing operations and so lower production costs. On the other hand, both the device and its dies are very expensive, so it is only cost-effective for high production volumes. Special attention has to be paid in order to reduce the presence of the porosity in the castings: the component produced by this process suffers from porosity and cannot be submitted to heat treatment, because blistering occurs. A vacuum pump decreases the pressure inside the die and the difference in pressure promotes entrance of the liquid metal in the die: not only air is removed but also the metal flow is less turbulent producing a more limited presence of gas inclusions. A metal mold is a good heat conductor and dissipates the heat more quickly and is also resistant enough to absorb the high levels of pressure caused by the injection of aluminium into the cavity. In addition, it facilitates the removal of the product once it is solidified. These products have already good surface finishing properties, avoiding the need for further machining operations. With shapes which have complex geometry, a reduction of the assembling step has been obtained. The benefits achieved by this technique make it one of the most reliable processes: it is dimensionally consistent, economical, environmentally acceptable, versatile and clean. This process is ideal for high-volume thin-walled castings production because of the fast cycle times. The development of gas due to the extremely turbulent liquid metal flow constitutes the main weakness of this method. The presence of gas porosity in the central part of the casting inhibits any heat treatments damaging the features of the overall casting product.

The final product quality depends not only on the process but other aspects have to be considered, i.e. the composition of the Al alloy that will be cast, the impurities or gas that can be retained in the metal and its temperature. Most Mg castings are made by high-pressure die casting [45-51].

1.2.4. Vacuum die casting and modified vacuum process

For parts which have to be submitted to further heat treatments, vacuum die casting proves to be of interest and constitutes an ideal method to produce high-quality structural automotive Al alloy components because it represents a solution for most of the problems associated with porosity. Through vacuum die casting, it is possible to produce high-quality thin-walled parts with expected and repeatable mechanical properties, with or without heat treatment or welding. Vacuum die casting was first used in Japan and it extended rapidly around the world. Vacuum die casting has some important advantages: the vacuum systems remove the air from the cavity reducing gas porosity. In addition, very thin sections can be casted easily; good surface finishing properties and appearance can be obtained with no need for further machining. Using this technique, casting defects are low and the rejection of the component is reduced. The general principle is the same as in low-pressure die casting. Depending on the alloy used, the required properties can be achieved in vacuum die casting even without additional heat treatment; but whenever such treatment is required, it will produce superficial defects in the presence of even minor gas porosity, which are usually not tolerated on the final product.

The ranges of application that have components manufactured by the vacuum process are almost the same as those produced by the conventional die casting process, but with enhanced quality. In addition, the components obtained by this method can be subjected to further heat treatments and welding processes without the appearance of new problems of porosity, as usually occurs with high-pressure die casting. This system improves overall component quality by reducing or eliminating the presence of porosity due to gases; but in the case of shrinkage, porosity it is not an effective solution, and other measures should be considered and adapted [37, 52-56].

For high-pressure die casting of Al alloys, two types of vacuum systems have been developed: (i) a complete vacuum system where the complete die casting system are fully enclosed and evacuated during casting and (ii) a vacuum assistance system where a vacuum valve is incorporated into a die to evacuate the air entrapped in the cavity. The system is relatively simple, cheap and stand-alone, but needs a complex set-up of the complete vacuum system [54, 57, 58] and requires fast, accurate control of the vacuum and precise timing of its cut-off.

The latest improvements on vacuum die casting are related to a modified vacuum process to produce components with structural applications, i.e. automobile chassis with very high quality. The ultra-high vacuum casting process involves lowering the pressure with respect to the ordinary vacuum system. In this process, air and gas content will be 5 to 20 cm^3 per 100 g of Al [54].

1.2.5. Squeeze casting process

The concept of squeeze casting was originally introduced in 1819 [59] and the first scientific experiment was scientifically carried out in Germany on Al-Si alloy [60]. This process can be considered among the best to be produced casting components, and it is based on the principle of pressurized solidification in which the final product can be produced in a single step from the molten metal to a final component using a re-usable die. A significant advantage of this process is the lack of a runner system, leading to high yield and a cost benefit. In the squeeze casting process the injection of the metal takes place vertically and at a slow speed, producing a high cooling rate due to the instantaneous contact between the molten alloy and the cavity. As in the high-pressure die casting process, the molten metal solidifies under pressure inside the cavity giving rise to cast components with finer microstructure and therefore better

http://dx.doi.org/10.5772/60769

mechanical properties approaching those of a wrought product. The components manufactured by this technique have wide range of applications because they meet higher requirements. This process is suitable to produce structural aluminium components to reduce vehicle weight and improve efficiency.

Squeeze casting has been widely used to produce non-ferrous components, i.e. Al, Mg and Cu components, besides some ferrous components with a relatively simple geometry. Squeeze casting is used for producing components with high mechanical properties and it is also suitable to produce structural Al components of relatively low weight, which are becoming more in demand in the automotive industry to reduce fuel consumption and improve vehicle efficiency. However, the quality of the parts depends on the same variables as the more traditional methods, i.e. alloy composition, design of gates and runners, the pressure and temperature of the metal when it fills the cavity, solidification time and lubricants. Pumps, steering column components, suspension links, transmissions components are just a few of the range of components casted by this method. The process is characterized by a low injection speed, with minimum turbulence and a high pressure maintained during cooling to produce heat-treatable, high-integrity components. In this process, the molten metal solidifies under pressure inside a cavity which is vertically positioned. Die halves are also closed by hydraulic presses that are pre-heated before the metal reaches the die to reduce thermal shock. The metal flows slowly into the die, providing instant contact between the die surface and itself and pressure is applied during the whole solidification. This produces a rapid heat transfer and results in a pore-free fine-grain casting with mechanical properties approaching those of a wrought product.

The squeeze casting process eliminates both types of porosities: shrinkage porosity is eliminated by using a relatively large gating area, while the other form of gas entrapment derived from dissolved hydrogen is reduced using a conventional degassing technique. Air entrapment is avoided by using a relatively slow in-gate velocity, maintaining a planar front filling of the die. The squeeze casting process provides a net shape with minimum porosity which allows secondary porosity sensitive processes, i.e. solution heat treating and welding. The most important advantage of a squeeze casting process is its ability to produce high-integrity castings with negligible porosity, by minimizing turbulence during the filling of the cavities, and by guaranteeing directional solidification to the gates.

1.2.5.1. Types of squeeze casting process

Two alternatives of this special process have been developed based on different approaches of metal movement, namely direct and indirect squeeze casting. In the direct process the pressure is applied by the mold itself as it closes, while in the indirect squeeze casting the mold is totally close before applying the pressure [61-64].

- Direct squeeze casting: during direct squeeze casting, also known as liquid metal forging, metal is poured into a pre-heated and lubricated die contained within a hydraulic press. It is then covered by the other die half which applies pressure gradually as it is placed over the lower half. The load is applied just after the metal starts to freeze, in order to prevent the beginning of solidification prior to pressurization, and it is maintained until solidification has been completed.

- Indirect squeeze casting: this is more similar to high-pressure casting: the metal is not poured directly into the mold, but into a cold chamber. As in the conventional method, the metal is then injected in the cavity and pressure is applied through the shot system throughout solidification. However, in this case the injection velocity is relatively lower and the cold chamber is vertical. As the metal is poured through the sidewall of the cold chamber with a vertical movement, it is tilted back before filling the die, thus turbulence and consequently also porosity are minimized. The slow velocity of about 0.5 m/s of the metal flow during the injection does not enhance waves or swirls, so entrapped air almost disappears. To reduce the production cycle, pouring is done while the die is being sprayed. The shot system applies pressure since it starts to force the metal into the die until the solidification phase is finished. Pressure is continuously maintained to reduce shrinkage porosity and enhance heat transfer to achieve a small grain size.

In squeeze casting, there are several parameters that must be carefully controlled to guarantee the quality of the casted components, i.e. volume of metal, the temperature of the metal and die, the time delay between the pouring of the molten metal and the instant in which pressurization starts, pressure - depending on the size and geometry of the component and on the required mechanical properties, lubrication - to guarantee the ejection of the cast once it has solidified.

1.3. Commonly used alloys for die casting

A wide variety of non-ferrous die casting alloys is available for a wide range of applications on the basis of the physical and mechanical properties requested. Generally, Al and Zn alloys, followed by Mg, Zn-Al (ZA) alloys, Cu alloys, Sn and Pb alloys are the materials most frequently used. On the basis of the melting temperature, these alloys can be classified as low melting point metals (less than 385°C); ZA alloys have a slightly higher melting range of 430-480°C. While Al and Mg alloys are considered to be moderate melting point alloys, casted in the range of 620-700°C, Cu alloys are considered to be high melting point alloys, with a casting temperature of over 900°C. In addition, low melting point alloys and high melting point alloys can be differentiated by the casting method used in order to process them: low melting range alloys are casted in hot-chamber machines while the higher ones are casted in cold-chamber machines. Aluminium die casting alloys are lightweight, offer good corrosion resistance, no difficulty as regards casting, good mechanical properties and good dimensional stability. Moreover, a variety of aluminium alloys can be die-casted from primary or recycled metal. Actually, at industrial level the most commonly used Al alloys are:

- A356 and A357 alloys (belonging to the AlSi7Mg system) and the A319 alloy (belonging to AlSi7Cu3Mg), usually processed by low-pressure die casting and gravity semi-permanent mold technology; they are employed to produce for cylinder heads, knuckle components, engine blocks, pistons;

http://dx.doi.org/10.5772/60769

- A360 alloys show higher corrosion resistance, superior strength at high temperatures, and slightly better ductility, but reveals some difficulties during casting;
- A380, A383 andA384 alloys offers the best combination of properties and production cost; they offer superior die filling but with a significant loss in mechanical properties, i.e. toughness;
- A390 alloys are used for special applications where high strength, fluidity and wear-resistance properties are required. They are employed for cylinder head and piston production;
- A413 is used for its maximum pressure tightness and fluidity [65].

The chemical compositions of the alloys mentioned earlier are reported in Table 1:

Chemical composition (wt%)										
Type of alloys	Elements									
	Si	Fe	Cu	Mn	Mg	Ni	Zn	Sn	Ti	Al
A356	6.5-7.5	0.2	0.2	0.1	0.25-0.45	-	0.1	-	0.1	bal.
A357	6.5-7.5	0.2	0.2	0.1	0.4-0.7	-	0.1	-	0.04-0.2	bal
A360	9-10	1.3	0.6	0.35	0.4-0.6	0.5	0.5	0.15	-	bal
A380	7.5-9.5	1.3	3-4	0.5	0.1	0.5	3.0	0.35	-	bal
A383	9.5-11.5	1.3	2-3	0.5	0.1	0.3	3.0	0.15	-	bal
A384	10.5-12	1.3	3-4.5	0.5	0.1	0.5	3.0	0.35	-	bal
A390	16-18	1.3	4-5	0.5	0.45-0.65	0.1	1.5	0.20	0.2	bal
A319	6.5-8	0.8	2.8-3.5	0.5	0.25-0.5	-	-	-	0.25	bal
A413	11-13	1.3	1.0	0.35	0.1	0.5	0.5	0.15	-	bal

Table 1. Composition of the main alloys used in aluminium die casting

1.4. Innovative manufacturing processes for light metal alloys - Semi-solid metal processing

Generally, during solidification, in traditional manufacturing process, volume contraction is observed. This is due to an inappropriate feeding system and/or gas development, as also illustrated previously, which in turn generates voids or cavities within a casting, which are then responsible for the presence of defects in the casting components. Commonly, defects such as interdendritic shrinkage pores, inclusions, enlarged secondary dendrite arm spacing are considered the most important reasons for crack growth. Independent of the loading conditions, they reduce the alloy's mechanical resistance and during severe stress conditions lead to the failure of the alloy.

For these reasons, the research community and manufacturing industries are giving a high level of attention to the development of innovative production procedures.

In this context, semi-solid metal processing is able to attain at least the same level of properties and performances as those obtained by conventional techniques. Despite half a century of evolution, there is still a serious need for additional and more detailed rheological data for the successful use of components developed through these methods. The real and important advantages of semi-solid metal processing are prevalently related to faster production times at lower cost. When fully developed, it is expected to represent a viable solution compared to conventional methods. The group of innovative manufacturing techniques, based on the thixotropic characteristic of the alloys in a non-dendritic structure in the semi-solid state, approaches its fourth decade, its starting point being the scientific studies of Spencer et al. (1972) [66]. The most important characteristic of semi-solid metal processing, known as slurry, which makes it superior to conventional casting processes, is the non-turbulent (or thixotropic flow) behaviour which occurs at the two-phase field of solid and liquid. In such conditions, materials have to be managed in a way analogous to solids, but flowing as liquids when a shear stress is applied and the viscosity decreases dramatically so that the alloy can be cut and spread like butter. The laminar flow and thixotropy of the semi-solid metal are directly connected to its microstructure: the semi-solid state consists of spheroids of solid phase enclosed in a liquid phase. Semi-solid metal processes produce the metal slurry at a semi-solid casting temperature with globular microstructure. Rheocasting and thixocasting processes belong to this approach. In the 'Rheo', one-step process, the alloy is introduced into a die with no transitional stage. As starting alloy, it is a regular and completely liquid alloy which is cooled to arrive at the chosen solid fraction and then casted. For the 'thixo', two-step process, the starting alloy is prepared externally and then the billet is cut to length, re-heated to arrive at the proper solid fraction and then casted [67-71].

Semi-solid casting is one of the most reliable technologies to cast metal components of complex shapes. To produce aluminium cast components, this latest patented technology is based on keeping the metal in the semi-solid phase between solid and liquid. Two main types of processes have been developed to cast aluminium components: thixocasting and rheocasting. The difference between these methods is the process to make the slurry. In thixocasting a cast billet, which has a dendritic microstructure, is heated to the liquidus temperature, and then injected into a mold cavity. Rheocasting consists of cooling the molten metal to the liquidus temperature and then injecting it into the sleeve of a conventional die casting machine. This is a considerable advantage over thixocasting because it results in less-expensive row material. Semi-solid processed components are heat-treatable, weldable and show superior mechanical properties. In addition, the wear of the die-casting machines is low due to the low injection speed. These processes also reduce traditional die casting process costs due to the elimination of the secondary process, the increase of productivity and the reduction of die maintenance [72-76].

The semi-solid concept fits very well with Mg alloys too; in fact, the use of Mg alloys produced in this way during the past decades comes rapidly of age as a reliable alternative to high-pressure die casting. In the case of Mg alloys, Thixomolding® process is preferred, adopted from the injection moulding of polymers. As a result of this process, there are a number of possible advantages, including high process control, dimensional constancy in produced parts, low porosity, capacity to mould parts with complex geometries, good surface finishing properties, and production of near-net-shape parts. Because of lower and predictable shrinkages, the component produced has good dimensional tolerances. The thixoformed parts are able to meet stringent dimensional tolerances [72, 77-83].

1.5. Advantages and disadvantages of traditional and innovative methods

Different processes have been used to produce lightweight components, including conventional and new technologies. Forging is evidently always the best way, but also the most expensive one. Casting technologies are more competitive from the economical point of view, while forging is able to guarantee the best performances, thanks to the high level of soundness. Castings are generally affected by defects. The new trend is to use a new manufacturing route to obtain a similar performance compared to the product obtained by traditional processes. Semi-solid processes can reduce the existing gap between casting and forging; and during such a process, better control of the defect level can be attained. The components manufactured by the semi-solid process acquire a very low content of defects (shrinkage or gas porosities, segregations, surface defects). Semi-solid processed components are heat-treatable, weldable and reveal superior mechanical properties. Productivity can be improved with a using semi-solid process because of the shorter production cycle time. Semi-solid slurries normally contain about 50% solid fraction before they come into the die cavity, so only about 50% of the latent heat is released within the die during the solidification stage. The lower heat content of semi-solid slurries reduces thermal shocks on the die, improving the life of the die. In addition, wear of the die-casting machines is lower than during the traditional manufacturing route due to the low injection speed connected to the reduction of die maintenance. The casting surfaces of the heat-treated components are free from formation of blisters. Another important advantage of the semi-solid process is related to its weight-saving capacity: all the metal introduced at the beginning of the process evolves to the development of the final component and risers are absent. Due to the excellent surface quality all finishing operations are reduced, saving additional time and cost.

2. Experimental study results

2.1. Influence of the Al-based alloy composition on the performance of the product manufactured by modified squeeze casting process: Real case study

2.1.1. Modified squeeze casting process - General approach

This technique combines the advantages of low-pressure die casting, one of the most efficient casting method, and of forging, till now the most efficient technique to obtain superior mechanical characteristics. The modified squeeze casting process produces near-net-shape components in a single step, obtaining materials with high quality and good mechanical properties. A schematic illustration of the modified squeeze casting process is given in Figure 5 [84]. In this process the liquid metal is injected bottom up from the furnace into the closed mould at very low velocity; once filling is complete, a high pressure is applied to the liquid material thus 'forging' the component. The high pressure is maintained throughout the component solidification producing a final component with very fine microstructure and no defects. This way a very interesting process was built for automotive industry safety and high-performance components such as suspension and engine components. Moreover, this process can use wrought alloys, high-performance alloys that until now could only be processed by forging. Modified squeeze casting technology is different from already established casting technologies, since the cavity of the die is filled due to the application of low pressure (about 0.5 bar); during the solidification phase the pressure rises to 120 MPa, due to an applied force of 1,100 tons on a large diameter cylinder, approximately 0.1 m^2, to realize the 'forging'. The working cycle is varied between 60 and 70 s (similar to the high-pressure die casting process), but all the finishing operations are reduced, due to the absence of risers and scraps from the feeding system. In this method no contact between liquid metal and the environment occurs and, if necessary, can be used protective gas, especially interesting for the production of Mg-based components. Compact and fine microstructure is obtained, due to the pressure applied on the liquid metal during the solidification phase. It is possible to obtain components with inserts or sand cores, quite isotropic components, with similar microstructural and mechanical properties in all directions.

A pre-quantified liquid metal, at 700-730°C inserted at very low pressure inside a pre-heated die (at about 200°C) die (Figure 2a) and then the power system is closed (Figure 2b). About 90 MPa pressure is applied (Figure 2c) on the metal, and it is applied until total solidification of the metal. The die is opened (Figure 2d) and the final shape is removed (Figure 2e). The continuous application of pressure during solidification generates a defect-free part.

Applying pressure during solidification develops a fine crystalline structure with a minimum of defects. The size of grains is considerably smaller than those obtained by other casting techniques and the developed microstructure is uniform. The presence of gas or shrinkage porosity is radically reduced (or even absent) resulting in superior mechanical properties. Although porosity is minimal, this process promotes grain boundary segregation and sometimes a non-uniform macrostructure can develop.

Despite these benefits, the process also has several drawbacks. One of them is related to productivity: at present the process does not reach the typical productivity of die casting process and its production capacity can be considered to be between high-pressure die casting and classic processes of sand casting. The complex tooling involves higher initial costs and complex shapes cannot be produced by this process. To compensate for the initial investment, a high production volume is needed to make this process a cost-effective technique.

Compared to conventional high-pressure die casting, in the modified advanced squeeze casting process, components are gated in a different casting position. In modified squeeze casting the gate is settled at the bottom of the side wall, while the plunger is guided horizontally in conventional high-pressure die casting machines, and in this way it fills the die at a high velocity; while in the modified squeeze casting there is no plunger to inject the metal in the cavity. The metal fills the die due to the pressure difference created by vacuum pumps. The holding furnace is placed at the bottom of the machine, right under the dies. The furnace and

http://dx.doi.org/10.5772/60769

the cavity are connected by the filling channel through which the metal fills the dies. This happens rather slowly and with a vertical movement, hence the die is filled without significant gas inclusions.

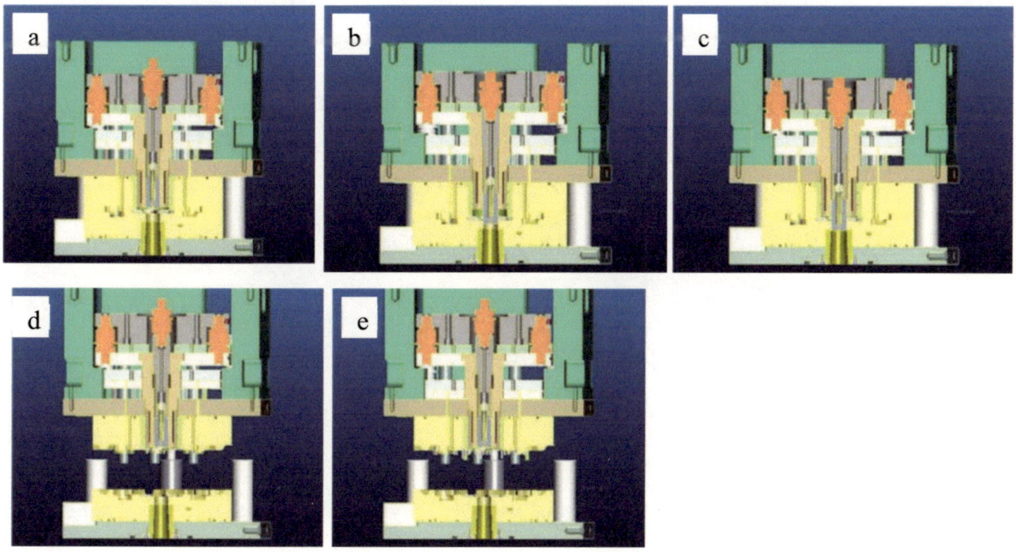

Figure 2. Modified squeeze casting cycles: a) low-pressure feeding, b) closure of the feeding gate, c) application of the pressure, d) opening of the die, e) rejection of the components

2.1.2. Modified squeeze casting process - Experimental results

In this section some results will be presented based on the study carried out on two components, using A356 and A380 (with the chemical compositions reported in Table 1) Al alloys, manufactured by advanced squeeze casting [85]. The components were T6 heat-treated (510°C for 6 h, water quenched to room temperature and aged at 200°C for 8 h) and their properties were evaluated to study how the alloy composition and the process parameters influence the performance of the final component. In addition, the feasibility and the efficiency of the modified squeeze casting process in a real industrial frame was evaluated. On one hand, the pressure applied (i.e. uniform distribution of the pressure on the whole pieces) and the period of the pressure applied can be considered the most important key process parameters analysing and comparing two types of samples, from two different sites (Figure 3). Sample A was extracted from the area where pressure was directly applied on the piece, while Sample B received lower pressure during production. The effect of the alloy composition was estimated, comparing the properties of Sample A and of Sample C (A380 Al alloy).

2.1.3. Microstructural analysis

Figure 4 reports the microstructures related to the composition of Samples A and B. Dendritic microstructures with comparable features were obtained, made of primary α-Al solid solution

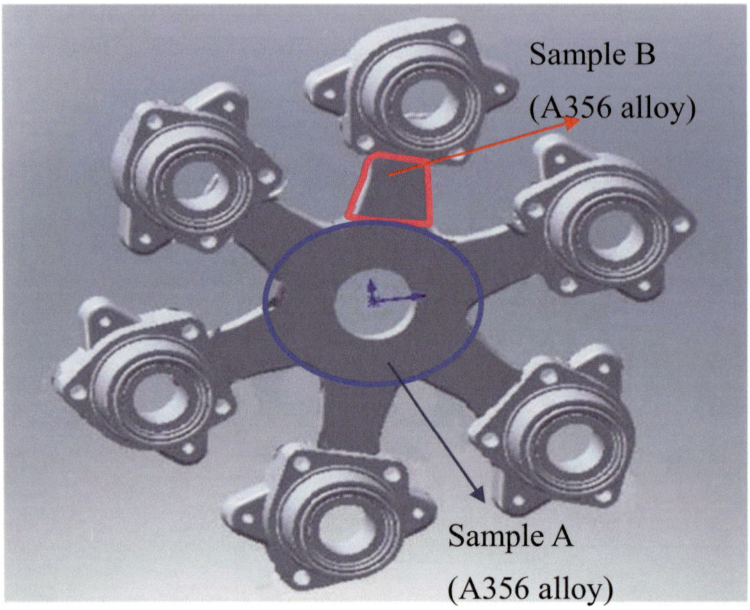

Figure 3. Scheme of the investigated component, in A356 and A380 alloy showing the initial position of samples before analysis

phase enclosed within the eutectic mixture of Al and Si. The size of the primary α-Al particles increased (Sample A, 30-50 μm) as a sign of a lower solidification rate (Sample B, 50-70 μm). Coarse-grained morphology developed in the case of the A380 (Figure 4). Figure 5 reports the microstructure of the A380 alloy sample. Apart from α-Al grain the presence of some other phases can be detected: one of these was chemically constituted by Al, Fe (about 14% wt), Mn and Si while the other consisted in block-like Al_2Cu particles (about 15 % wt of Cu). The presence of Chinese script intermetallic compounds are more accentuated in the case of the A380 composition controlling matrix stiffness and contributing to the brittleness of the material.

2.1.4. Mechanical property evaluation and fracture surface analysis

Evolution of hardness was evaluated, starting from the border of samples A, B and C and going toward the core along a direction normal to their edges. The results are reported in Table 2. Higher hardness values were achieved for the samples subjected to higher pressure, caused by the growth of hardening phases. Tensile properties of both alloys were determined at room temperature, according to the UNI EN 10002 standard and the results are reported in Table 2. Actually, the samples did not show an outstanding mechanical performance, but an indication of the positive effect of pressure during solidification can be deduced.

Figure 6 reveals some characteristics of the fractured surfaces: for both alloy compositions and for both regions generally ductile fractures occurred as illustrated in Figure 6. A380 alloy revealed the occasional presence of some shrinkage porosities (Figure 6c), and due to the higher

http://dx.doi.org/10.5772/60769

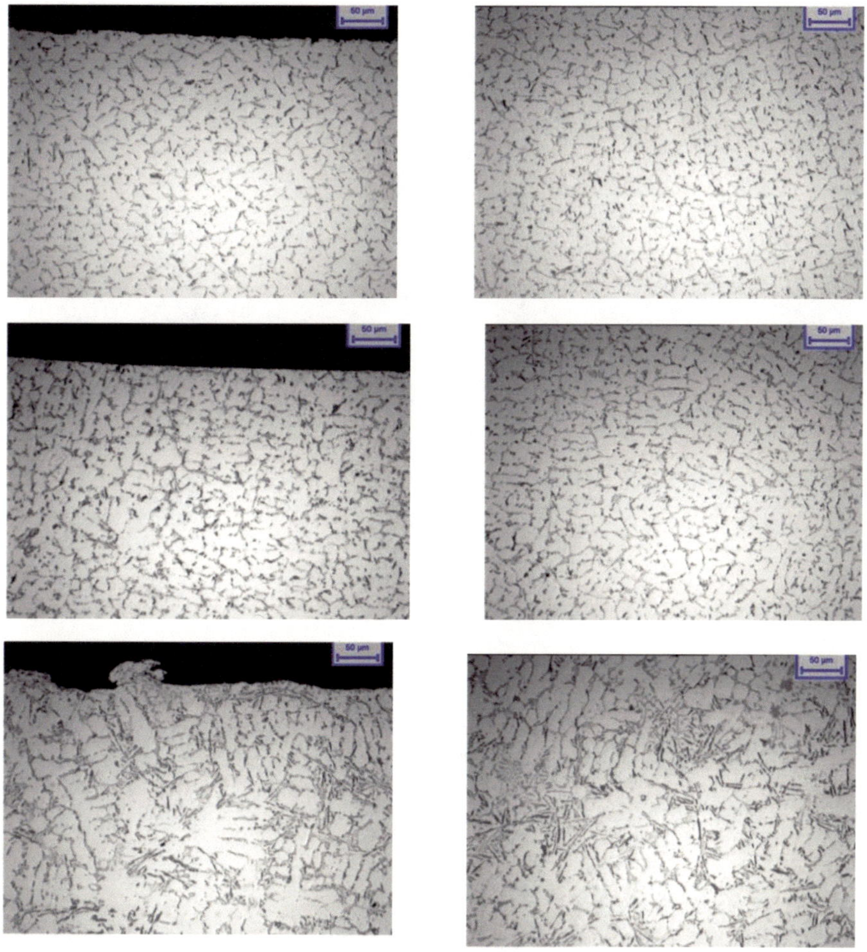

Figure 4. Microstructure of the samples obtained by optical microscopy, showing the differences between the samples and between the edge and the central part of the samples

presence of some intermetallic fragile phases some crack development was observed (Figure 6d).

	Sample A	Sample B	Sample C
YS [MPa]	240 ± 20	170 ± 20	182 ± 20
UTS [MPa]	418 ± 20	386 ± 20	374 ± 20
HV	71 ± 1	67 ± 0.5	96 ± 1.5

Table 2. Average mechanical properties of samples

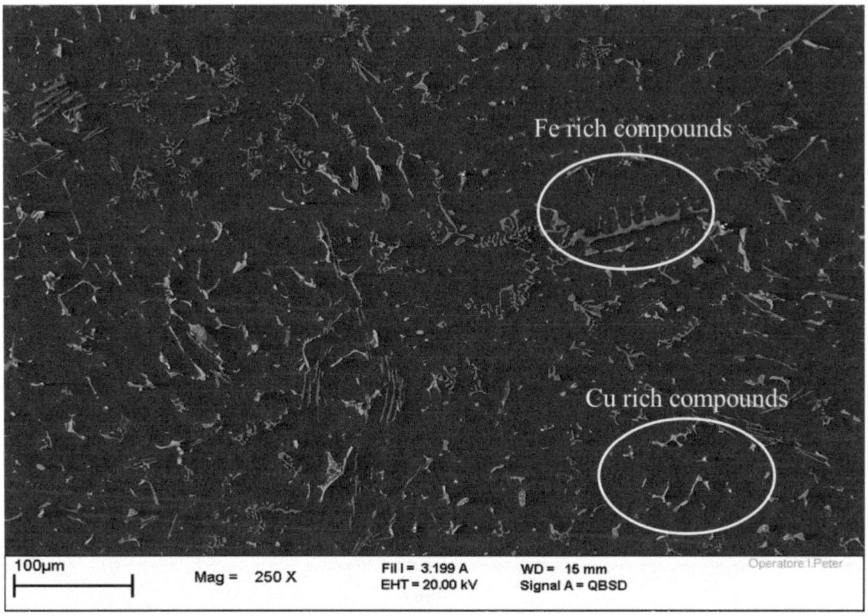

Figure 5. Microstructure of the A380 alloy obtained by scanning electron microscopy

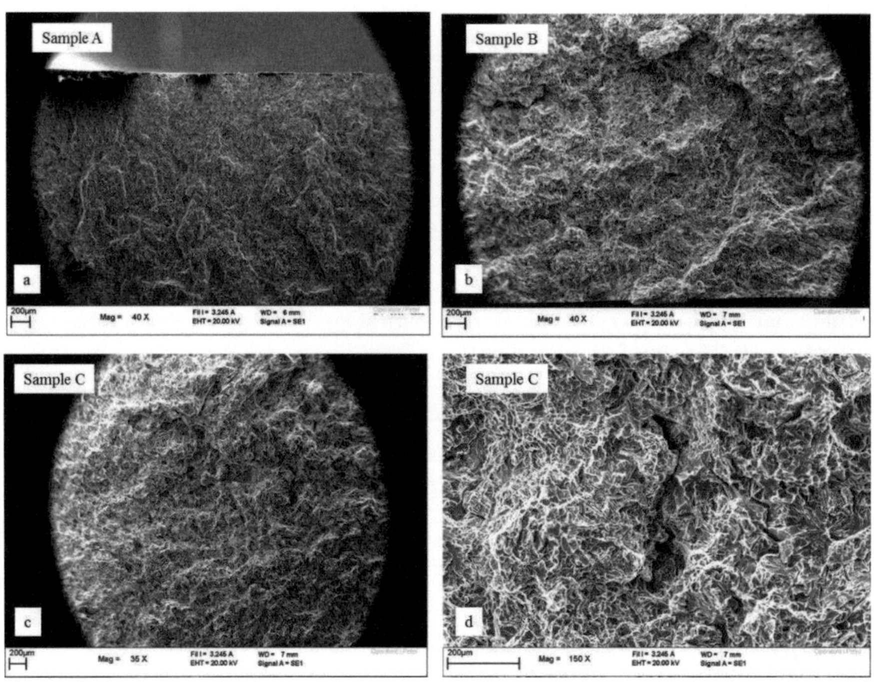

Figure 6. Fracture surface analysis: scanning electron microscopy images showing the nature of the fracture and some details on the fractured samples

http://dx.doi.org/10.5772/60769

In accordance with the data in literature, the experimental analysis presented demonstrates that the modified squeeze casting process is suitable for the production of automotive components using A356 and A380 aluminium alloys. The positive effect of the pressure applied during solidification was observed on both the microstructure and on the mechanical properties of the modified squeeze casted components. Among the two considered compositions, the A356 alloy appears to be more appropriate for the production of automotive components, in particular suspension lever arms, through modified squeeze casting process.

2.2. Feasibility of a new rheocasting process - Weldability of T6 heat-treated Al alloy parts manufactured by new rheocasting process - Real case study

2.2.1. New rheocasting process - General approach

To confirm why automotive component designers consider the semi-solid state forming process as an alternative to traditionally employed processes, some highly stressed parts (flanges for truss) for structural applications and automotive space frame component were produced by innovative rheocasting process at ATS Company (Lugo Ravenna, Italy):

1. The structural part is characterized by a relatively heterogeneous geometry. Microstrucral analysis and mechanical characterization was performed. Because of the particular shape of the rib arms, it was not possible to obtain samples for tensile tests, so the most suitable opportunity was to machine bars, with a cross-section of 10 × 5 mm and about 7 mm long to perform three point bending tests.

2. The automotive component characterized by quite an intricate shape with 2 mm maximum thickness and a length in the range of 300 mm was considered. Two different aluminium alloys were investigated B356.2 and B357.2, and in combination two different kinds of heat treatment, in particular T5 and T6 on B356.2 and T5 only on 357.2 parts. From these parts two different series of dog-bone-shaped flat samples for tensile tests were obtained, the first one obtained directly as an appendix of the casting, with a transverse section of about 6 × 22 mm and a total length of about 200 mm; henceforth, these samples will be identified with the bold letter A: appendix. The second series was machined from the parts using three different zones where it was possible to extract dog-bone-shaped samples with a rectangular section of about 2 × 5 or 10 mm dimension and 100 mm total length: hereafter, these samples will be identified by the bold letter T: thin section.

An important aspect related to this process is the possibility of obtaining quite a wide range of thicknesses, starting from 2.5 mm.

Two types of heat treatments (HT) were used after production: T5 HT (by water quenching out of die followed by ageing at 165°C for 6 h) and T6 HT (520°C for 6 h and then ageing at 165°C for 6 h); and following the encouraging results obtained in the case of the T6 HT structural parts, their weldability has demonstrated. In this section, only the results on the welded parts are presented and more details on the microstructural and mechanical behaviour concerning the un-welded samples can be found in references [86, 87]. Welding was done with the tungsten inert gas (TIG) procedure with Peraluman Al alloy (Table 3) as filler material.

For innovative rheocasting process a TCS vertical hydraulic press Rotorone 400 tons model was employed. The press has an injection piston of 180 mm diameter; the closure power is 400 tons with an injection power of 320 tons. Under the lower level of the press, a turning table moves with the containers set up at 120°: the first contains the slurry to be injected after rotation; the second one has the evacuation of the biscuit; the third is lubricated ready to inject a new quantity of slurry from the ladle. The two upper and lower half dies are heat-controlled by oleodynamic panel controls. When the right injection temperature is reached (between 577 and 590°C, depending on the employed alloy), the piston pushes the slurry very slowly through the in-gate until the filling of the cavity is complete. After a very short time, just to maintain the pressure so that the cycle can finish, the piston comes down carrying the biscuit cuts off the in-gate pieces. The press opens and the upper part goes up allowing the piece to come out, helped by the ejector. This part is immediately quenched into water when T5 heat treatment is requested. The cycle is ready to continue after die lubrication [86, 87].

Elem.	Si	Fe	Cu	Mn	Mg	Zn	Ti	Ni	Impurity	Al
Wt (%)	0.4	0.4	0.1	0.4-1	4-4.9	0.25	0.15	0.05	0.05	Balance

Table 3. Chemical composition (%wt) of the filler metal

2.2.2. New rheocasting process - Welded component characterization and mechanical behaviour of the automotive component

The welded component for structural applications and the automotive space frame component are shown in Figure 7a and 7c, respectively. The shape of automotive component produced had quite a complex shape with 2 mm maximum thickness and a length of about 300 mm. Figure 6a reports the photo of the welded element. Following some mechanical stresses, failure of the component occurred, but the failure only affected the area external to the welding zone (Figure 7b), showing that the welding was successful and the fracture of the sample was independent from the welding.

Figure 7. Photograph of the massive welded element: (a) position of the fracture, (b) automotive components, (c) produced by an innovative rheocasting process [86]

http://dx.doi.org/10.5772/60769

Figure 8 shows the microstructure of the welded samples prior and post T6 heat treatment, with the evidence of some porosity. The base metal without T6 heat treatment contained α-Al particles and Al-Si eutectic phases with a coarse microstructure (on the right-side bottom Figure 8). After T6 thermal treatment the microstructure became slightly finer. The weld fusion zone resulted in a finer dendrite structure due to the high cooling velocity during the solidification phase. A heat-affected zone was developed into a combined structure with non-uniform grain size, as a sign of the influence of heat from the welding.

The tensile properties of the samples prior and after welding and T6 heat treatment are reported in Table 4. The first sign from these measurements confirms the expected results: (i) a slightly higher tensile property was obtained; (ii) an improvement of the ductility of the alloy was reached. This latter point is also confirmed by the fracture surface analysis. Additionally, the fracture of the welded and T6 heat-treated samples always takes place external to the welding area, an indication that there was adequate welding which amplified the whole component's mechanical resistance. Prior to T6 heat treatment, a heat-affected zone was directly involved in the alloys failure. The presence of some small-sized brittle particle had no influence on the mechanical failure of the alloy, demonstrating that the weldability of the A356 alloy was successful and the welded component present promising mechanical properties. This aspect is important at industrial level when joining of different parts produces the final component.

Figure 8. Weld profile of A356 Al alloy with microstructural details of the involved areas prior and after T6 heat treatment

Table 5 reports the tensile test results obtained for the automotive component: a slightly higher performance was obtained on the thin samples (T) with respect to the thick samples (A) obtained as an appendix of the castings. Probably, the difference is due to the different feeding system used: feeding of the appendix was carried out as a separate branch from the main feeding gate; in that branch the flow of the semi-solid slurry favoured a higher concentration

Samples	T6 A356 alloy	Welded T6 A356 alloy
YS [MPa]	200 ± 10	230 ± 10
UTS [MPa]	280 ± 10	300 ± 10
Elongation [%]	7 ± 1	12 ± 1

Table 4. Tensile properties of the samples considered in the A356 alloy

of oxides particles and impurities introduced during mechanical stirring to produce the slurry. In fact, although observation of the fracture morphology of the type A samples showed the typical ductile fracture with dimples areas that follows the microstructure details like silicon particles, inside some dimples the presence of very small hard particles can be detected. These very small particles probably originated during the mechanical stirring of the semi-solid slurry. As expected, the tensile strength and ductility of T6 heat-treated samples were better than in the T5 state; moreover, the B357.2 alloy confirmed its slightly superior level of resistance.

Sample	A: 356 (Thick)		T: 356 (Thin)		A: 357 (Thick)	T: 357 (Thin)
Heat treatment	**T5**	**T6**	**T5**	**T6**	**T5**	**T5**
$\sigma_{0.2}$ [MPa]	132	190	140	200	160	165
UTS [MPa]	220	260	240	280	225	260
Elongation %	3	6	4	7	2	4

Table 5. Tensile properties of samples obtained from automotive parts

Our research demonstrates that it is appropriate to use the innovative rheocasting casting process to produce structural components and automotive space frame components. The positive effect of T6 heat treatment on A356 Al alloys was verified. Additionally, the feasibility of the welding was demonstrated: an adequate welding strengthened the mechanical resistance of the final component.

2.3. Study of the development of an alternative solution to the heat-treatable A356/A357 Al alloy - Self-hardening Al alloys: real case study

Self-hardening aluminium alloys (Al-Zn-Si-Mg alloys) represent an innovative class of light aluminium alloys. They exhibit excellent mechanical properties, which make them suitable for many applications in different industrial fields, especially in the transport industry. The most important and relevant feature of these alloys is their good performance without the need of any heat treatment: they are subjected to a natural ageing phenomenon at room temperature after a storage period of about 7-10 days. The possibility to avoid heat treatment represents an important benefit, contributing to considerably reduce both the production cost of some components and the amount of energy involved in the manufacturing process. Furthermore, without heat treatment the risk of the deformation of some component during the production is eliminated [88].

http://dx.doi.org/10.5772/60769

The goal of this study was to find an alternative solution to the T6 heat-treated A356/A357 alloys currently used for automotive component production. The feasibility of the development of a knuckle suspension component, starting from the modified self-hardening alloy with an increased Mg content, was studied and investigated [89]. Comparison of the alloy properties, also considering the solidification rate, was carried out with the aim to find an optimal composition and condition for good mechanical performance and good corrosion resistance achievements. The alloys were produced using the permanent mould casting technique by Teksid Aluminium Srl (Carmagnola, Torino, Italy) and then investigated.

Since the maximum solubility of Mg in Al is at 17.4% wt. under equilibrium conditions, the Mg content does not usually exceed 5% in wrought alloy and 10% in cast alloy, respectively. For these reasons, the Mg content is maintained in this range. The alloys produced, with an increasing Mg content, were labelled as AlZn10Si8Mg, AlZn10Si8Mg1 and AlZn10Si8Mg3, as reported in Table 5. To analyse the effect of the solidification rates on the microstructure of the casting, a special geometry, presented in Figure 9, was developed. The weight of the aluminium alloy casting without a runner system is about 0.5 kg.

Elements	Si	Fe	Cu	Mn	Mg	Zn	Ti	Al
AlZn10Si8Mg	7,5-9,5	0,30	0,10	0,15	0,2-0,5	9,0-10,5	0,15	bal.
AlZn10Si8Mg1	7,5-9,5	0,30	0,10	0,15	1,0	9,0-10,5	0,15	bal.
AlZn10Si8Mg3	7,5-9,5	0,30	0,10	0,15	3,0	9,0-10,5	0,15	bal.

Table 6. Chemical composition (wt. %) of the three alloys produced

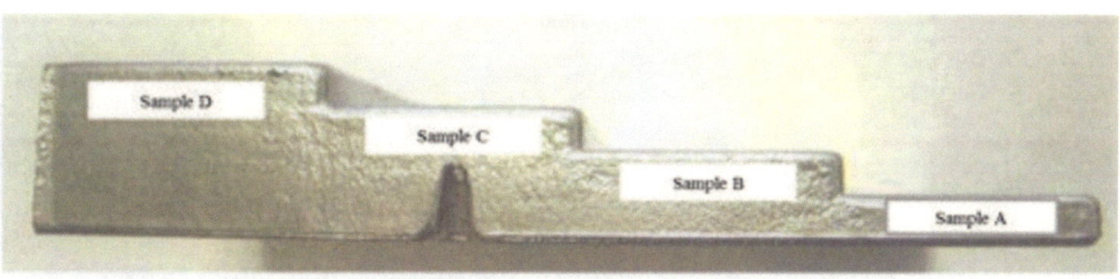

Figure 9. Step casting geometry

2.3.1. Microstructure analysis

Solidification is critical for the development of the microstructure in casting and can be considered as responsible for the development of mechanical properties. A typical casting microstructure made of α-Al dendrites and of Al-Si eutectic phase is reported in Figure 10. At comparable positions among the castings, no considerable variation was evidenced. As expected, higher values of cooling rate in all cases gave a finer microstructure. The precipitation of large Mg-based intermetallics, with a Chinese script morphology, is favoured by the high content of Mg as well as by high values of cooling rate. Mg_2Si hardening precipitates were developed as illustrated in Figure 11 and their level was directly associated to the Mg as

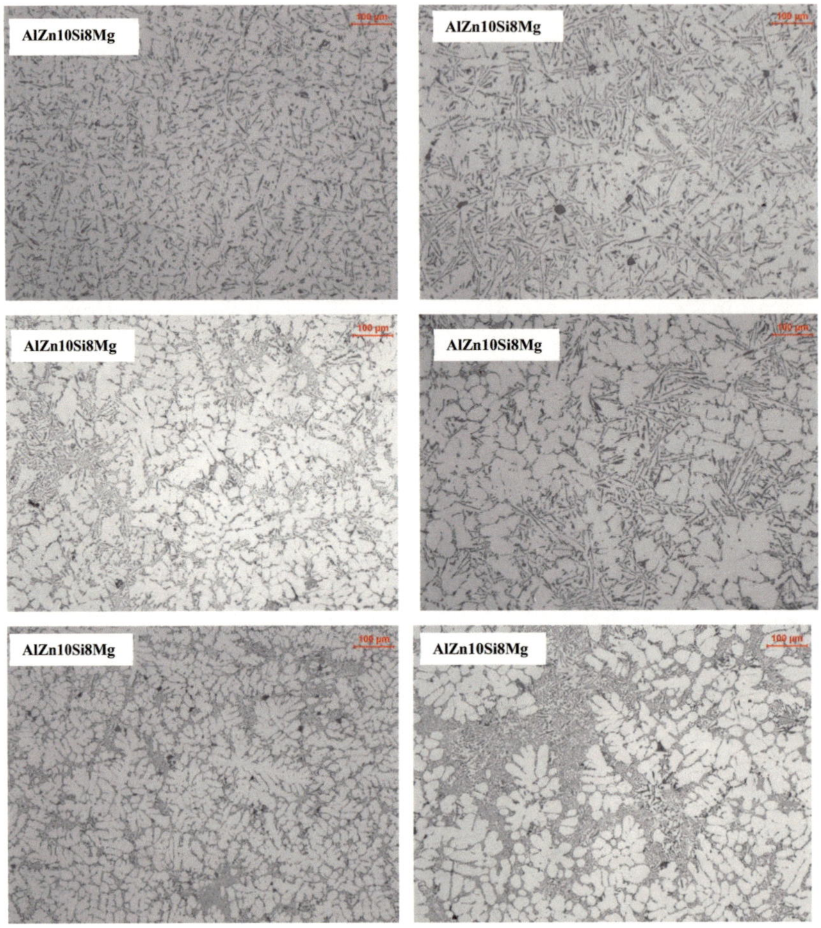

Figure 10. Microstructure of the alloys considered as function of Mg content and of solidification rate, left-side samples type A, right-side samples type D

reported in Figure 11a, b and c. Simultaneously, the solidification rate controlled the growth of the intermetallic particles: in the case of samples extracted from zone D with lower solidification rate, the largest precipitates were developed (in Figure 11 moving from d to f). A relatively equivalent condition was obtained with high Mg content (Figure 10c) and low solidification rate (Figure 11f).

The presence of Zn-based and the hexagonal α-AlFeMnSi intermetallic particles was detected by SEM observation and EDS analysis. They randomly interrupt the microstructure of the alloy for all compositions. The distribution of Zn-based compounds in the as-cast condition (Figure 12a) was a little bit different with respect to the situation after 20 days of natural ageing (Figure 12b), due to the diffusion of Zn within the Al-based matrix. After 20 days, no important differences were detected and stabilization of the microstructure was reached. A dense α-AlFeMnSi intermetallic particles were developed in the presence of a small amount of Fe which led to higher ductility and superior corrosion resistance. Higher Mg content favoured the development of $MgZn_2$ particles as reported in Figure 13. Usually, the growth of $MgZn_2$ intermetallics was close to the AlZn intermetallics. Migration of Zn occurred within the grain boundary and a diffusion of this element to produce $MgZn_2$ intermetallics was observed.

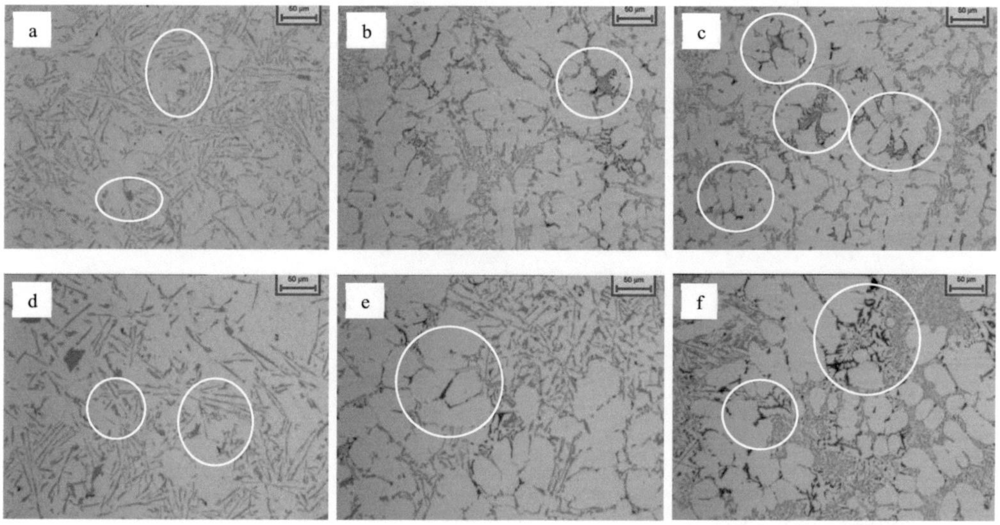

Figure 11. Microstructure of AlZn10Si8Mg alloy zone B (a) and zone D (d), AlZn10Si8Mg1 alloy zone B (b) and zone D (e), AlZn10Si8Mg3 zone B (c) and zone D (f).

The use of the modified self-hardening Al alloy avoided any heat treatment which in turn contributed to important energy saving during manufacturing, especially in terms of gas and electricity consumption, important features for the environment.

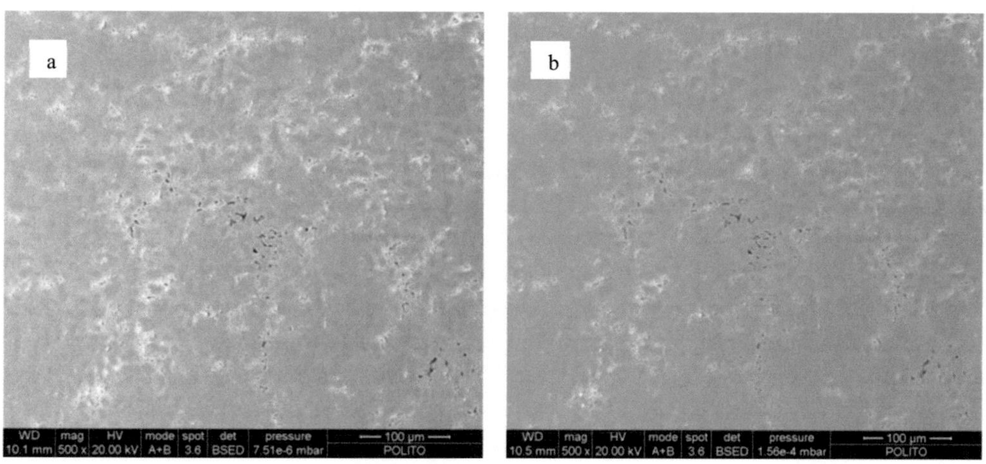

Figure 12. SEM microstructure showing the dispersion of the Zn-based intermetallics in as-cast sample (a) and after 20 days of natural ageing (b)

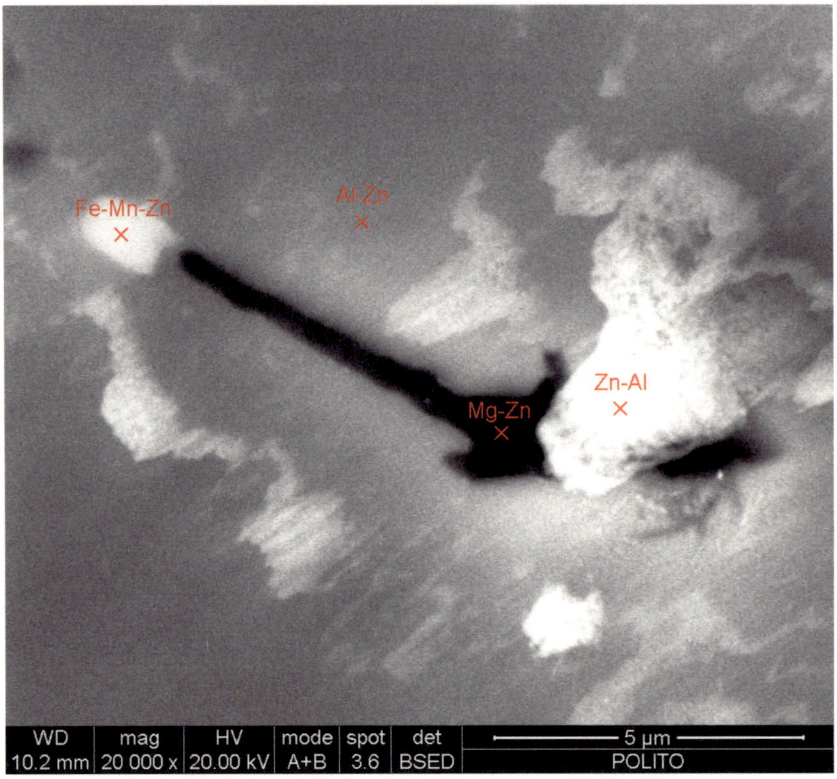

Figure 13. SEM microstructure showing the presence of the intermetallic particles in the as-cast sample

3. Conclusions

In this chapter, an overview of the actual state of the manufacture and use of light alloys has been presented. It includes a short history and description of state-of-the art techniques with special attention both to large-scale applications and to special use. The general description based on data available in the recent and past literature are integrated with some results of the current research carried out by the authors. It was pointed out that, in such critical areas of alloy development, detailed knowledge of the alloys' properties represents a key factor. Continuous interaction, data and know-how exchange between materials scientists, metallurgical engineers and industrial partners constitute a key-issue for the production of high-quality components based on innovative processes. The results presented within the research performed by the authors in collaboration with different foundries demonstrates the effectiveness of such strong interaction between industrial players and universities, e.g., the use of the developed modified self-hardening Al alloy can be considered an important energy-saving example, especially in terms of gas and electricity consumption. The increasing use of light alloys in different applications requires production of high integrity and improved perform-

http://dx.doi.org/10.5772/60769

ance components, similar to those obtained by traditional processes, also taking into account the economical aspect of the production. In fact, commonly used production processes used in a conventional way by the industry can be enhanced by developing and making available innovative production procedures. In this context, the development of new industrial processes, namely modified squeeze casting and innovative rheocasting processes for light-weight alloy production represents a valid support to enhancing the life of engineering components and their reliability. Aluminium has proven to be the ideal light-weighting material allowing weight savings also in mass production within reasonable cost limits and without compromising safety. When considering various development methods, it may be helpful to consider the advantages and drawbacks of using a specific production process. Once the correct alloy for the production has been chosen, there are numerous factors to be considered, including the type of casting process to use and the intended purpose of the produced component.

Acknowledgements

The authors wish to thank FOMT Spa, Teksid Aluminium Srl and ATS Company for their collaboration and for the effort of implementing the new production processes providing the samples at different stages of the study and the whole components.

Author details

Ildiko Peter* and Mario Rosso

*Address all correspondence to: ildiko.peter@polito.it

Politecnico di Torino, Department of Applied Science and Technology, Institute of Science and Engineering of Materials for Innovative technologies, Corso Duca degli Abruzzi, Torino, Italy

References

[1] Horst E, Mordike, BL. Magnesium Technology, Metallurgy, Design Data, Applications, Friedrich, 2006, XXII, 677.

[2] Darel E. Hodgson, Properties and selection: nonferrous alloys and special-purpose materials, ASM Handbook, Vol. 2; 1990, Materials Park, OH, ASM International.

[3] St John D, Caceres C, Zhang D, Edwards G, Taylor J, Schaffer G. Aluminum alloys for cast automotive components, 2nd Conference and Exhibition on Materials in the

Automotive Industry, IMMA, Conference Proceedings, Melbourne, Victoria, Australia, April 1996, pp.140-3.

[4] Spada T. In search of light-weight components: automotive's cast aluminum conversion. Eng Cast Solut 2002:4(2):28-31.

[5] Hatch JE (ed). Aluminum: properties and physical metallurgy, 5, 153, 177, 348; 1984, Materials Park, OH, American Society for Metals.

[6] Davis JE (ed). Aluminum and aluminum alloys, ASM Specialty Handbook; 1993, Materials Park, OH, ASM International.

[7] Fridlyander IN, Sister VG, Grushko OE, Berstenev VV, Sheveleva LM, Ivanova LA. Aluminum alloys: promising materials in the automotive industry. Met Sci Heat Treatment 2002;44(9-10):365-70.

[8] Dorum C, Laukli HI, Hopperstad OS, Langseth M. Structural behaviour of Al-Si die-castings: experiments and numerical simulations. Eur J Mechanics A/Solids 2009;28:1-13.

[9] Campbell J. Thin wall castings. Mater Sci Technol 1988;4(3):194-204.

[10] Aluminium in cars, EAA brochure, Brussel, 2007-Internet: www.alucars.org.

[11] Polmear IJ. Metallurgy of Light Alloys. New York; Toronto: J. Wiley & Sons, 1996.

[12] Jianhui LI, Wenbo DU, Shubo LI, Wang Z. Effect of aging on microstructure of Mg-Zn-Er alloys. J Rare Earths 2009;27(6):1042-945.

[13] Avedesian MM, Baker H. ASM Specialty Handbook, Magnesium and magnesium alloys, 1999, ISBN 0-87170-657-1.

[14] Kainer KU (ed). Magnesium alloys and technology, DMG, 2003, WILEY-VCH, ISBN 3-527-30256-5.

[15] Hartman D. Magnesium into the Next Millenium, International Magnesium Association, Rome, Italy, 1999, P.J.

[16] Magers D, Willekens J. In: Mordike B, Kainer K. (eds.) Magnesium Alloys and their Applications, Werkstoff Informationsgesellschaft mbH, Wolfsburg, Germany, 1998, p. 105.

[17] Edgar R. Magnesium into the Next Millenium, International Magnesium Association, Rome, Italy, 1999, p. 21.

[18] Joseph RD. Aluminum and Aluminum Alloys Handbook, ASM S ASM International, 1993.

[19] Liu G, Zhang GJ, Ding XD, Sun J, Chen KH. Modeling the strengthening response to aging process of heat-treatable aluminum alloys containing plate/disc- or rod/needle-shaped precipitates. Mater Sci Eng: A 2003;344(1-2):113-24.

[20] Filatov YA, Yelagin VI, Zakharov VV. New Al-Mg-Sc alloys. Mater Sci Eng: A 2000;280(1):97-101.

[21] Dahle AK, Arnberg L. Development of strength in solidifying Aluminium alloys. Acta Mater 1997;45:547-59.

[22] Arnberg L, Chai G, Backerud L. Determination of dendritic coherency in solidifying melts by rheological measurements. Mater Sci Eng A 1993;173A:101-3.

[23] Gourlay CM, Laukli HI, Dahle AK. Defect band characteristics in Mg-Al and Al-Si high-pressure die castings. Metall Mater Trans A 2007;38A:1833-44.

[24] Dahle AK, St John DH. Rheological behaviour of the mushy zone and its effect on the formation of casting defects during solidification. Acta Mater 1999;47:31-41.

[25] Dahle AK, Lee YC, Nave MD, Schaffer PL, St John DH. Development of the as-cast microstructure in magnesium-aluminium alloys. J Light Met 2001;1(1):61-72.

[26] Edward JV. High Integrity Die Casting Processes. John Wiley & Sons- Editore, 16 ottobre 2002.

[27] Terzi S, Salvo L, Suery M, Dahle AK, Boller E. Internal melting and coarsening of liquid droplets in an Al-Cu alloy: a 4-D experimental study. J Mater Sci 2013;48(21): 742234, doi:10.1007/s10853-013-7557-5.

[28] Pedersen L, Arnberg L. The effect of solution heat treatment and quenching rates on mechanical properties and microstructures in Al-Si-Mg foundry alloys. Metall Mater Trans A 2001;32A(3):525-32.

[29] William A. Die Cast Engineering: A Hydraulic, Thermal, and Mechanical Process, Marcel Dekker Ed. New York 2005.

[30] Rosindale I, Davey K. Transient thermal model for the hot chamber injection system in the pressure die casting process. Appl Math Model 1999;23(4):255-77.

[31] Luo AA, Sadayappan K. Technology for Magnesium Castings, American Foundry Society, Schaumburg, IL, 2011. pp.29-47.

[32] Avedesian MM, Baker H. Magnesium and Magnesium Alloys, ASM International, Materials Park, OH, 1999.

[33] Freuling JW, Hanawalt JD. Transactions of the American Foundrymen's Society 77 (1969) 159e164.

[34] Alan AL. Magnesium casting technology for structural applications. J Magnesium Alloys 2013;1:2-22.

[35] Damjan K, Janez T. Thermal stresses in aluminium alloy die casting dies. Comput Mater Sci 2008;43(4)1147-54.

[36] Street A. The Die Casting Books, Portcullis Press Ltd., 1977, pp.3-16, pp.625-41.

[37] Kaye A, Street A. Die Casting Metallurgy, Butterworths, London, 1982, pp.231-5.

[38] Zhang B, Maijer DM, Cockcroft SL. Development of a 3-D thermal model of the low-pressure die-cast (LPDC) process of A356 aluminum alloy wheels. Mater Sci Eng A 2007;464:295-305.

[39] Esdaile RJ, Nguyen TT, de Looze GR, Murray MT (eds). Proceedings of the First International Non-Ferrous Processing and Technology Conference, TMS International, Warrendale, PA, 1997, pp.213-8.

[40] Muller W, Feikus, Manufacturing - Casting methods FJ. AFS Trans 1996;104:1111-7.

[41] George E, Totten GE, MacKenzie DS (eds). Handbook of Aluminum Vol. I Physical Metallurgy and Processes. 2003.

[42] George E, Totten GE, MacKenzie DS (eds). Handbook of Aluminum: Volume 2: Alloy Production and Materials Manufacturing. 2003.

[43] Laws KJ, Gun B, Ferry M. Large-scale production of Ca65Mg15Zn20 bulk metallic glass samples by low-pressure die-casting. Mater Sci Eng A 2008;475:348-54.

[44] Miller AE, Maijer DM. Investigation of erosive-corrosive wear in the low pressure die casting of aluminum A356. Mater Sci Eng A 2006;435-436:100-11.

[45] Otarawanna S, Laukli HI, Gourlay CM, Dahle AK. Feeding mechanisms in high-pressure die castings. Feeding mechanisms in high-pressure die castings. Metallurg Mater Trans A 2010;41(7):1836-46. doi:10.1007/s11661-010-0222-6.

[46] Otarawanna S, Gourlay CM, Laukli HI, Dahle AK. Microstructure formation in high pressure die casting. Trans Indian Inst Metals 2009;62(4-5):499-503. doi:10.1007/s12666-009-0081-2.

[47] Bainbridge IF, Taylor JA, Dahle AK. Surface formation on VDC casting. In: Grandfield JF, Eskin DG. (ed) Cast Shop for Aluminum Production, pp.783-8, Hoboken, NJ, USA: John Wiley & Sons. doi:10.1002/9781118647783.ch98.

[48] Otarawanna S, Dahle AK. Casting of aluminium alloys. In Roger L. (ed.) Fundamentals of Aluminium Metallurgy: Production, Processing and Applications, pp.141-54, Oxford, U.K.: Woodhead.

[49] Skrotzki B, Gottsein G. Aluminium Alloys: Their Physical and Mechanical Properties, Volume 1(2008). Jürgen H (ed.), ISBN-13: 978-3527323678.

[50] Heinz A, Haszler A, Keidel C, Moldenhauer S, Benedictus R, Miller WS. Recent development in aluminum alloys for aerospace applications [J]. Mater Sci Eng A 2000;280:102–9.

[51] Barresi J, Chen Z, Davidson C, Murray MT, Nguyen T, St John DH, Thorpe WR. Castings of Aluminium Alloy Components, Materials Forum (Australia), 1996;20:53-70.

http://dx.doi.org/10.5772/60769

[52] Woithe H, Achwab W. Producing heat treatable Al die castings: Part II, Die Casting Management (October 1993) 29-31.

[53] Pennington JN. Re-engineered casting combines, metal ¯flow, vacuum technologies. Modern Metals 1996;July:48B±48G.

[54] Brevick JR, Cheng PS. Evaluating the quality of vacuum die castings. Trans North American Die Casting Association 1995;pp.145-8.

[55] Akira U, Susumu M, Koji Y, Tomoyuki S. Fatigue behavior of die casting aluminum alloys in air and vacuum. Proc Eng 2010;2:1937-43.

[56] Sheng-yong L, De-jiang L, Xiao-qin Z, Wen-jiang D. Microstructure and mechanical properties of Mg–6Gd–3Y–0.5Zr alloy processed by high-vacuum die-casting. Trans Nonferrous Met Soc, China 2014;24:3769–76.

[57] Niu XP, Hu BH, Pinwill I, Li H. Vacuum assisted high pressure die casting of aluminium alloys. J Mater Process Technol 2000;105:119-27.

[58] Baumgartner K. Vacuum high technology for top quality pressure die castings. In: Proceedings of the 15th International Pressure Die Casting Conference, Montreaux, Switzerland, May 22, 1996.

[59] Hollinggrak J. Casting Metals, UK Patent 4371 (1819).

[60] Chernow DK. Reports on the Imperial Russian Metallurgical Society, 1878.

[61] Ghomashchi MR, Vikhrov A. Squeeze casting: an overview. J Mater Process Technol 2000;Volume 101: pages 1-9.

[62] Yue TM, Chadwick GA. Squeeze casting of light alloys and their composites. J Mater Process Technol 1996;58:302-7.

[63] Maleki A, Niroumand B, Shafyei A. Effects of squeeze casting parameters on density, macrostructure and hardness of LM13 alloy. Mater Sci Eng A 2006;428:135-40.

[64] Luo SJ, Hu LX, Li HJ. The effect of plastic flow on the strengthening and toughening of alloy during liquid-metal forming under pressure. J Mater Process Technol 1995;49:425-9.

[65] ASM Handbook Vol.15 Casting, ASM International 2008.

[66] Spencer DB, Mehrabian R, Flemings MC. Rheological behavior of Sn-15Pb in the crystallization range. Metallurg Trans 1972;3:1925-32.

[67] Flemings MC. Behavior of metal alloys in the semisolid state. Metallurg Trans 1991;22B:269-93.

[68] Hall K, Kaufmann RH, et al. Detailed processing and cost considerations for new rheocasting of light metal alloys. Proceedings of the 6th International Conference of Semi-solid Processing of Alloy and Composites, pp.23-28, 2000, Turin Italy.

[69] Haga T, Suzuki S. Casting of Aluminum Alloy Ingots for Thixoforming Using A Cooling Slope. J Mater Process Technol 2001;118:169-72.

[70] Xie SS, Yang HQ, et al. Numerical simulation of semi-solid Magnesium alloy in continuous roll-casting process. International Conference on Semi-Solid Processing of Alloys and Composites, Trans Tech Publications, Switzerland, Solid State Phenomena, 2006;116-117:583-6.

[71] Xie SS, He YF, et al. Study on Semi-solid Continuous Roll-casting Strips of AZ91D Magnesium Alloy. Proceedings of the 10th International Conference of Semi-solid Processing of Alloy and Composites, Solid State Phenomena, 2008;141-143:469-73.

[72] Rosso M, Peter I. New frontiers for thixoforming. Int J Microstructure Mater Properties 2013;8(1-2)113-24; ISSN: 1741-8410.

[73] Gourlay CM, Dahle SAK, Nagira T, Nakatsuka N, Nogita K, Uesugi K, Yasuda H. Granular deformation mechanisms in semi-solid alloys Acta Materialia 2011;59:4933-43.

[74] Haga T, Kapranos P. Simple rheocasting processes. J Mater Process Technol 2001;130-131:594-8.

[75] Santella ML, Engstrom T, Storjohann D, Pan TY. Effects of rheocasting and heat treatment on microstructure and mechanical properties of A356 alloy. Scripta Materialia 2005;53:201-6.

[76] Atkinson HV, Liu D. Microstructural coarsening of semi-solid aluminium alloys. Mater Sci Eng A 2008;496(1-2)(25):439-46.

[77] Barnes H. Thixotropy - a review. J Non-Newtonian Fluid Mechanics 1997;70:1-33.

[78] Cetin A, Kalkanli A. Investigation of microporosity formation mechanisms in A356 aluminium alloy castings. Int J Microstructure Mater Prop 2009;4(3):377-85. Curle UA, Govender G. Semi-solid rheocasting of grain refined aluminum alloy 7075. Transaction of Nonferrous Metals. Society of China 2010;20:832-6.

[79] Atkinson HV. Modelling the semisolid processing of metallic alloys. Prog Mater Sci 2005;50:341-412.

[80] Flemings MC. Behavior of metal alloys in the semisolid state. Metall Trans B 1991;22A:957-81.

[81] Ji S, Fan Z, Bevis MJ. Semi-solid processing of engineering alloys by a twin-screw rheomoulding process. Mater Sci Eng A 2001;299:210-7.

[82] Birol Y. Semi-solid processing of the primary aluminium die casting alloy A365. J Alloys Compd 2009;473:133-8.

http://dx.doi.org/10.5772/60769

[83] Favier V, Atkinson HV. Micromechanical modelling of the elastic-viscoplastic response of metallic alloys under rapid compression in the semi-solid state. Acta Materialia 2011;59:1271-80.

[84] Peter I, Rosso M, Bivol C. Microstructure and mechanical behaviour of Al-based alloy obtained by liquid forging technique. Metallurgia International 2009;XIV(2):15-9, ISSN 1582-2214.

[85] Rosso M, Peter I, Chiarmetta G, Gatteli I. Development of industrial components by improved rheocasting system. Int J Mater Forming 2010;3(1):787-90, DOI 10.1007/s12289-010-0888-8.

[86] Rosso M, Peter I, Chiarmetta G, Gatelli I. Extremely light weight rheocast components for auto motive space frame. Solid State Phenomena 2013;192-193:545-50, doi: 10.4028/www.scientific.net/SSP.192-193.545.

[87] Rosso M, Peter I. Defect control on Al castings for excellent quality and improved performances through novel Rheocasting processes, TMS Annual Meeting Volume 2, 2012, pp.403-410, 141st Annual Meeting and Exhibition, TMS 2012; Orlando, FL; United States; 11 March 2012 through 15 March 2012.

[88] Tillova E, Durinikova E, Chapulova M. Structural analysis of secondary AlZn10Si8Mg cast alloy. Acta Metallurgica Slovaca 2011;17(1):4-10.

[89] Peter I, Rosso M, Castella C, Molina R. Self-hardening alloys for automotive application. Mater Sci Forum 2014;794-796:1221-6.

Chapter 2

Alloy Development through Rapid Solidification for Soft Magnetic Application

Rajat K. Roy, Ashis K. Panda and Amitava Mitra

Additional information is available at the end of the chapter

http://dx.doi.org/0.5772/60772

Abstract

This chapter describes different rapidly solidified processing routes of soft magnetic alloys and their properties and applications in different areas. Section-2 explains the details of process mechanism. The functions of different alloying elements are discussed with the alloy design of soft magnetic alloys in section-3. The structure-property correlation is described in section-4. Section-5 highlights different types of rapidly solidified soft magnetic alloys, like high permeability alloys, high induction alloys, Fe-6.5 wt% Si steel and GMI alloys. In the last section-6, the applications of different types of soft magnetic alloys are mentioned.

Keywords: Rapid solidification, soft magnetism, alloy design, saturation induction, Fe-6.5 wt% Si steel, GMI alloys

1. Introduction

In recent years, amorphous and nanostructured soft magnetic alloys have gained considerable interest owing to their superior properties. Consequently, it catered to a broad area of applications such as transformer cores, magnetic field sensors, sensors for non-destructive evaluation of materials, motors, electric vehicles, etc. The bulk use of rapidly solidified iron-based magnetic materials is in distribution transformers due to its low core loss as compared to cold rolled grain orient (CRGO) Si steel. However, the use of amorphous alloy is limited only to low power transformers as the saturation induction of such alloy is low (~1.6T) as compared to CRGO Si steel (~ 2.1T). To overcome this problem the current trend is to work on high induction alloy so that the materials can be used for high frequency power transformers. Beside the distribution transformer, the high permeability amorphous or nanostructured magnetic alloys are used in switched mode power supply, switchgear, etc. In nanostructured alloys or nanocomposites which typically comprise nanocrystallites dispersed in amorphous matrix, the magnetic properties are controlled by the exchange of magnetic coupling between nanocrystallites through intergranular amorphous matrix. In these nanostructured materials, the extrinsic properties like coercivity and permeability are controlled through crystallite size and their distribution, whereas the intrinsic properties, like saturation magnetic induction (B_S) and Curie temperatures (T_C), are the functions of alloy chemistry of amorphous and nano-crystallite phases. The ultimate advancement of alloys is determined by the optimization of these intrinsic and extrinsic properties. Therefore, both designing of alloy composition and processing of alloys are equivalently important for attaining the ultimate goal of applications. In this chapter, the processing of different soft magnetic alloys through rapid solidification has been discussed, and the alloys are categorized according to their properties and applications.

2. Rapid solidification processing routes

2.1. Melt spinning technique

Amongst different rapid solidification processing routes, melt spinning is the most common technique for yielding soft magnetic metallic ribbons in large quantity. Sometimes this technique is also described as chill block melt spinning (CBMS). Basically, the ribbons are synthesized as the stream of molten alloy from a quartz crucible is purged through argon gas pressure on a rapidly rotating wheel made from the metals like pure Cu, Cu-Be alloy and stainless steel (Fig. 1). As shown in Fig. 1, the ribbon is typically maintaining a contact of sticking distance ($d_{t1}<d_{t2}<d_{t3}$) with the wheel surface before flying out of the wheel surface by centrifugal force. The adhesion time of ribbon at wheel surface is important for the ribbon preparation and depends on the process parameters and alloys used. It is observed that sticking distance of $Fe_{40}Ni_{40}B_{20}$ is less than that of $Fe_{81.5}B_{14.5}Si_4$ amorphous alloy on the Cu wheel [1]. Moreover, the ribbon thickness is quite important for controlling the ribbon properties, which is basically controlled by the rotating wheel speed, ejection pressure, nozzle slot size and crucible-wheel gap.

Huang et al. described the ribbon thickness effect on the core loss and excitation for straight and toroidal samples [2]. The optimum ribbon thickness of 30 μm has been found the most suitable for low core loss as well as low excitation (Fig. 2). For thin ribbon (<30 μm), both straight and toroidal samples show similar core loss and excitation, but for thick ribbons (>30 μm), core loss and excitation in toroidal samples are higher than the other. The magnetic domain structure is also responsible for better soft magnetic properties for the ribbons having thickness 25–30 μm as in such cases magnetic domains become parallel to ribbon thickness, whereas the ribbon with lower and higher thickness than above leads to transverse domain structure.

http://dx.doi.org/0.5772/60772

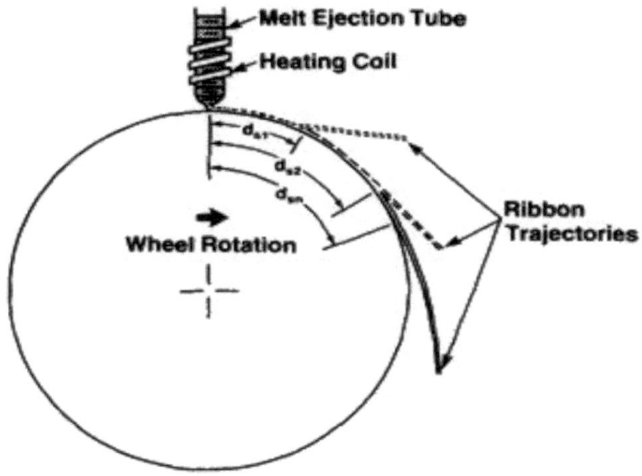

Figure 1. Schematic diagram showing ribbon preparation by single roller melt spinning technique and ribbon-wheel sticking distance at adhesion times t_1, t_2 and t_3 [1].

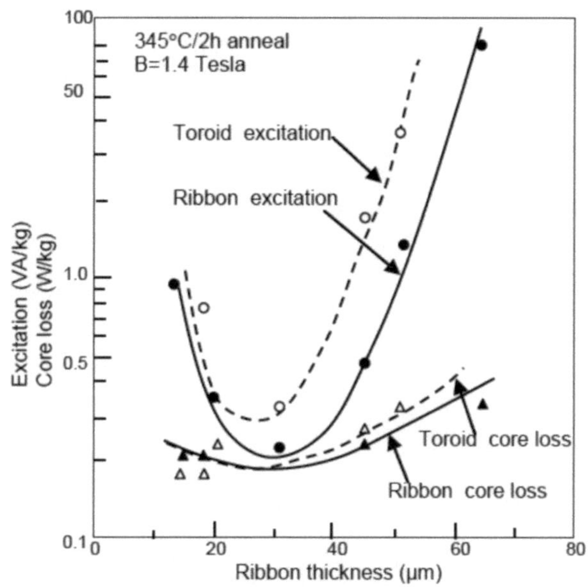

Figure 2. Ribbon thickness dependence of core loss and excitation for straight ribbons and 7.5 cm toroids, annealed at 345°C/2h, B = 1.4T [2].

2.2. In-rotating water quenching technique

Another way of preparing rapidly solidified materials is the processing of microwires (~100 μm diameter) by in-rotating water quenching technique. Unlike melt spinning, the amorphous wire preparation through in-rotating water quenching technique involves rapid quenching of molten material inside the water of a rotating drum (Fig. 3). This processing route involves a stable melt impingement and minimum melt stream break into water surface and suitably carried along with the water's centrifugal velocity to acquire continuous wires. Therefore, in addition to the glass forming ability of alloys, the molten jet stability prior to quenching and high stability of water surface layer are prerequisite conditions for obtaining defect-free, continuous wires. Heyder et al. has reported that a non-viscous fluid jet may break into different steps like dropping, varicose breakup, sinuous breakup and atomization during the wire processing [3] and causes the powder formation in the quenching water. The increasing jet velocity can result in the rapid jet disintegration. The ratio between melt jet velocity (V_j) and cooling water velocity (V_w) is an important parameter to achieve straight wire. The ratio (V_j/V_w) has been found to be 0.9 for $Fe_{75}Si_{10}B_{15}$ metallic wires [3]. However, Olofinjana et al. have reported that the optimized ratio is in the range of 1.14–1.20 for the same aforementioned alloy and 1.16–1.18 for $Fe_{77.5}Si_{7.5}B_{15}$ alloy for continuous wire formation [4]. Therefore, a little difference in alloy composition and processing route may change the optimum ratio for achieving continuous wires. Additionally, other parameters, such as crucible nozzle to water level separation (N_W), melt ejection angle (θ), casting temperature, ejection pressure and nozzle orifice diameter (N_D), have similar importance in maintaining the stability of melt jet and solidification rate of the wire and, subsequently, in obtaining good quality wires [5]. The crucible of large orifice diameter (N_D) tends to produce powders due to destabilized melt jet with lower jet velocity impinging into the water. In contrast, the lower limit of N_D is related to higher pressure which is required to overcome the surface tension for the melt ejection. The melt jet is more likely to resist, through surface tension forces, being flattened by the dynamic pressure or drag force of the cooling water. The low melting point alloy is instantaneously solidified on ejecting the melt alloy because of the proximity of cooling water. Thus it is observed empirically that an optimal dwell time before casting is essential to avoid stream disturbance at high temperature melt and nozzle blockage at low temperature melt [5].

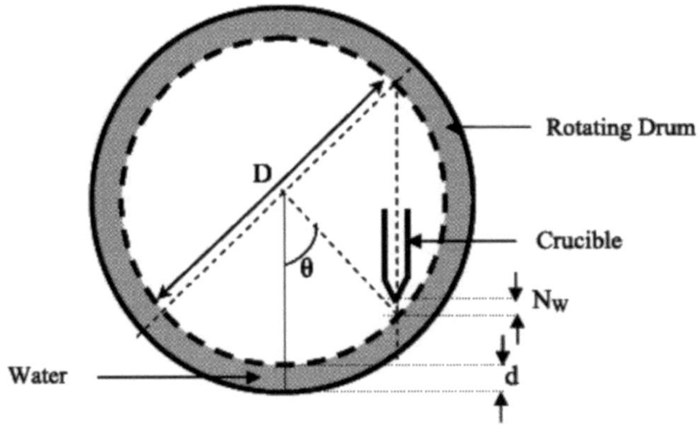

Figure 3. Schematic presentation of in-rotating-water quenching instrument (D: diameter of drum; N_W: separation between nozzle and water surface; d: depth of water level; :ejection angle of the melt) [5].

3. Design of alloy composition

Nanostructured soft magnetic alloys are designed on the basis of refinement of grain size to tailor intrinsic and/or extrinsic magnetic properties. The intrinsic magnetic properties (magnetic flux density, Curie temperature, magnetocrystalline anisotropy, magnetostrictive coefficient) are dependent on alloy chemistry, whereas extrinsic magnetic properties (coercivity, permeability) are influenced by the microstructures. Accordingly, a nanostructured soft magnetic alloy is basically designed with one ferromagnetic and three different non-ferromagnetic elements which are categorized as glass formers, grain growth inhibitors and nucleating agents, as explained schematically in Fig. 4. The stoichiometric variation between these elements lead to a competition between amorphization and crystallinity for obtaining critical ribbon thickness in glassy alloy systems like Fe-Si-B and Co-Si-B (Fig. 5) [6]. It is noteworthy to mention that smaller amount metalloids are required for achieving amorphization and smaller thickness ribbons in Co-based system compared to Fe-based system.

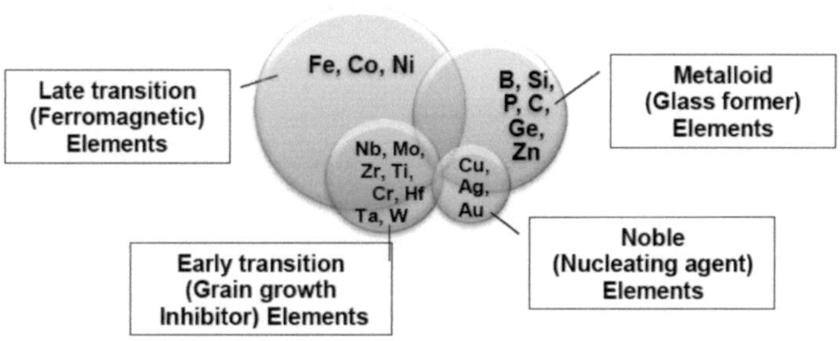

Figure 4. Schematic diagram showing alloy design of soft magnetic alloys (circle diameter representing approximate elemental percentage in alloy).

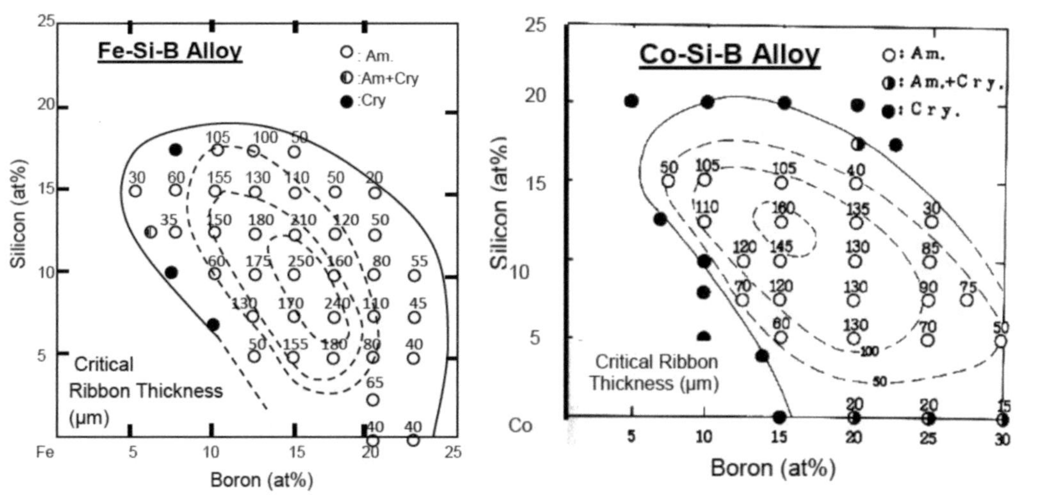

Figure 5. Compositional dependence of amorphization and crystallinity in (a) Fe-Si-B and (b) Co-Si-B systems [6].

Although the elemental composition of nanocrystalline and amorphous alloys may vary within wide ranges, a typical chemical formula may be explained as F_xM_y, where F = ferromagnetic elements (e.g., Fe, Co, Ni), M = non-ferromagnetic elements (e.g., B, Si, P, Nb, Mo, Zr, Cu, etc.), x = 70–90 at% and y = 10–30 at%. According to the compositional ranges of ferromagnetic elements and consequent property developed, nanostructured alloys may be classified as FINEMET (Fe-M-Nb-Cu, where M = B, Si), NANOPERM (Fe-B-Zr-Cu) and HITPERM (Fe-Co-B-Zr-Cu). The composition and magnetic properties of these alloys are explained in Table 1. Therefore, the intrinsic magnetic properties of these alloys are basically the function of ferromagnetic elements, and the extrinsic properties depend on the non-ferromagnetic elements along with the effect of nanocrystallinity. Accordingly, there are three factors to be emphasized during alloy design which are (i) desired magnetic properties, (ii) high thermal stability and (iii) good glass forming ability. Thus, controlled chemistry of nanostructured alloys can cater to different soft magnetic applications pertaining to either ultralow coercivity or high permeability or high Curie temperature or high induction.

Alloy Name	Typical Composition	Saturation Magnetization (T)	Coercivity (A/m)	Permeability	Curie Temp. (T_c), °C
FINEMET	$Fe_{73.5}Si_{13.5}B_9Nb_3Cu_1$	1.24	0.53	10^5	370
NANOPERM	$Fe_{88}Zr_7B_4Cu_1$	1.64	4.5	$3.4X10^4$	770
HITPERM	$Fe_{44}Co_{44}Zr_7B_4Cu_1$	1.6-2.1	-	-	>965

Table 1. Compositions and magnetic properties of three types of amorphous alloys [7]

4. Structure-property correlation

Saturation magnetization and Curie temperature are the intrinsic magnetic properties, depending on alloy chemistry and crystal structure, whereas soft magnetic properties like coercivity and permeability are extrinsic properties, relating to nanocrystallinity of alloys and preferred arrangement of magnetic domains. The soft magnetism of any material is monitored through the coercivity (H_c) and/or permeability (μ), which are inversely related to each other. The excellent soft magnetism is achieved with the tailoring of alloy chemistry and optimizing the microstructure [7].

According to Herzer diagram [8], the relation of soft magnetism with grain size can be divided into two segments, (i) nanocrystalline and (ii) polycrystalline (Fig. 6). For large grain size (D> 0.1–1μm), such as polycrystalline alloys, the coercivity is inversely proportional to D, i.e., H_c 1/D. So, in the segment II, the soft magnetism increases towards right side. In other words, for polycrystalline alloys, the magnetic hardness of fine grained material is larger than that of coarse grained materials. For small grain size (segment-I, D<100nm), such as amorphous and nanostructured alloys, the coercivity rapidly decreases with grain size, and it follows the relation H_c D^6.

In nanostructured materials, the fine (~15 nm diameter) ferromagnetic nanocrystallites are impregnated in the amorphous matrix, and thus magnetic coupling of nanocrystallites exchanges through intergranular amorphous matrix. In other words, the grain sizes in the order of 10–15 nm lowers the magnetocrystalline anisotropy proportionally, resulting in superior soft magnetic properties [8]. Due to higher Curie temperature of nanocrystallites than that of amorphous matrix, the intergranular coupling rapidly decreases while the measuring temperature approaches Curie temperature of amorphous matrix [9]. Since the soft magnetic properties are related to intergranular exchange interaction, Curie temperature of amorphous matrix also plays a dominant role in controlling soft magnetic properties in nanocomposites. If the alloy is designed in such a way that the amorphous matrix has high Curie temperature, the alloy can be used for high temperature applications. Thus a new series of nanostructured alloy named HITPERM was developed which contain equal at% of Fe and Co along with glass former B, nucleating element Cu and the grain growth inhibitor Zr.

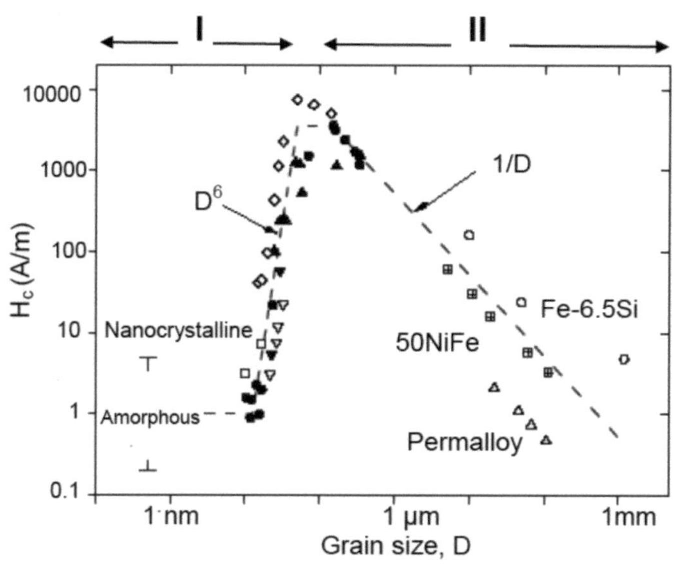

Figure 6. Relationship between coercivity and grain size in soft magnetic alloys as proposed by G.Herzer [8].

In case of HITPERM alloys, the composition variation in amorphous matrix during the nanocrystallization process is a complex function of the enrichment in B, and Nb and increase in the Co/(Fe+Co) ratio. The magnetic properties are thus correlated with nanocrystallites formation in amorphous matrix [10]. Although the Co concentration is homogeneous throughout the amorphous matrix and nanocrystals, Fe is relatively enriched in the amorphous matrix and expels Nb and B to the matrix during crystallization process [11]. The enrichment of matrix with Nb and B stabilizes the residual amorphous phase. Moreover, due to low diffusivity and solubility of Nb in Fe(Co) phases, Nb acts as a diffusion barrier located at the primary nanocrystal/ amorphous matrix interface. The thickness of the diffusion barrier increases with increasing Nb percentage for fixed B containing alloys. Consequently, it hinders primary nanocrystal grain growth and also restricts Fe enrichment in Fe(Co) phases, resulting into an increase in Fe content in the amorphous matrix and retaining its higher Curie temperature.

5. Different soft magnetic alloys and their properties

Rapid solidification reduces the magnetocrystalline anisotropy, increases the solubility of alloying element and also produces the materials in the form of thin sheet, and hence rolling and forging process can be avoided for many brittle materials. Composition of the materials also plays a vital role in developing soft magnetic materials. A small variation in composition can change intrinsic magnetic property like magnetostriction which in turn alters the magnetoelastic anisotropy of the system and thus, modifies the soft magnetic behaviour of alloys. Alloy compositions have also a strong influence on the crystallization behaviour of rapidly solidified materials. This section will be discussed on the influence of alloy composition for the soft magnetic properties.

5.1. High permeability alloys

The high permeability amorphous alloys may be categorized as FINEMET type alloys with a nominal composition of $Fe_{76.5-x}Si_{13.5}B_9M_xCu_1$ (where M = Nb, Ta, Mo, W, V, Cr and x = 0–3 at%) [12–16]. Many attempts have been made for the modifications of FINEMET alloys, as explained in Table 2. The primary crystallization onset of these alloys approximately 435–500°C, varying with alloy composition. Upon annealing within 50° above primary crystallization temperatures, the α-Fe(Si) nanocrystallites of 10–15 nm grain size are embedded in a residual amorphous matrix, which may be called as nanocomposites. Annealing near or above 600°C leads to the precipitation of boride compounds (Fe_2B or Fe_3B) [17] which is detrimental for soft magnetism of nanostructured materials. The Curie temperature and saturation induction of α-Fe nanocrystallites are about twice than that of amorphous phase for Fe-Si-B-Nb-Cu alloys [18]. However, both magnetic properties decrease with Si percentage in $Fe_{96-z}Si_xB_{z-x}Nb_3Cu_1$. Panda et al. have examined the effect of Si/B ratio on magnetic properties for high temperature annealed alloys $Fe_{73.5}Nb_3Cu_1Si_{22.5-X}B_X$ (X = 5, 9, 10, 11.25, 19) [19]. The magnetic softening has been found maximum at Si/B ratio of 1.5, that is, for the alloy $Fe_{73.5}Nb_3Cu_1Si_{13.5}B_9$, with a coercivity of 1.56A/m and susceptibility of 1.35×10^5 while the alloys is annealed at 527°C. For other Si/B ratio, the magnetic softening is deteriorated before 470°C annealing. In Co added FINEMET alloys, Si has a detrimental effect on saturation magnetization due to the formation of $(Fe,Co)_3Si$ phase than that of (Fe,Co) phase in Si-free alloy, causing less localized moment between Fe and Co atoms in former alloy [20]. On the other hand, Co-based amorphous alloys with small additions of Fe reveal nearly zero λ_s, resulting in good soft magnetic behaviour. Their application is limited due to the fact that their saturation induction is considerably lower compared to that of Fe-based alloys [17].

Alloy Composition	T_{x1} (°C)	T_{x2} (°C)	B_s (T)	Hc (A/m)	μ	Reference
$Fe_{73.5}Si_{13.5}B_9Nb_3Cu_1$	-	-	1.24	0.53	10^5	[21]
$Fe_{73.5}Si_{13.5}B_9Nb_3Cu_1$	497	633	1.23	1.0	35×10^3	[22]
$Fe_{72.5}Si_{13.5}B_9Nb_3Cu_1Al_1$	480	623	1.09	0.74	33×10^3	[22]
$Fe_{73.5}Si_{13.5}B_9Mo_3Cu_1$	487	-	-	-	7×10^3	[23]
$Fe_{73.5}Si_{13.5}B_9V_3Cu_1$	467	577	-	-	750	[23]
$Fe_{36.75}Co_{36.75}Si_{13.5}B_9Mo_3Cu_1$	460	560	-	-	-	[24]
$Fe_{73.5}Si_{17.5}B_5Nb_3Cu_1$	435	615	0.55	1	-	[25]
$Fe_{72.7}V_{1.8}Si_{13.5}B_9Nb_2Cu_1$	-	-	1.21	-	81×10^3	[26]
$Fe_{71.5}Si_{13.5}B_9Nb_3Cu_1Al_2$	-	-	1.3	0.32	-	[27]
$Fe_{62.8}Co_{10}Si_{10}B_{13.5}Nb_3Cu_{0.7}$	572	662	1.35	3	-	[28]

Table 2. Thermal and magnetic properties of some FINEMET alloys

(T_{x1} = Primary crystallization temperature onset, T_{x2} = secondary crystallization temperature onset, B_s = saturation induction, H_c = coercivity and μ = permeability)

5.2. High induction alloys

In high permeability alloys, the distribution of fine nanocrystallites in amorphous precursor is a key factor for the improvement of soft magnetism. This is attributed to the presence of high (>20 at%) metalloids by replacing ferromagnetic elements, like Fe, Co, Ni, etc. As a result, alloys like FINEMET have superior soft magnetic properties, but the saturation induction of these alloys is lower than that of Si steel. To fill up this dearth, many high (>1.24T) induction alloys have been developed in recent years (Table 3).

Alloy Composition	B_s (T)	H_c (A/m)	μ	Reference
$Fe_{83.16}P_{4.95}C_{3.96}B_{3.96}Si_{2.97}Cu_1$	1.74	4.9	22.9×10^3	[29]
$Fe_{82.65}B_{14}Si_2Cu_{1.35}$	1.85	10	-	[30]
$Fe_{83.7}Nb_{6.6}B_{8.6}P_1Cu_{0.1}$	1.5	5	48×10^3	[31]
$Fe_{81.7}P_2B_7Si_9Cu_{0.3}$	1.56	7	-	[32]
$Fe_{85}P_4B_8Si_2Cu_1$	1.85	6.1	27×10^3	[33]
$Fe_{91}Zr_7B_2$	1.7	7.2	14×10^3	[34]
$Fe_{89}Zr_7B_3Cu_1$	1.64	4.5	34×10^3	[35]
$Fe_{86}Zr_7B_3Si_4$	1.54	-	10^4	[36]
$Fe_{53.3}Co_{28.7}Si_{2.8}B_{11.2}Nb_3Cu_1$	1.54	-	-	[37]
$Fe_{55.25}Co_{29.75}Si_{2.8}B_{11.2}Cu_1$	1.73	-	-	[37]
$Fe_{54.6}Co_{29.4}Si_{2.8}B_7P_{4.2}Nb_1Cu_1$	1.62	-	-	[38]

Table 3. Saturation induction (B_s), coercivity (H_c) and permeability (μ) of some amorphous alloys

The design of high induction alloys may be based on Slater-Pauling curve [42], which explains the high magnetic induction achievable in Fe-Co alloys compared to either Fe- or Co- based alloys (Fig. 7). The Curie temperature is also high in these alloys than Fe-rich alloys. The crystalline alloys of $Fe_{65-70}Co_{35-30}$ exhibit maximum saturation induction and Curie temperature of 2.4T and 1000°C, respectively [39]. These values are deteriorated in amorphous or nano-crystalline alloys due to the additions of metalloids and other elements. However, many Fe-based alloys have been explored with a saturation of 1.85T rather than Fe-Co based alloys (Table 3). Although the Curie temperature of former alloys may be lower than later alloys. The minimum magnetocrystalline anisotropy is investigated at concentration of x = 0.5 for $Fe_{1-x}Co_x$ system [39]. The magnetostriction coefficient (λ_s) is also significant in equiatomic composition of this alloy system. Since Fe is positively magnetostrictive, addition of negatively magnetostrictive Co to Fe reduces the overall magnetostriction λ_s of $Fe_{80-x}Co_xB_{20}$ system [39]. Similarly addition of Co to Fe raises the saturation induction as depicted in Slater Pauling curve with the maximum saturation magnetization value reported for $Fe_{70}Co_{10}B_{20}$ alloy [41].

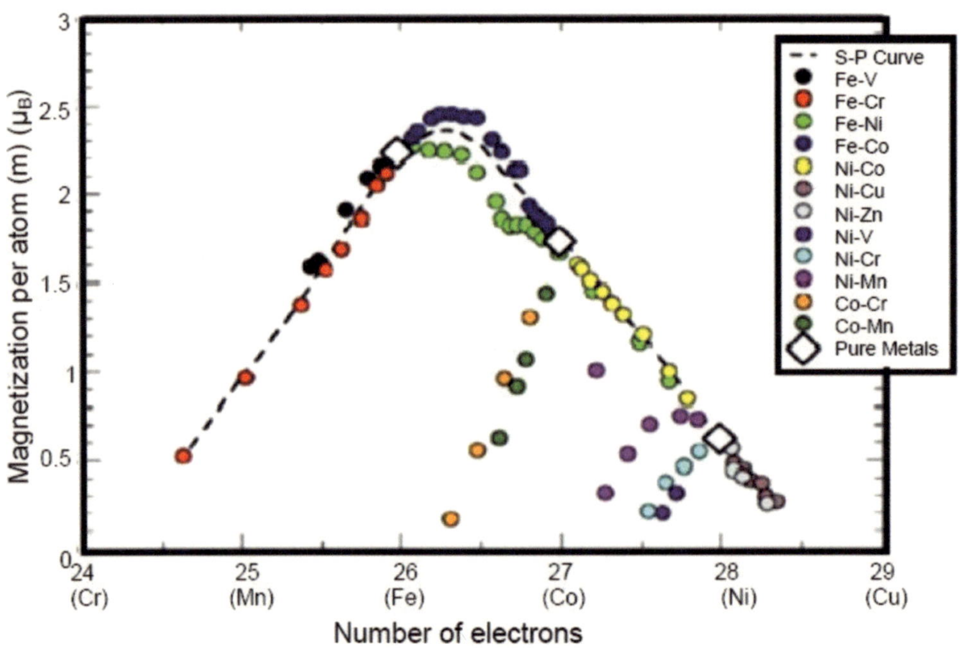

Figure 7. Slater-Pauling curve of late transition elements [42].

5.3. Fe-6.5 wt% Si steel

The development of high silicon (Si) steels bears technological as well as commercial importance not only due to their superior soft magnetic properties but also due to low cost. The 3.5

http://dx.doi.org/0.5772/60772

wt% Si steel has a major application in distribution transformers owing to their excellent soft magnetic properties and low core losses [43]. This material is being produced in the form of cold rolled grain oriented (CRGO) or non-grain oriented (CRNO) Si steel. Increasing Si content beyond 3.5% in iron enhances the electrical resistivity of the alloy, resulting in low eddy current component. Si also decreases the magnetostriction constant which becomes minimum at 6.5 wt%, leading to improved soft magnetic properties and low core losses [43]. However, the workability of alloy deteriorates with enhancement in Si content due to the formation of some ordered phases B2 and DO_3 (see Fig. 8), coarse grains and impurities on grain boundary, etc. [44, 45]. Therefore, the fabrication of strips with high ~6.5 wt% silicon through conventional rolling process is difficult owing to inherent brittleness of the material [47]. The brittleness is due to the formation of ordered phases, coarser grains and impurities on grain boundary, etc. A breakthrough in the improvement of ductility by suppression of ordered phases may be was achieved while processing the alloy through several rapid solidification techniques like chemical vapour deposition (CVD), gas atomization, splat quenching and melt spinning [48–50]. The products of these processes are subjected to various annealing treatments to control the grain sizes and ordered phases for achieving superior soft magnetic properties [51]. Amongst these methods, the development of 6.5 wt% Si steel ribbons by melt spinning technique has drawn an intense research interest due to its potential for production in bulk quantities demand as a high-efficiency material for magnetic components [52, 53].

In the Fe-Si alloys, mainly one disordered structure A2 and two types of ordered phases, viz., B2 (FeSi type) and DO_3 (Fe_3Si type) structures are observed within 5.5 to 10 wt% Si contents [54]. The ordered phases B2 and DO_3 are superstructures of A2. They form from A2 phase by unlike-atom pairing of first and second nearest neighbor place in BCC lattice, respectively [53]. It is observed that melt spun ribbons comprise mainly A2 disordered phases owing to the suppression of ordered phases [52]. Subsequent heat treatment can bring about second order transformation of A2 to B2 phase, as shown in the Fe-Si phase diagram (Fig. 8). However, B2 to DO_3 transformation which is of first order can take place simultaneously, leading to the co-existence of both these phases [55]. The grain size becomes finer and the ductility is also improved in the rapidly solidified ribbons. Therefore, the rapid solidification route emerges as a potential technique for the fabrication of Fe-6.5 wt % Si alloys [50, 55]. During the course of annealing treatment, the ordered phases B2 and DO_3 form and affect the magnetic properties.

It is reported that the as-cast 6.5 wt% Si steel ribbon shows low permeability ($\mu_m \sim 1.13\times10^{-3}$) and high coercivity ($H_c$ ~150A/m) owing to large internal stress and grain refinement caused by rapid solidification (Fig. 9) [56]. With progress of annealing, the maximum permeability continuously increases with annealing temperature and gets saturate after 800ºC. Conversely, the coercivity decreases linearly with the annealing temperature and gets the lowest value after annealing at 850ºC. Therefore, the soft magnetic properties are quite improved by controlling the distribution and size of ordered phases [58]. In this regard, the post annealing treatment has an influential effect on the morphologies of ordered phases [59].

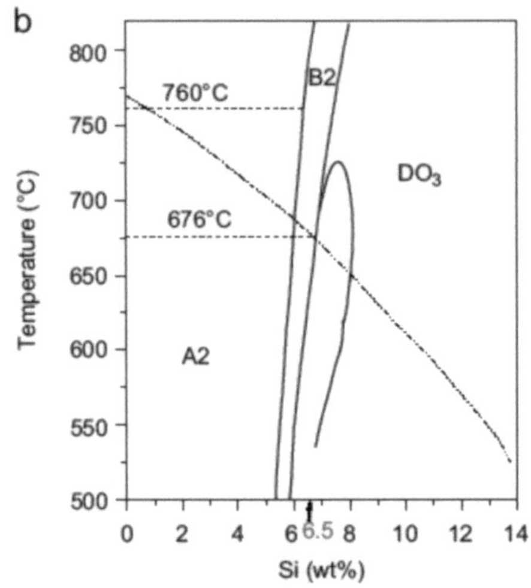

Figure 8. Fe-Si Phase diagram [54]

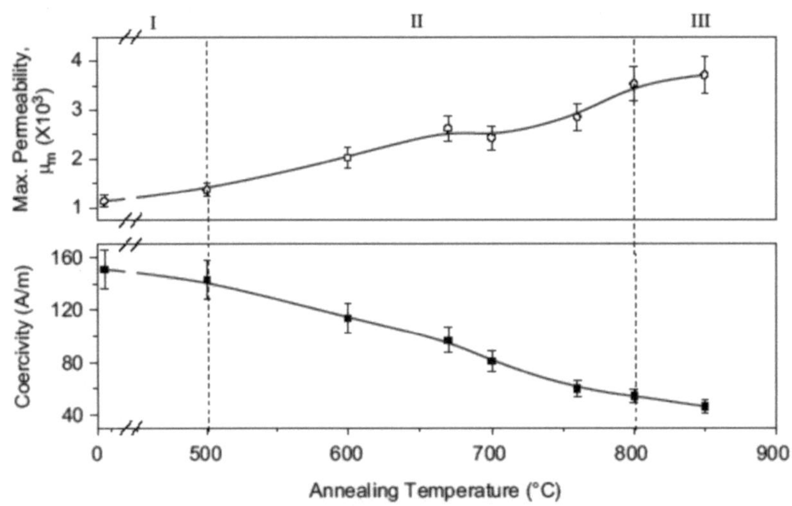

Figure 9. Variation of maximum permeability and coercivity as a function of annealing temperature [57].

5.4. Alloys with giant magnetoimpedance properties

An interesting phenomenon called giant magneto-impedance (GMI) where large magnetic field dependence of electrical impedance (Z) is observed in ferromagnetic wire prepared by in-rotating water quenching technique (discussed in section-2.2). Unlike the well-known phenomena of GMR where large change in material resistance takes place upon application of magnetic field, complex impedance suffers drastic changes as a function of external magnetic field in GMI materials within the frequency range of 100 kHz to few MHz. GMI behaviour is also observed in ribbon-shaped materials, but the value in wire-shaped material exhibits much higher value. The influence of external magnetic field on the skin depth is the major cause for large change in magnetoimpedance of rapidly solidified wire. The skin depth of the materials is expressed as

$$\delta = \frac{c}{\sqrt{4\pi^2 f \sigma \mu}},$$

where μ is the material permeability, c is the speed of light, σ is the electrical conductivity and is frequency of the ac current used for measuring GMI voltage.

The circular permeability is a function of external magnetic field, temperature and also stress. Thus the impedance in rapidly solidified wire changes not only with the external magnetic field but also with temperature and stress. Hence, the GMI materials have potential application as field, temperature and stress sensors. Anisotropy in rapidly solidified materials is predominantly magnetoelastic in origin. Thus the value of saturation magnetostriction constant plays major roll in controlling the circular permeability and hence the skin depth. The materials with low magnetostriction constant exhibit higher response in impedance under the influence of applied magnetic field. Thus the materials should be designed in such a way that their saturation magnetostriction constant should be very low. Usually, Co-based amorphous alloys have negative magnetostriction (λ_s ~ -3×10^{-6}), whereas Fe-based alloys have positive magnetostriction value (λ_s ~ 25×10^{-6}). An alloy with the combination of Fe and Co will provide low saturation magnetostriction and expected to be the better GMI materials.

The GMI materials are characterized by GMI ratio (Z/Z) which is the normalized value of magnetoimpedance with respect to high magnetic field and expressed as:

$$\frac{\Delta Z}{Z}(\%) = \frac{Z(H)\text{-}Z(H_{MAX})}{Z(H_{MAX})} \times 100\%$$

where Z(H) is the impedance of the material with the applied magnetic field H and Z(H_{max}) is magnetic field saturating the impedance [60].

Moreover, the Fe-, Co- and FeCo-based alloys reveal different GMI properties depending on the core–shell domain structure of the wires [60]. The Fe-based microwires (e.g., Fe-Si-B) possess cylindrical domain with longitudinal magnetization and radially closure domains. Such domain patterns attribute to a large Barkhausen effect (LBE) in the wire [61]. The LBE causes a square loop in positively magnetostrictive FeSiB systems, leading to a perspective use in switching devices [62, 63]. On the other hand, the Co-based wires (e.g., Co-Si-B) have bamboo shaped circular domains in the wire outer surface, resulting in large Barkhausen effect (LBE) and axial pattern in inner core. Such differences in domain pattern in Fe-based and Co-based alloys arise due to their positive and negative magnetostriction, respectively. Garcia et al. has reported a large GMI ratio of 1200% at 14.2 MHz for the zero magnetostrictive amorphous microwire $Co_{68}Fe_{4.35}Si_{12.5}B_{15}$ [64]. However, other researchers have reported a lower (600%) GMI ratio for the same composition [65, 66]. Recently, the Cr addition has revealed the improvement of GMI effect as well as field sensitivity in amorphous wires with nominal composition of $(Co_{0.5}Fe_{0.5})_{78-x}Cr_xSi_8B_{14}$ (X = 0 to 12 at%), where saturation magnetostriction of the materials changes with the Cr [67]. Fig. 10 shows the change of GMI ratio with the saturation magnetostriction at driving current (I_{ac}) = 10mA and the driving frequency 400 kHz. The variation of maximum GMI ratio (GMI_{max}) with Cr for as-quenched and annealed wires is shown in Fig. 11.

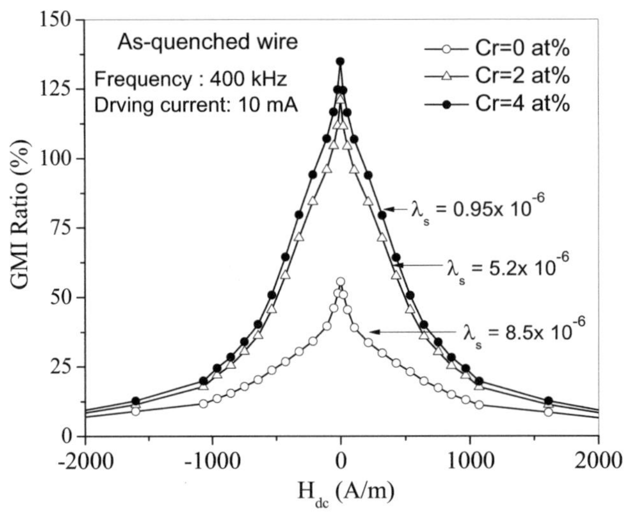

Figure 10. GMI plots of Cr-0, Cr-4, Cr-7 for as-quenched wires showing variation of GMI ratio with the saturation magnetostriction constant of the materials $(Co_{0.5}Fe_{0.5})_{78-x}Cr_xSi_8B_{14}$ (X = 0, 4, 7 at%)

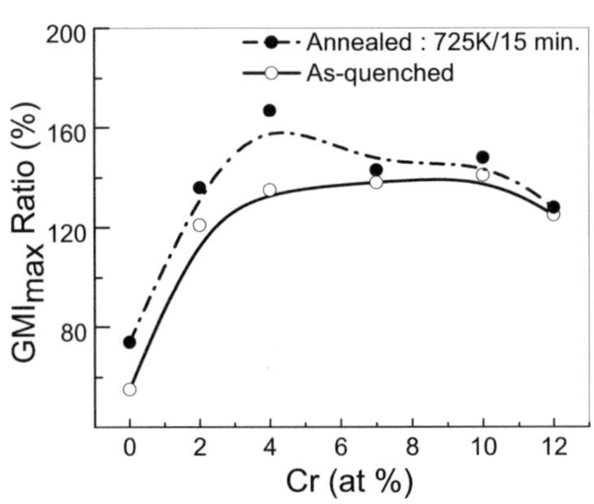

Figure 11. Effect of annealing on GMI_{max} with Cr content in $(Co_{0.5}Fe_{0.5})_{78\_x}Cr_xSi_8B_{14}$ (x = 0 to 12 at%) wires

6. Applications

The uniqueness of rapidly solidified amorphous and nanocrystalline materials is that the structural correlation (grain size) is much smaller than the ferromagnetic correlation length (domain wall with) [17]. It causes the lowering of magnetocrystalline anisotropy, and there is a tendency of vanishing magnetostriction for certain alloy compositions. As a consequence, nanocrystalline alloys achieve superior soft magnetic properties. Moreover, the improved other magnetic properties, such as higher saturation induction, Curie temperature, thermal stability of some FeCo-based amorphous and nanocrystalline alloys (HITPERM) envisage the reduction in size and weight of magnetic components, and therefore, it can be affective for the versatile applications of these alloys.

Accordingly, the nanocrystalline soft magnetic alloys may be applicable in the following field such as

i. Core material for power transformer and magnetic choke coil

ii. Inductive components for switched mode power supplies

iii. Magnetic sensors

The performance of magnetic choke coil made using FINEMET alloys is improved than that of Mn-Zn ferrite based choke coil [12–13]. The NANOPERM choke coil is also effective to prevent signal distortion in reactor elements of the phase modifying equipment [68]. Owing to high saturation induction, good thermal stability and low core loss, the NANOPERM alloys can also be applicable as a core material for high frequency power transformer [35]. Some FeCo-based amorphous alloys (HITPERM) have a demand as core materials and winding wires for audio and radio frequency transformers in space power system [7]. These materials may also be applicable as a rotor assembly in More-Electric Aircraft Integrated Power Unit (MEA-IPU).

Beside the applications in the field of transformer, rapidly solidified materials are widely used as core materials for sensors and transducers. A wide range of magnetic sensors, such as induction sensors, fluxgate sensors (FGS), Hall Effect sensors, giant magnetoresistive sensors and superconducting quantum interference device (SQUID) gradiometers are available. A magnetic sensor directly converts the magnetic field into some measurable parameters and the field sensitivity of the sensor plays a key role in determining its operating regime and potential applications. High permeability of amorphous and nanostructured materials are rapidly replacing their crystalline counterpart to achieve better performance of the sensing device.

The GMI effect has attracted considerable scientific and technological interest especially because of its applicability in magnetic field sensing and as an additional tool to investigate soft magnetic properties of other materials, for example, ferritic steel. Magnetic sensors based upon the GMI effect offer several advantages over conventional magnetic sensors. Amorphous metallic materials cast in the form of wire having diameter of the order of 100 μm are generally used as the core of this type of magnetic sensors. These materials have their superior mechanical, electrical and magnetic properties which originate from the absence of long range order. The formation of amorphous state depends on the alloy composition as well as on the processing condition. The microwires with large and positive magnetostriction, exhibit bistable behaviour with magnetization reversal through a giant Barkhausen jump originating in the propagation of a single-domain wall. On the other hand, microwires with very low or vanishing magnetostriction show excellent GMI effect. Both the types show natural ferromagnetic resonance (NFMR) at microwave frequencies.

The GMI materials can be used as the magnetic field and current sensors because GMI changes as a function of external dc magnetic field or applied dc/ac currents, respectively. The sensing elements of GMI sensors are in the form of amorphous wires [69], thin films [70] or ribbons [71]. The Aichi Steel Corporation of Japan has designed and developed a variety of GMI sensors using amorphous wires for a wide range of technological applications [72]. Since GMI effect is dependent on the applied stress, it is possible to design and develop stress and magnetoelastic sensors using Co-based amorphous ribbons and wires, respectively [73, 74]. Recently, GMI sensors have been projected for structural health monitoring of industrial components [75]. Additionally, many GMI sensors have been proposed for car traffic monitoring, antitheft system, electronic compass, non-destructive crack detection, detection of ferromagnetic dust inside human body, tyre pressure monitoring system, etc. [60].

GMI phenomena attracted a great attention for the sensor applications owing to the large sensitivity of the electrical impedance to the DC magnetic field of soft magnetic conductor [76]. The ultra-high sensitivity of GMI to external dc magnetic field (down to 10^{-4} A/m) can be used for magnetic field sensors and other sensors based on the change of a local magnetic field.

All these novel properties make amorphous magnetic microwires very attractive for exceptional technological applications, and also provide opportunities for fundamental micromagnetic studies mainly due to their unique magnetic domain structure. The sensitivity of GMI sensors is found to be high and have improved characteristics among the micromagnetic sensor families (Table 4) [77, 78].

Sensor element	**Hall**	**Giant magnetoresistive sensors**	**Flux gate sensor (FGS)**	**Giant magnetoimpedance (GMI) sensors**
Materials	In-Sb, Ga-As	Multilayer thin film	Fe-Ni	Amorphous wire/glass coated
Parameter	Hall voltage	Magnetoresistance	Magnetic flux	Magnetoimpedance
Field detection (A/m)	$0.1–2\times10^{5}$	10–20k	$10^{-4}–10^{-2}$	$10^{-4}–10^{-2}$
Frequency	0–5 kHz	0–10MHz	0 –10kHz	0.1–10MHz
Head size (mm)	0.1	0.01–0.1	3–20	0.1–3
Resolution (A/m)	10^{-2}	10^{-2}	10^{-4}	10^{-4}
Sensitive axis	Perpendicular	Parallel	Parallel	Parallel
Power consumption	1W	10mW	1W	10mW

Table 4. Micromagnetic sensor families and properties of sensor elements

http://dx.doi.org/0.5772/60772

Such type of magnetic field sensor can be extended for non-invasive way of monitoring the structural health of steel components which are in-service and intended to use for an extended period during which various damages can be developed such as residual stress, fatigue, creep or the formation of magnetic phase in non-magnetic steel in non-destructive way.

7. Conclusion

Rapidly solidified soft magnetic materials have been placed in special class of functional materials due to their excellent soft magnetic properties. They can be used as a transformer core or sensing element depending on composition. Due to low core loss and subsequently less heating effect, amorphous metal core transformers are slowly replacing the conventional Si steel–based distribution transformer. However, these materials are not suitable for power transformer as the saturation induction is lower than the crystalline counterpart due to the presence of non-magnetic metalloid. Extensive research is now going on to enhance the saturation induction by increasing the ferromagnetic components. The bulk production of rapidly solidified materials for transformer are mainly carried out by Hitachi Metals, Japan; Metglas, USA; Advanced Technology and Materials Co. Ltd (AT&M), China; POSCO, Korea which are the major players for large scale production of amorphous materials for transformer core where about 90 mm wide ribbons are required. Nanostructured materials are slowly replacing the core for small scale transformer like Switched Mode Power Supply (SMPS), circuit breaker and also for various types of sensor development. Nanostructured magnetostrictive wires are excellent element for pulse generating sensor application due to their large Barkhausen effect. Nearly zero magnetostrictive wires are excellent materials for GMI-based sensor element used for magnetic field determination.

Author details

Rajat K. Roy*, Ashis K. Panda and Amitava Mitra

*Address all correspondence to: rajat@nmlind.org, rajatroy.k@gmail.com

MST Division, CSIR-National Metallurgical Laboratory, Jamshedpur, India

References

[1] H. H. Liebermann, *Ribbon- substrate adhesion dynamics in chill block melt spinning processes*, Metall. Trans. B, 15 (1984) 155–161.

[2] Huang et al., *Rapidly solidified amorphous and crystalline alloys*, Ed. by B.H. Kear, B.C.Giessen, M. Cohen, Material Research Society, 1981, pp. 211–216

[3] B. Heyder, G.Frommeyer, *Direct casting of continuous fibres and wires by in-rotating-liquid spinning,* Mater. Sci. Eng., A, 133 (1991) 667–670.

[4] A. O. Olofinjana, J. H. Kern, H. A. Davies, *Effects of process variables on the multi-strand casting of high strength sub-millimeter metallic glass wire* J. Mater. Process Technol., 155–156 (2004) 1344–1349.

[5] P. Sarkar, R. K. Roy, A. K. Panda, A. Mitra, *Optimization of process parameters for developing FeCoSiB amorphous microwires through in-rotating-water quenching technique,* App. Phys. A: Mater. Sci. Process, 111 (2013) 575–580.

[6] L. Schultz, *Formation of amorphous metals by mechanical alloying,* Ed. by M. Von Allmen, Proc. MRS Europe Meet. on Amorphous Metals and Non-Equlibrium Processing, 1984, p. 135.

[7] M. E. McHenry, M. A. Willard, D. E. Laughlin, *Amorphous and nanocrystalline materials for applications as soft magnets,* Prog. Mater Sci., 44(1999) 291–433.

[8] G. Herzer, *Nanocrystalline soft magnetic alloys,* in Handbook of magnetic materials, Ed. by K. H. J. Buschow, vol. 10. Amsterdam: Elsevier Science BV; 1997.

[9] A. Mitra, A. K. Panda, S. R. Singh, V. Rao, P. Ramachandrarao, *Magnetic and structural behaviours of nanocrystalline Fe70.8Nb3.7Cu1Al2.7Mn0.7Si13.5B7.6 alloy.* Philos. Mag., 83 (2003) 1495–1509

[10] M. E. McHenry, F. Johnson, H. Okumura, T. Ohkubo, A. Hsiao, V. R. V. Ramanan, D. E. Laughlin (invited), *The kinetics of nanocrystallization and microstructural observations in FINEMET, NANOPERM and HITPERM nanocomposite magnetic materials,* Scripta Mater., 48(2003) 881.

[11] Y. Zhang, J. S. Blázquez, A. Conde, P. J. Warren, A. Cerezo, *Partitioning of Co during crystallisation of Fe–Co–Nb–B(–Cu) amorphous alloys,* Mater Sci. Eng., A, 353 (2003) 158.

[12] Y. Yoshizawa, S. Oguma, K. Yamauchi, *New Fe-based soft magnetic alloys composed of ultrafine grain structure,* J. Appl. Phys., 64(1988) 6044.

[13] Y. Yoshizawa, K. Yamauchi, T. Yamane, H. Sugihara, *Common mode choke cores using the new Fe-based alloys composed of ultrafine grain structure,* J. Appl. Phys., 64 (1988) 6047.

[14] Y. Yoshizawa, K. Yamauchi, J. Jpn. Inst. Met., 53(1989) 241.

[15] Y. Yoshizawa, K. Yamauchi, Mater Trans., JIM, 31(1990) 307.

[16] Y. Yoshizawa, K. Yamauchi, *Magnetic properties of Fe-Cu-M-Si-B (M = Cr, V, Mo, Nb, Ta, W) alloys,* Mater Sci. Eng., A,133(1991) 176.

[17] G. Herzer, *Modern soft magnets: amorphous and nanocrystalline materials,* Acta Mater., 61 (2013) 718–734.

[18] G. Herzer, IEEE Trans Magn., MAG-25 (1989) 3327.

[19] A. K. Panda, B. Ravikumar, S. Basu, A. Mitra, *Crystallization and soft magnetic properties of rapidly solidified Fe* $_{73.5}$ *Nb3Cu1Si* $_{22.5\text{-}X}$ *B* $_X$ *(X=5, 9, 10, 11.25, 19) alloys,* J. Magn. Magn. Mater., 260 (2003) 70–77

[20] M. A. Willard, M. Daniil, K. E. Kniping, *Nanocrystalline soft magnetic materials at high temperatures: A perspective,* Scripta Mater., 67 (2012) 554.

[21] Y. Zhang, K. Hono, A. Inoue, T. Sakurai, *Partitioning of Si in a Fe87Zr7Si4B2 nanocrystalline soft magnetic alloy,* Appl. Phys. Lett., 69(1996) 2128.

[22] S. H. Kim, W. K. Pi, T. H. Noh, H. J. Kim, I. K. Kang, *Effects of Al on the magnetic properties of nanocrystalline Fe73.5Cu1Nb3Si13.5B9 alloys,* J. Appl. Phys., 73(1993) 6591.

[23] X. Z.Zhou, A. H. Morrish, D. G. Naugle, R. Pan, *Mössbauer study of amorphous and nanocrystalline Fe73.5Cu1Nb3Si13.5B9,* J. Appl. Phys., 73(1993) 6597.

[24] J. Wang, Z. Wang, Y. Jia, R. Shi, Z. Wen, *High temperature soft magnetic properties of (FexCo1−x)73.5Cu1Mo3Si13.5B9 (x=0.5,1) alloys,* J. Magn. Magn. Mater., 328 (2013) 62–65

[25] T. Kulik, A. Hernando, M. Vazquez, *Correlation between structure and the magnetic properties of amorphous and nanocrystalline Fe73.5Cu1Nb3Si22.5−x Bx alloys,* J. Magn. Magn. Mater., 133(1994) 310.

[26] A. R. Yavari, D. Negri, Nanostruct. Mater., 8(1997) 969.

[27] L. K. Varga, A. Lovas, L. Pogany, L. F. Kiss, J. Balogh, T. Kemeny, *The role of nucleating element additives in the crystallization and soft magnetic properties of Fe-Zr-B based amorphous alloys,* Mater Sci. Eng., A226–228(1997) 740.

[28] B. Shen, A. Inoue, J. Mater. Res., 19(2004) 2549–2552.

[29] M. Zhang, A. Wang, W. Yang, B. Shen, *Effect of Fe to P concentration ratio on structures, crystallization behavior, and magnetic properties in (Fe0.79+xP0.1-xC0.04B0.04Si0.03)99Cu1 alloys,* J. Appl. Phys., 113, 17A337 (2013)

[30] M. Ohta, Y. Yoshizawa, Appl. Phys. Lett., *Magnetic properties of nanocrystalline Fe 82.65 Cu 1.35 Si x B 16−x alloys(x=0–7), 91, (2007) 062517.*

[31] A. Makino, M. Bingo, T. Bitoh, K. Yubuta, *Improvement of soft magnetic properties by simultaneous addition of P and Cu for nanocrystalline FeNbB alloys,* A. Inoue, J. Appl. Phys., 101, 09N117(2007)

[32] A. Makino, H. Men, K. Yubuta, T. Kubota, *Soft magnetic FeSiBPCu hetero-amorphous alloys with high Fe content,* J. Appl. Phys., 105, 013922(2009)

[33] T. Kubota, A. Makino, A. Inoue, *Low core loss of* $Fe_{85}Si_2B_8P_4Cu_1$ *nanocrystalline alloys with high* B_s *and* B_{800}, J. Alloys Compd., 509 (2011) S416-S419.

[34] K. Y. Kim, T. H. Noh, I. K. Kang, T. Kang, *Microstructural change upon annealing Fe-Zr-B alloys with different boron contents,* Mater Sci. Eng. A,179/180 (1994) 552.

[35] A. Makino, A. Inoue, T. Masumoto, Mater. Trans., JIM, 36(1995) 924.

[36] A. Inoue, Y. Miyauchi, A. Makino, T. Masumoto, Mater. Trans., JIM, 37(1996) 78.

[37] R. K. Roy, S. J. Kernion, S. Shen, M. E. McHenry, *Crystallization behavior and high temperature magnetic phase transitions of Nb-substituted FeCoSiBCu nanocomposites,* App. Phys. Lett., 99 (2011) 192506.

[38] R. K. Roy, S. Shen, S. J. Kernion, M. E. McHenry, *Effect of P addition on nanocrystallization and high temperature magnetic properties of low B and Nb containing FeCo nanocomposites,* J. App. Phys., 111, (2012) 07A301.

[39] F. Pfeifer, C. J. Radelo€, *Soft magnetic Ni-Fe and Co-Fe alloys - some physical and metallurgical aspects.,* J. Magn. Magn. Mater., 19(1980) 190.

[40] R. Boll, H. R. Hilzinger, H. Warlimont, The magnetic, chemical and structural properties of glassy metallic alloys, Ed. by R. Hasegawa, Boca Raton, FL: CRC Press; 1983.

[41] R. C. O'Handley, R. Hasegawa, R. Ray, C.P. Chou., *Ferromagnetic properties of some new metallic glasses,* Appl. Phys. Lett., 1976; 29(6): 330–332.

[42] J. C. Slater, *The Ferromagnetism of Nickel. II. Temperature Effects,* Phys. Rev. 49 (1936) 931.

[43] L. Pauling, *The Nature of the Interatomic Forces in Metals,* Phys. Rev. 54 (1938) 899.

[44] K. Narita, M. Enokizono, *Effect of nickel and manganese addition on ductility and magnetic properties of 6.5% silicon-iron alloy,* IEEE Trans Magn., MAG-14 (1978) 258

[45] D. Bouchara, M. Fagot, J. Degauque, J. Bras, *Ordering influence on magnetic properties of rapidly quenched Fe-6.5 wt% Si,* J. Magn. Magn. Mater., 83 (1990) 377.

[46] B. Viala, J. Degauque, M. Barico, E. Ferrara, F. Fiorillo, *Study of the brittle behaviour of annealed Fe-6.5 wt%Si ribbons produced by planar flow casting,* Mater. Sci. Eng., A, 212 (1996) 62.

[47] H. Haiji, K. Okada, T. Hiratani, M. Abe, M. Ninomiya, *Magnetic properties and workability of 6.5% Si steel sheet,* J. Magn. Magn. Mater., 160 (1996) 109.

[48] P. R. Swann, L. Grånäs, B. Lehtinen, Met. Sci., 9 (1975) 90–96.

[49] K. Narita, Y. Enokizono, N. Teshima, Y. Mori, *Magnetic properties of 6.5% silicon-iron ribbon produced by splat-cooling technique,* J. Magn. Magn. Mater., 19 (1980) 143.

[50] Y. Takada, J. Appl. Phys., 64 (1988) 5367.

[51] Y. Sato, T. Sato, Y. Okazaki, *Production and properties of melt-spun Fe-6.5wt.%Si ribbons,* Mater. Sci. Eng., 99 (1988) 73.

[52] X. F. Bi, Y. Tanaka, K. Sato, *Effect of microstructure on the magnetic properties of 6.5% Si-Fe alloy,* J. Magn. Magn. Mater., 112 (1992) 189.

[53] K. Narita K, N. Teshima, Y. Yamashiro, Y.J. Shin, Y. Yoshida, *Magnetic properties of rapidly quenched silicon-iron ribbons,* J. Magn. Magn. Mater., 41 (1984) 86.

[54] M. Enokizono, N. Teshima, K. Narita, IEEE Trans Magn., MAG-18 (1982) 1007.

[55] O. Kubaschewski, Iron Binary Phase Diagram, Springer, Berlin, 1982, p.136.

[56] H. R. Hilzinger, *Recent advances in rapidly solidified soft magnetic materials,* J. of Magn. Magn. Mater., 83 (1990) 370.

[57] S. K. Das, C. F. Chang, L. Richard, *Heat treatment of rapidly quenched Fe-6.5 wt% Si ribbon,* US Patent No 4865657 (1989).

[58] R. K. Roy, A. K. Panda, M. Ghosh, A. Mitra, R. N. Ghosh, *Effect of annealing treatment on soft magnetic properties of Fe–6.5 wt% Si wide ribbons,* J. Magn. Magn. Mater., 321 (2009) 2865.

[59] F. E. Luborsky, J. L. Walter, *Stability of amorphous metallic alloys,* J. Appl. Phys. 47 (1976) 3648–3650.

[60] J. E. Wittig, G. Frommeyer, *Deformation and Fracture Behavior of Rapidly Solidified and Annealed Iron-Silicon Alloys,* Metall. Trans. A, 39A, (2008) 252.

[61] M. H. Phan, H. X. Peng, *Giant magnetoimpedance materials: Fundamentals and applications,* Prog. Mater. Sci., 53 (2008) 323-420.

[62] P. Ripka, Magnetic Sensors and Magnetometers, Artech House Publishers, Boston, London, 2001.

[63] H. Chiriac, T.A. Ovari, *Amorphous glass-covered magnetic wires: Preparation, properties, applications,* Prog. Mater Sci., 40 (1996) 333.

[64] M. Vazquez, A. Hernando, J. App. Phys., 299 (1996) 39.

[65] D. Garcia, V. Raposo, O. Montero, I. J. Inguez, *Influence of magnetostriction constant on magnetoimpedance–frequency dependence,* Sens. Actuators, A, 129 (2006) 227–230.

[66] M. Vazquez, M. Knobel, M. L. Sanchez, R. Valenzuela, *Giant magnetoimpedance effect in soft magnetic wires for sensor applications,* Sens. Actuators, A, 59 (1997) 20–29.

[67] M. Vazquez, *Giant magneto-impedance in soft magnetic "Wires",* J. Magn. Magn. Mater., 226-230 (2001) 693.

[68] P.Sarkar, A. Basu Mallick, R. K. Roy, A. K. Panda, *Structural and Giant Magneto-impedance properties of Cr-incorporated Co–Fe–Si–B amorphous microwires,* A. Mitra, J. Magn. Magn. Mater., 324 (2012) 1551–1556

[69] I. Navarro, A. Hernando, *Magnetic and structural properties of nanocrystalline Fe77B19Cu1Nb3 alloy,* J. Magn. Magn. Mater., 133(1994) 306.

[70] K. Mohri, T. Uchiyama, P. V. Panina, *Recent advances of micro magnetic sensors and sensing application,* Sens. Actuators, A, 59 (1997) 1–8.

[71] S. Yabukami, H. Mawatari, N. Horikoshi, Y. Murayama, T. Ozawa, K. Ishiyama, et al., *A design of highly sensitive GMI sensor,* J. Magn. Magn. Mater., 290–291(2005) 1318.

[72] K. Nesteruk, M. Kuzminski, H. K. Lachowicz, Sen. Trans Mag., 65 (2006) 515.

[73] Y. Honkura, *Development of amorphous wire type MI sensors for automobile use,* J Magn Magn Mater, 249 (2002) 375.

[74] M. Tejedor, B. Hernando, M. L. Sanchez, V.M. Prida, M. Vazquez, *Magneto-impedance effect in amorphous ribbons for stress sensor application,* Sens. Actuators, A, 81 (2001) 98.

[75] A. F. Cobeno, A. Zhukov, J. M. Blanco, V. Larin, J. Gonzalez, *Magnetoelastic sensor based on GMI of amorphous microwire,* Sens. Actuators, A, 91 (2001) 95.

[76] P. Sarkar, R.K. Roy, A.K. Panda, A. Mitra, *Development of a GMI Sensor for Evaluating-Microstructural Degradation in Ferromagnetic Materials*, Inter. Conf. proceeding of IEEE Xplore, (2012) 108–111.

[77] V. Panina and K. Mohri, *Magneto-impedance effect in amorphous wires,* Appl. Phys. Lett., 65, (1994), 1189.

[78] T. Meydan, Application of amorphous materials to sensors, J. Magn. Magn. Mater., 133 (1994) 525–32.

Section 2

Mechanism of Self Diffusion in Alloys

Chapter 3

Self-Diffusion in Alloys

Kazu-masa Yamada and Nobuaki Matsuhashi

Additional information is available at the end of the chapter

http://dx.doi.org/10.5772/60993

Abstract

It has been successfully provided that in Fe, Co, Ni, Cu, Zn, Al, Ga, Cr, and Mn, alloy has been done to obtain reliable values of diffusion coefficient particularly with Arrhenius relationship graphic plotter tool. In the presented work, the Arrhenius plots of self-diffusions and other diffusion mechanisms have been exemplified. It is an aim to summarize diffusion coefficients in Arrhenius relations that are important for physical constant values in specified materials via free-of-charge Web-based diffusion coefficient diffusion database.

Keywords: Diffusion coefficient, Arrhenius relation, Co, Ni, Cu, Zn, Al, Ga, Cr, Mn, Metal and alloy

1. Introduction

There has been considerable important work to investigate that seems to be a reliable value of diffusion coefficient and temperature dependences of diffusivity in all around alloy and composite because it would be an essential physical constant value in specified materials and vitally useful for material development [1,2]. Particularly coefficients for self-diffusion are the most essential and have shown to be a good base element for thermal property in bulk-forming alloy. But it is difficult to measure the self-diffusivity in materials and alloys basically because the measurement is impossible other than using radioisotope tracer. In the present work, the use of a drawing tool with Arrhenius relation plots and data analysis function has been applied to determine the relations of thermal property regarding numerical activation energies and pre-exponential factors (frequency factors) and to evaluate whether it represents several Arrhenius relation platforms focusing on the developing materials [3]. Additionally, Web-based diffusion coefficient database presented the NIMS, National Institute for Materials Science, Japan, on October 10, 2014, including 8,925 diffusion data and 4,242 references which needed to be registered. They said that the diffusion database aims to cover all the basic diffusion data that mainly targeted metallic and inorganic materials and substantially contains information of pure metals, alloys, semiconductors, ceramics, and intermetallics [4].

The main objective of this research is to provide a diffusion data in alloys as well as a usage of Web-based diffusion database platform from all over the world to present diffusion research results and development activities in materials science. Additionally, to clarify a self-diffusion among alloys to develop for explorer thermal property using the process of plotting diffusion coefficient and temperature dependence, Arrhenius relations in alloy and composite all around the world focusing on the activation energy and pre-exponential factor discussion by using Web-based diffusion coefficient database-presented NIMS have been shown clearly in specially using freeware GP.exe plotting tool [3]. This discussion focusing on activation energy for diffusion coefficient in a relationally atomic diffusivity was able to investigate perspectives regarding discussed numerical values. Moreover, in activation energies for diffusion coefficient within all alloys, a quantity alloy development in materials has been discussed with the use of total relationship plots in Arrhenius relations that depend on diffusion temperature.

2. Procedure

Suitable for Arrhenius relation plots and data analysis, even a spreadsheet software and database relationally atomic diffusivity including the MIMS are good procedures among references of treatise for Arrhenius relation plot data and diffusion coefficients. Consequently, in the failure of searching the database, the term of an activation energy narrowing can prevent the error and be able to avoid limitation of MIMS database owing to be less than 100 results. Using freeware GP.exe plotting tool is a respectable way to discuss the activation energy and pre-exponential factor of Arrhenius relation diffusivity in alloys.

In Figure 1, the schematic diffusion coefficient tendency of 84 data alloys is related with the diffusion Web database list of MIMS, especially in Fe alloy system and with diffusant of Fe through handmade relational data-based processing by using the so-called presented work AWK-GP-PDF drawing system with GP.exe [5, 6, 7] where PDF means the Portable Document Format which the Adobe Systems Incorporated (ADBE) developed. It was found that using the AWK-GP drawing system made clear the relations between the T-inverse and T-linear value. Additionally, the D shows the extrapolated D_0 strongly related among the Q and T; diffusion mechanism and thermodynamics easily show the nearly neighbored equilibrium alloy state even if it does not understand the diffusivity in objective-based alloy. The certain overall atoms in an around alloy have a rule in the tendency of this AWK-GP-PDF drawing Arrhenius plot rather than in without the extrapolated D_0 relation. Subsequently symbol meanings are given below:

D: diffusion coefficient (m^2/s)

D_0: diffusion constant (pre-exponential factor, frequency factor) (m^2/s)

Q: activation energy (kJ/mol), (1 eV=96.5 kJ/mol)

http://dx.doi.org/10.5772/60993

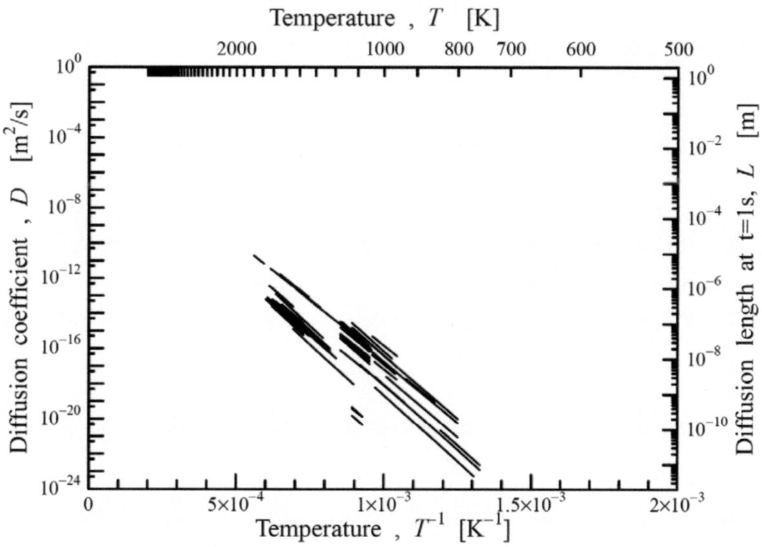

Figure 1. Schematic illustrations of Arrhenius plots for picked-up 84 data alloys between the activation energy of diffusion coefficient from 251 to 300 kJ/mol described with lower horizontal axes of temperature inversed, upper horizontal of linear temperature, left perpendicular axis of logarithm diffusion coefficient, and right perpendicular of logarithm diffusion length at time t=1 s, respectively

R: gas constant=8.31446 (J/mol K)

T: absolute temperature (K)

t : diffusion time (s)

And regarding Figure 1 diffusion data, in the minimum and maximum range of *T* during the diffusion process, the temperature dependence of diffusivity *D* available among references of treatise is shown below:

$$D = D_0 \exp\left(-\frac{Q}{RT}\right) \tag{1}$$

And in diffusion length [2], *L* means in general as

$$L = 2\sqrt{Dt} = 2\sqrt{D}, \; at \; t = 1s. \tag{2}$$

In alloy development, the characteristics of the objective alloy from analysis of neighboring information of nearly alloy systems and diffusant can be predicted. Because it is difficult to obtain new experimental diffusivity, the superior study by analogy with well-known data can be modified.

It may be concluded that the AWK-GP-PDF system with NIMS diffusion database presented one of the superior level prediction processes in the world using the nearest-neighbor diffusion characteristics for user objective developing alloys.

2.1. Process with AWK: An interpreted programming language

AWK [8] which was created at Bell Labs in the 1970s is an interpreted programming language design of ASCII, abbreviated from American Standard Code for Information Interchange, for data processing and typically used as a data extraction and reporting tool. It is now presented in Unix-like operating systems, although its platform has that of Windows OS, Mac OS, and Linux OS unfluctuating on Android OS.

Result: 67

F1: D=D0*exp(-Q/RT)

Material	Diffusant	D_0 [m^2/s]	Q [kJ/mol]	Tmin [K]	Tmax [K]
Fe-Ni	Fe;Ni	2.00E-05	264	1123	1699
Fe-Ni	Fe;Ni	1.50E-05	263	978	1699
Fe-Ni	Fe;Ni	1.74E-05	272	1383	1699
Fe-Ni	Fe;Ni	2.60E-05	262	978	1699
Fe-Ni	Fe;Ni	3.00E-05	259	978	1699
Fe-Ni	Fe;Ni	3.80E-05	257	978	1699
Fe-Ni	Fe;Ni	4.10E-05	255	978	1699
Fe-Ni	Fe;Ni	5.60E-05	255	978	1699
Fe-Ni	Fe;Ni	7.10E-05	256	978	1699
Fe-Ni	Fe;Ni	6.30E-05	255	978	1699
Fe-Co	Fe;Co	1.50E-07	219	1273	1673
Fe-Co	Fe;Co	7.85E-08	198	1273	1673
Fe-Co	Fe;Co	2.90E-07	215	1273	1673
Fe-Co	Fe;Co	2.07E-07	205	1273	1673
Fe-Co	Fe;Co	4.40E-07	212	1273	1673
Fe-Co	Fe;Co	5.80E-07	216	1273	1673
Fe-Co	Fe;Co	7.00E-07	215	1273	1673
Fe-Co	Fe;Co	8.80E-07	217	1273	1673
Fe-Co	Fe;Co	1.15E-06	218	1273	1673
Fe-Co	Fe;Co	1.20E-06	218	1273	1673
Fe-Co	Fe;Co	1.31E-06	219	1273	1673

Figure 2. Schematic search tendency of 67 data alloy jointed diffusant list; relational database for alloy diffusivity using method limiter by activation energy values, e.g. material, Fe-based alloy; diffusant, Fe. Reference from NIMS database, using clipboard pasted and related with spreadsheet software, e.g., MS Excel would be highly user friendly

In Figure 2, the schematic search tendency of color-coded 67 data (only the top 21 are illustrated in Figure 2) jointed diffusant list (column 3), pre-exponential factor D_0 (column 4), activation energy Q (column 5), and minimum and maximum temperature for Arrhenius relation's linear function span (columns 6 and 7), respectively. It should be rearranged in formula F1 (in Figure

http://dx.doi.org/10.5772/60993

2) as $D=D_0$ exp $(-Q/RT)$; then in MS Excel formula, "=[cell#3]*EXP(-1*[cell#4]*1000/8.31429/[cell#5]) " and "=[cell#3]*EXP(-1*[cell#4]*1000/8.31429/[cell#6])," D_{min} and D_{max}, would be adapted, respectively.

As shown later summarized afterward AWK script make into the 3 lines of reformation CSV (comma-separated values) or space-separated value (3 lines cycled) formation for optimize into the GP.exe data format, as shown in Table 1.

The AWK, process in Figure 3, a sample AWK script for calculation and reforming suitable for GP.exe data format as filename data01.TXT is shown in Table 1. Now for adequate usage to be a reasonable AWK script, it should be named with filename ex2gp.awk and then a command line that is executable in circumstances and command as gawk –f ex2gp.awk exceldata.txt > data01.TXT should be used. For example, it is the Windows OS GNU that is a Unix-like computer operating system developed by the GNU Project tool of gawk.exe for interpreting awk script as a multi-byte version of GNU awk 3.1.5 modified for Windows OS including interactive pipe and Internet correspondence with supporting character code Shift_JIS, EUC-JP, and UTF-8. On the other hand, in Mac OS and Linux, replacement of the only gawk name should be able to bring effect on the above command line script.

Regarding the before-mentioned "exceldata.txt," in Figure 4, a typical numerical example for copied-and-pasted text file for Arrhenius relations plots datasheet is shown. In Figure 4, [tab] means a Tab key (abbreviation of tabulator key or tabular key) on a computer keyboard. Meanwhile, on the computer screen, [tab] would be usually invisible. It is only necessary for the display of the Arrhenius relation plots of awk fields 1, 2, 3, and 4 as to be $1, $2, $3, and $4. But additionally, it would be useful for the other field of so-called code in awk $0 that means fully one line information from the start to the end.

Additionally in Figure 5, a sample batch file script for the AWK script exaction is shown. For adequate usage to be a reasonable script, the filename should be ex2gp.bat in Windows OS. After the main processing in ex2gp.bat, e.g., in the second half, a text editor Terapad.exe should be used for recognition. Other free text editors should be replaced, for example, the Emacs, etc.

2.2. Process with GP.exe

As AWK exploited technicalities to process the data, data01.TXT shown in Table 1 has been created. Then the next would be plotting the Arrhenius relation graph as horizontal axis of temperature T inverse and vertical axis of logalism diffusion coefficient D via diffusion mechanism for discussion infinity T of D_0.

For plotting the Arrhenius relationship, the freeware in Tohoku University, by Prof. K. Edamatsu, GP.exe that was designed until 1999 to make smart graphs for publication with powerful data analysis ability such as numerical complex differentiation and comparison was used. And now it is shown that the GP.exe has been useful for genuine data processing even in the year 2015. Fortunately, GP.exe is now supported with DOS, Disk Operating System, emulator and being executed GP.exe on it. Presented tutorials show a freeware DOSBox that is DOS emulator enabled on platform of Windows OS, Mac OS, and Linux OS including Android OS. After it has been difficult in general to calculate and plot numerical T inverse and

D logalism between any kinds of diffusion data and temperature, the freeware GP.exe tutorial to short-course calculation and plotting method will be provided in this session.

x-axis data y-axis data	
A main title of presented Graph	
A title of x-axis	
A title of y-axis	
#Comment No.01	
956	**3.63977E-17**
1041	**4.94613E-16**
#Comment No.02	
956	**8.45635E-18**
1041	**1.59639E-16**
#Comment No.03	
800	**8.75231E-22**
992	**2.53967E-18**
#Comment No.04	
1386	**2.67551E-15**
1528	**2.07494E-14**
.	.
.	.
.	.
.	.

Table 1. Typical numerical example for Arrhenius relation plots and lines as the special suitable format for GP.exe as data filename data01.TXT. It is necessary for instructions to include the filename within length of 8 and 3, because of the software of legacy-type DOS. The header of 3 lines are the main title, x-axis title, and y-axis title, respectively. In addition, more than 1 blank line makes an effect of snapping regarding the continuous line of GP.exe drafting

In Figure 6, a schematic illustration of DOSBox of DOS emulator and executed GP.exe as platform on its DOSBox is shown. The left and right windows are the prompt and main frame of DOSBox emulator, respectively. GP.exe users have to add DOSBox configuration descriptions as in Figure 7 for GP.exe executable circumstances via DOSBox application menu for configurations. Additionally, GP.exe have to read firstly the initial file of INIT.GPR file as Figure 8 for easy reading the data file data01.TXT and further adding useful extra properties.

Furthermore, Figure 7 has shown the menu of "DOSBox 0.74 Options," a sample configuration script for [autoexec] area; it should be necessary to add the MOUNT and Change-Directry and then execute the GP.exe. If the user needs to use the Japanese keyboard, then the line "keyb jp" should be added and also its module. In case of English keyboard, it is not needed. In the case of GP.exe, the current directory might be C:/prog/gp/GP.exe.

2.3. Plot confirmation and characterization with GP.exe

If the cases that the [autoexec] area execution might be started, or in DOSBox command line "gp" followed "enter" key in to the graph plot tool GP.exe start, it would be started GP.exe opening. In Figure 8, the standard INIT.GPR file for GP.exe was shown, and one point modified description included as colored red and underlined "*.TXT". For example, if the user needs to use a "data01.TXT" in the presented case, the user firstly should change from "*.xy" to "*.TXT" in the [Path and Directories] DataPath of INIT.GPR that is a good way to easy mounting data such as "data01.TXT".

Meanwhile, in Figure 9, Arrhenius relationship plot profile file is shown in detail, and descriptions of Figure 9 are explained below.

For example, on the other "data01.TXT" as shown in Table 1, 4 kinds of linear Arrhenius relations are conformed; the user can display computer graphics on graph plot tool GP.exe, finally resulting as in Figure 10 through high-resolution PostScript and PDF format.

On graph plot tool GP.exe, first of all, it is best that the user of the Arrhenius plot use not "INIT.GPR" but "ARRHEN.GPR" in the beginning as shown in Figure 9. In this figure, the use of tool extraction of freeware df.exe and schematic illustration of differences between "ARRHEN.GPR" and "INIT.GPR" for executable parameters on graph plot tool GP.exe were shown. The "INIT.GPR" is completely similar as in the list in Figure 8. On the other hand, "ARRHEN.GPR" has a file of "data01.txt" that have 4 groups of data as shown in Table 1 and 4 groups of linear line in Arrhenius relationship plotting on temperature T inverse and legalism D value as shown in Figure 10.

Regarding GP.exe plot confirmation and characterization in Figure 11, GP.exe schematic illustrations for searching the plots and their points, which plots for 4 groups of linear line in Arrhenius relationship plotting on temperature inverse and legalism D value, are represented. That is, there are 8 edges of the right and left on the 4 linear lines. The graph plot tool GP.exe has the superior function that can show the accurate value of data as shown in Figure 11 of green-colored cross-grid. Data points from relational database for alloy diffusivity using clipboard pasted and related with spreadsheet software were concluded, and then data were delivered on GP.exe by suitable optimized processing using AWK into the GP.exe format.

2.4. Process with GP.exe into postscript file

In Figure 12, schematic illustrations on the graph plot tool GP.exe, for creating the high-resolution PostScript picture as shown in Figure 11, which file of 01.ps for common forms of Arrhenius plots using GP.exe. If the user wants to reproduce the similar frame of Arrhenius plots but with another diffusion data, the data should be replaced with (filename from the data01.TXT to another filename, e.g., data02.TXT) the *.GPR graph parameter file. Meanwhile, the user can transform precisely from 01.ps to 01gw.pdf (PDF: Portable Document Format) using the freeware command line tool Ghostscript.

2.5. Process with PDF graphic file

Using the freeware command line tool Ghostscript, the user can transform PS to PDF. Then the user can use another freeware, Adobe Reader or Adobe Acrobat Reader. In Figure 13, Adobe reader schematic illustrations for creating the high-resolution GIF (Graphics Interchange Format) picture as all pictures shown are presented. When using those of freeware PDF reader, the user opens the PDF of 01gw.pdf, sets the magnitude to "400 %," activates "Take a Snapshot," chooses the "Select All," and finally chooses the "Copy." All through the process, the user could copy the graphic data onto the Windows OS, Mac OS, and Linux OS clipboard.

2.6. Process with GIF, JPEG, PNG, etc., graphic file

In Figure 14, it will be a pair of image processing software schematic illustrations for creating the high-resolution GIF (Graphics Interchange Format) picture as all pictures shown are presented in this paper. For example, using the freeware "IrfanView," the user opens the menu "Save Picture As" of clipboard picture data and pastes it by "Paste Ctrl+V," and the user can copy and paste through the graphic Windows OS clipboard examples. Finally, almost 94 kByte of compact-size and high-resolution GIF file was created via the software "IrfanView." This high-resolution GIF file of around 94 kByte would be user friendly for making documentation with graphic pictures. Also the other standard graphic file formats of JPEG, PNG, BMP, TIF, etc., are able to apply in a similar procedure the high-performance software "iView."

```
# ex2gp.awk
BEGIN{
# adding 3 lines for GP.exe standard graph configuration format as *.GPR
printf("A main title of presented Graph\n");
printf("A title of x-axis \n");
printf("A title of y-axis \n");
}

{
i++;  # Data numbering
printf("#%s\n",$0); # Marking data and comment one by one for
relationship between excel and GP text data.
printf("%s  %s\n",$1,$3);
printf("%s  %s\n\n\n",$2,$4); # It was acquired two empty lines following
diffusion data for easiness in seeing.
#$1 is Tmin [K]
#$3 is D(Tmin)  [m^2/s]
#$2 is Tmax [K]
#$4 is D(Tmax)  [m^2/s]
}

END{
printf("%d  are now re-arranged.\n",i)>"/dev/stderr";
}
```

Figure 3. A sample AWK script for calculation and reforming suitable for GP.exe data format as filename data01.TXT shown in Table 1. The AWK script should be a name as filename ex2gp.awk and then command line executable circumstances as gawk –f ex2gp.awk exceldata.txt > data01.TXT. For example, it is the Windows OS GNU that is a Unix-like computer operating system developed by the GNU Project tool of gawk.exe for interpreting awk script as multibyte version of GNU awk 3.1.5 modified for Windows OS including interactive pipe and Internet correspondence with supporting character codes Shift_JIS, EUC-JP, and UTF-8

http://dx.doi.org/10.5772/60993

```
956[tab]1041[tab]3.63977E-17[tab]4.94613E-16[tab][tab]Fe[tab][tab]Fe55;Fe59[tab]2.75E-
        03[tab]254[tab]956[tab]1041[tab][tab]3.64E-17[tab]4.95E-16
956[tab]1041[tab]8.45635E-18[tab]1.59639E-16[tab][tab]Fe[tab][tab]Fe55;Fe59[tab]3.58E-
        02[tab]286[tab]956[tab]1041[tab][tab]8.46E-18[tab]1.60E-16
800[tab]992[tab]8.75231E-22[tab]2.53967E-18[tab][tab]Fe[tab][tab]Fe59[tab]6.80E-
        04[tab]274[tab]800[tab]992[tab][tab]8.75E-22[tab]2.54E-18
1386[tab]1528[tab]2.67551E-15[tab]2.07494E-14[tab][tab]Fe[tab][tab]Fe59[tab]1.00E-
        05[tab]254[tab]1386[tab]1528[tab][tab]2.68E-15[tab]2.07E-14
   •          •
   •          •
```

Figure 4. Typical numerical example for copied-and-pasted text file as name exceldata.txt Arrhenius relations plots and lines MS Excel. In this figure, the [tab] means a Tab key (abbreviation of tabulator key or tabular key) on a keyboard. It is only necessary for the display of the Arrhenius relation plots of awk fields 1, 2, 3, and 4 as to be $1, $2, $3, and $4

```
gawk -f ex2gp.awk exceldata.txt > data01.TXT
pause
TeraPad\TeraPad.exe data01.TXT
exit
```

Figure 5. A sample Batch File script for the AWK script exaction. Filename should be ex2gp.bat. After the main processing, in the second half, it should be used with a text editor as Terapad.exe after recognition. Other free text editors should be replaced, for example, the Emacs

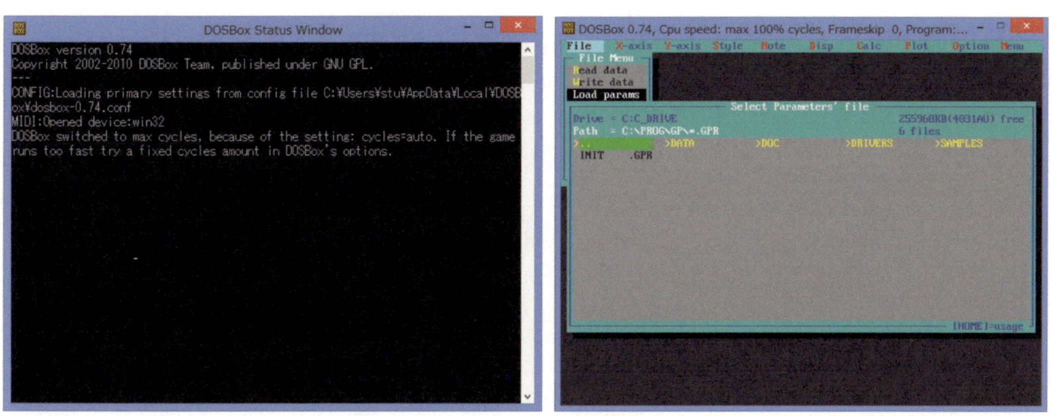

Figure 6. Schematic illustration of DOSBox of DOS emulator and executed GP.exe as platform on its DOSBox. The left and right windows are the prompt and main frame of DOSBox emulator, respectively. GP.exe users have to add configuration descriptions as in Figure 7 for GP.exe executable circumstances. Additionally, GP.exe has to read firstly the initial file of INIT.GPR file as in Figure 8 for easy reading the data file data01.TXT and furthermore adding useful extra properties

```
[autoexec]
# Lines in this section will be run at startup.
# You can put your MOUNT lines here.
@ECHO OFF
MOUNT c C:\
c:
CD \prog\gp
keyb jp
gp
```

Figure 7. The menu of "DOSBox 0.74 Options," a sample configuration script for [autoexec] area, should be necessary to add the MOUNT and Change-Directry and then execute the GP.exe. If the user needs to use the Japanese keyboard, then the line "keyb jp" should be added and also its module. In the case of GP.exe, the current directory might be C:/prog/gp/GP.exe

```
GraphP parameters file
version 4.31

[Remarks]
*** You can put remarks here
***


[Hardware Parameters]

PrinterType
 2
HdcopyType
 1
PlotterPort
 3
RSparams
 2 7 2 2 0 0
ParallelPort
 2
PlotSpeed
 0
PlotColors
 8
PlotterDriver
DRIVERSWPS.DLL

Beep
 0


[Text Colors]

TextColorSet
 1


[Graphic Colors]

GraphicColorSet
 1
WindowPaneColor
 2
AnalogColors ($GRB)
 $000 $FFF $1F1 $00F $F22
$FF2 $0BF $F0F $333 $999 $0A
$00A $A00 $AA0 $0AA $A0A


[File Selecter Switches]

DisplaySwitch
W

SortSwitch
N


[Special Menus]
```

```
XaxisTrans
 0


[Data Sizes]

MaxNData
 1000
MaxNFiles
 9
NScales
 5
NNotes
 10
NCaptions
 9
NArrows
 10


[Data Form]

TitleLineNo
 1
XlabelLineNo
 2
YlabelLineNo
 3
DataHeadLineNo
 4
ColumnStrings
X, Y, YE
X, Y, YE
X, Y, YE
X, Y, YE
X, Y, YE
X, Y, YE
X, Y, YE
X, Y, YE
X, Y, YE


[Path and Directories]

DataPath
*.TXT

ParamPath
*.GPR

PlotPath
*.PS

DriverPath
*.DLL

LogPath
*.LOG
```

```
[Data Files]

NumberOfFiles
 0
FileNames


[Overlay Files]

NumberOfFiles
 0
FileNames


[Plot File Parameters]

PlotterPort
 3
PlotFileName
01.PS


[Plot Sizes]

PlotForm
 2
FormSize
 0 210 0 297
PlotArea
 30 170 20 120
PlotMag
 100 100


[Axis Parameters]

AxisSw
 1 1 0 0 2 2
AxisLabel
X-AXIS
Y-AXIS



AxisFunction
 0 0 0 0 0 0
AxisPower
 1 1 1 1 1 1
AxisMin
 0 0 0 0 0 0
AxisMax
 10 10 10 10 10 10
AxisPosition
 0 0 0 0 0 0
AxisAutoScale
```

http://dx.doi.org/10.5772/60993

```
 1 1 1 1 0 0


[Axis Scales]

AxisTick
 1 0 0 0 0 1 0 0 0 0 1 0 0
0 1 0 0 0 0 0 0 0 0 0 0 0 0 0
0
AxisSubTick
 0 0 0 0 0 0 0 0 0 0 0 0 0
0 0 0 0 0 0 0 0 0 0 0 0 0 0 0
0
AxisNum
 2 0 0 0 0 2 0 0 0 0 2 0 0
0 2 0 0 0 0 0 0 0 0 0 0 0 0 0
0
AxisGrid
 0 0 0 0 0 0 0 0 0 0 0 0 0
0 0 0 0 0 0 0 0 0 0 0 0 0 0 0
0
AxisMin
 0 0 0 0 0 0 0 0 0 0 0 0 0
0 0 0 0 0 0 0 0 0 0 0 0 0 0 0
0
AxisMax
 10 0 0 0 0 10 0 0 0 0 10 0
0 0 10 0 0 0 0 0 0 0 0 0 0
0 0 0
AxisNumStyle
 0 0 0 0 0 0 0 0 0 0 0 0 0
0 0 0 0 0 0 0 0 0 0 0 0 0 0 0
0


[Aixs LogScales]

LogTick
 1 1 1 1 1 1
LogNum
 2 2 2 2 2 2
LogGrid
 0 0 0 0 0 0
AxisNumStyle
 0 0 0 0 0 0


[Axis Options]

FrameColor
 1 1 1 1 1 1
TickColor
 1 1 1 1 1 1
TickLength
 2 2 2 2 2 2
TickAngle
 90 -90 -90 90 90 -90
NumColor
 1 1 1 1 1 1
NumFont
 0 0 0 0 0 0
NumDistance
 0 0 0 0 0 0
NumAngle
 -90 90 90 -90 -90 90
NumWidth
 1 1 1 1 1 1
NumHeight
 4 4 4 4 4 4
NumSpace
 0 0 0 0 0 0
LabelColor
 1 1 1 1 1 1
LabelFont
 0 0 0 0 0 0
LabelDistance
 0 0 0 0 0 0
LabelAngle
 -90 90 90 -90 -90 90
LabelWidth
 1 1 1 1 1 1
LabelHeight
 5 5 5 5 5 5
LabelSpace
 0 0 0 0 0 0
LabelPosition
 50 50 50 50 100 100
GridColor
 1 1 1 1 1 1
GridLength
 100 140 100 140 0 0
GridAngle
 90 -90 -90 90 90 -90
GridStyle
 2 2 2 2 2 2
GridPeriod
 1 1 1 1 1 1


[Data Operators]

DataZero
 0 0 0 0 0 0 0 0 0 0 0 0 0 0
0 0 0 0
DataPower
 1 1 1 1 1 1 1 1 1 1 1 1 1 1
1 1 1 1
DataMag
 1 1 1 1 1 1 1 1 1 1 1 1 1 1
1 1 1 1
DataShift
 0 0 0 0 0 0 0 0 0 0 0 0 0 0
0 0 0 0
DataMin
 -3.4E+38 -3.4E+38 -3.4E+38
3.4E+38 -3.4E+38 -3.4E+38 -
3.4E+38 -3.4E+38 -3.4E+38 -
3.4E+38 -3.4E+38 -3.4E+38 -
3.4E+38 -3.4E+38 -3.4E+38 -
3.4E+38 -3.4E+38 -3.4E+38
DataMax
 3.4E+38 3.4E+38 3.4E+38
3.4E+38 3.4E+38 3.4E+38
3.4E+38 3.4E+38 3.4E+38
3.4E+38 3.4E+38 3.4E+38
3.4E+38 3.4E+38 3.4E+38
3.4E+38 3.4E+38 3.4E+38
DataSmooth
 0 0 0 0 0 0 0 0 0 0 0 0 0 0
0 0 0 0
DataAxis
 0 0 0 0 0 0 0 0 0 0 0 0 0 0
0 0 0 0
DataClip
 3 3 3 3 3 3 3 3 3 3 3 3 3 3
3 3 3 3


[File Operators]

FileOprator
 0 0 0 0 0 0 0 0 0
FileObject
 0 0 0 0 0 0 0 0 0


[MathFunction Operators]

CalcSwitch
 0 0 0 0 0 0 0 0 0
AutoRange
 1 1 1 1 1 1 1 1 1
ParamRange
 0 10 1 0 10 1 0 10 1 0 10 1
0 10 1 0 10 1 0 10 1 0 10 1 0
10 1
Xfunction
x
x
x
x
x
x
x
x
x

Yfunction
y
y
y
y
y
y
y
y
y


[Math Variables]


[Fitting Parameters]

FitFunctions
```

```
InitialParams
 0

Range
 -3.4E+38 3.4E+38 -3.4E+38
3.4E+38

[Differentiation Parameters

Orders
 0 0 0 0 0 0 0 0 0

[Integration Parameters]

FileNumber
 0
Range
 -3.4E+38 3.4E+38

[Sort Parameters]

Orders
 0 0 0 0 0 0 0 0 0

[Mark Styles]

MarkColor
 1 2 3 4 5 6 7 8 0
MarkType
 0 0 0 0 0 0 0 0 0
MarkSize
 2 2 2 2 2 2 2 2 2

[Error-bar Styles]

ErrorBarColor
 1 2 3 4 5 6 7 8 0
ErrorBarType
 1 1 1 1 1 1 1 1 1
ErrorBarSize
 1 1 1 1 1 1 1 1 1

[Bar Styles]

BarColor
 1 2 3 4 5 6 7 8 0
BarType
 0 0 0 0 0 0 0 0 0
BarSize
 0 0 0 0 0 0 0 0 0

[Line Styles]

LineColor
 1 2 3 4 5 6 7 8 0
LineType
 1 2 3 4 5 6 7 1 1
LineSize
 0 1 2 3 10 15 15 0 0
Spline
 0 0 0 0 0 0 0 0 0

[Arrow Styles]

ArrowColor
 1 1 1 1 1 1 1 1 1 1
ArrowType
 0 0 0 0 0 0 0 0 0 0
ArrowSize
 2 2 2 2 2 2 2 2 2 2
ArrowX1
 0 0 0 0 0 0 0 0 0 0
ArrowY1
 90 80 70 60 50 40 30 20 10
ArrowX2
 10 10 10 10 10 10 10 10 10
10
ArrowY2
 90 80 70 60 50 40 30 20 10
LineType
 1 1 1 1 1 1 1 1 1 1
LineSize
 0 0 0 0 0 0 0 0 0 0

[Note Styles]

NoteColor
 1 1 1 1 1 1 1 1 1 1
NoteFont
 0 0 0 0 0 0 0 0 0 0
NoteRefPoint
 0 0 0 0 0 0 0 0 0 0
NoteX
 10 10 10 10 10 10 10 10 10
10
NoteY
 90 80 70 60 50 40 30 20 10
NoteWidth
 1 1 1 1 1 1 1 1 1 1
NoteHeight
 4 4 4 4 4 4 4 4 4 4
NoteSpace
 0 0 0 0 0 0 0 0 0 0
NoteDirection
 0 0 0 0 0 0 0 0 0 0
NoteStrings

[Caption Styles]

CaptionColor
 0 0 0 0 0 0 0 0 0
CaptionFont
 0 0 0 0 0 0 0 0 0
CaptionRefPoint
 0 0 0 0 0 0 0 0 0
CaptionX
 0 0 0 0 0 0 0 0 0
CaptionY
 150 145 140 135 130 125 120
115 110
CaptionWidth
 1 1 1 1 1 1 1 1 1
CaptionHeight
 4 4 4 4 4 4 4 4 4
CaptionSpace
 0 0 0 0 0 0 0 0 0
CaptionDirection
 0 0 0 0 0 0 0 0 0
CaptionStrings

[End of file]
```

Figure 8. The standard INIT.GPR file for GP.exe. If you need to use a data01.TXT, you should change from *.xy to *.TXT in the [Path and Directories] DataPath of INIT.GPR

http://dx.doi.org/10.5772/60993

```
ARRHEN.GPR
00001  |GraphP parameters file version 4.31
00002  |
00003  |[Remarks]
00004  |*** You can put remarks here. ***
00005  |
00006  |
       |
       |
       |
       |
       |
       |
       |
       |
       |
       |
       |
       |
       |
       |
       |
       |
       |
       |
       |
       |
       |
       |
       |
       |
       |
       |
       |
       |
       |
       |
       |
       |
       |
       |
       |
       |
       |
       |
       |
       |
       |
       |
       |
       |
       |
       |
       |
       |
       |
00007  |[Data Sizes]
00008  |
00009  |MaxNData
00010 L| 16
00011  |MaxNFiles
00012  | 9
00013  |NScales
00014  | 5
00015  |NNotes
00016  | 10
00017  |NCaptions
00018  | 9
00019  |NArrows
00020  | 10
00021  |
00022  |
```

```
INIT.GPR
||00001  |GraphP parameters file version 4.31
||00002  |
||00003  |[Remarks]
||00004  |*** You can put remarks here. ***
||00005  |
||00006  |
||00007 R|[Hardware Parameters]
||00008 R|
||00009 R|PrinterType
||00010 R| 2
||00011 R|HdcopyType
||00012 R| 1
||00013 R|PlotterPort
||00014 R| 3
||00015 R|RSparams
||00016 R| 2 7 2 2 0 0
||00017 R|ParallelPort
||00018 R| 2
||00019 R|PlotSpeed
||00020 R| 0
||00021 R|PlotColors
||00022 R| 8
||00023 R|PlotterDriver
||00024 R|DRIVERS₩PS.DLL
||00025 R|
||00026 R|Beep
||00027 R| 0
||00028 R|
||00029 R|
||00030 R|[Text Colors]
||00031 R|
||00032 R|TextColorSet
||00033 R| 1
||00034 R|
||00035 R|
||00036 R|[Graphic Colors]
||00037 R|
||00038 R|GraphicColorSet
||00039 R| 1
||00040 R|WindowPaneColor
||00041 R| 2
||00042 R|AnalogColors ($GRB)
||00043 R| $000 $FFF $1F1 $00F $F22 $FF2 $0BF $F0F $333 $999
||00044 R|
||00045 R|
||00046 R|[File Selecter Switches]
||00047 R|
||00048 R|DisplaySwitch
||00049 R|W
||00050 R|
||00051 R|SortSwitch
||00052 R|N
||00053 R|
||00054 R|
||00055 R|
||00056 R|[Special Menus]
||00057 R|
||00058 R|XaxisTrans
||00059 R| 0
||00060 R|
||00061 R|
||00062  |[Data Sizes]
||00063  |
||00064  |MaxNData
||00065 R| 1000
||00066  |MaxNFiles
||00067  | 9
||00068  |NScales
||00069  | 5
||00070  |NNotes
||00071  | 10
||00072  |NCaptions
||00073  | 9
||00074  |NArrows
||00075  | 10
||00076  |
||00077  |
```

```
00023  |[Data Form]                                           ||00078  |[Data Form]
00024  |                                                      ||00079  |
00025  |TitleLineNo                                           ||00080  |TitleLineNo
00026  | 1                                                    ||00081  | 1
00027  |XlabelLineNo                                          ||00082  |XlabelLineNo
00028  | 2                                                    ||00083  | 2
00029  |YlabelLineNo                                          ||00084  |YlabelLineNo
00030  | 3                                                    ||00085  | 3
00031  |DataHeadLineNo                                        ||00086  |DataHeadLineNo
00032  | 4                                                    ||00087  | 4
00033  |ColumnStrings                                         ||00088  |ColumnStrings
00034  |X, Y, YE                                              ||00089  |X, Y, YE
00035  |X, Y, YE                                              ||00090  |X, Y, YE
00036  |X, Y, YE                                              ||00091  |X, Y, YE
00037  |X, Y, YE                                              ||00092  |X, Y, YE
00038  |X, Y, YE                                              ||00093  |X, Y, YE
00039  |X, Y, YE                                              ||00094  |X, Y, YE
00040  |X, Y, YE                                              ||00095  |X, Y, YE
00041  |X, Y, YE                                              ||00096  |X, Y, YE
00042  |X, Y, YE                                              ||00097  |X, Y, YE
00043  |                                                      ||00098  |
00044  |                                                      ||00099  |
00045  |                                                      ||00100  |
00046  |[Path and Directories]                                ||00101  |[Path and Directories]
00047  |                                                      ||00102  |
00048  |DataPath                                              ||00103  |DataPath
00049 L|*.TXT                                                 ||00104 R|*.txt
00050  |                                                      ||00105  |
00051  |ParamPath                                             ||00106  |ParamPath
00052  |*.GPR                                                 ||00107  |*.GPR
00053  |                                                      ||00108  |
00054  |PlotPath                                              ||00109  |PlotPath
00055  |*.PS                                                  ||00110  |*.PS
00056  |                                                      ||00111  |
00057  |DriverPath                                            ||00112  |DriverPath
00058  |*.DLL                                                 ||00113  |*.DLL
00059  |                                                      ||00114  |
00060  |LogPath                                               ||00115  |LogPath
00061  |*.LOG                                                 ||00116  |*.LOG
00062  |                                                      ||00117  |
00063  |                                                      ||00118  |
00064  |                                                      ||00119  |
00065  |[Data Files]                                          ||00120  |[Data Files]
00066  |                                                      ||00121  |
00067  |NumberOfFiles                                         ||00122  |NumberOfFiles
00068 L| 1                                                    ||00123 R| 0
00069  |FileNames                                             ||00124  |FileNames
00070 L|DATA01.TXT                                            ||       |
00071  |                                                      ||00125  |
00072  |                                                      ||00126  |
00073  |                                                      ||00127  |
00074  |[Overlay Files]                                       ||00128  |[Overlay Files]
00075  |                                                      ||00129  |
00076  |NumberOfFiles                                         ||00130  |NumberOfFiles
00077  | 0                                                    ||00131  | 0
00078  |FileNames                                             ||00132  |FileNames
00079  |                                                      ||00133  |
00080  |                                                      ||00134  |
00081  |                                                      ||00135  |
00082  |[Plot File Parameters]                                ||00136  |[Plot File Parameters]
00083  |                                                      ||00137  |
00084  |PlotterPort                                           ||00138  |PlotterPort
00085  | 3                                                    ||00139  | 3
00086  |PlotFileName                                          ||00140  |PlotFileName
00087  |01.PS                                                 ||00141  |01.PS
00088  |                                                      ||00142  |
00089  |                                                      ||00143  |
00090  |                                                      ||00144  |
00091  |[Plot Sizes]                                          ||00145  |[Plot Sizes]
00092  |                                                      ||00146  |
00093  |PlotForm                                              ||00147  |PlotForm
00094  | 2                                                    ||00148  | 2
00095  |FormSize                                              ||00149  |FormSize
00096  | 0 210 0 297                                          ||00150  | 0 210 0 297
00097  |PlotArea                                              ||00151  |PlotArea
00098  | 30 170 20 120                                        ||00152  | 30 170 20 120
00099  |PlotMag                                               ||00153  |PlotMag
00100  | 100 100                                              ||00154  | 100 100
00101  |                                                      ||00155  |
00102  |                                                      ||00156  |
00103  |[Axis Parameters]                                     ||00157  |[Axis Parameters]
00104  |                                                      ||00158  |
00105  |AxisSw                                                ||00159  |AxisSw
00106 L| 1 1 1 1 2 2                                          ||00160 R| 1 1 0 0 2 2
00107  |AxisLabel                                             ||00161  |AxisLabel
00108 L|Temperature , Wf[2]TWf[0]^{-1} [K^{-1}]               ||00162 R|X-AXIS
00109 L|Diffusion coefficient , Wf[2]DWf[0]  [m^2/s]          ||00163 R|Y-AXIS
```

```
00110 L|Temperature , Wf[2]TWf[0]  [K]                    ||00164 R|
00111 L|Diffusion length at t=1s, Wf[2]LWf[0]  [m]        ||00165 R|
00112  |                                                  ||00166  |
00113  |                                                  ||00167  |
00114  |                                                  ||00168  |
00115  |AxisFunction                                      ||00169  |AxisFunction
00116 L| 0 1 2 1 0 0                                     ||00170 R| 0 0 0 0 0 0
00117  |AxisPower                                         ||00171  |AxisPower
00118 L| -1 1 -1 1 1 1                                   ||00172 R| 1 1 1 1 1 1
00119  |AxisMin                                           ||00173  |AxisMin
00120 L| 0 1.0E-24 3.4E+38 2.0E-12 0 0                   ||00174 R| 0 0 0 0 0 0
00121  |AxisMax                                           ||00175  |AxisMax
00122 L| 0.002 1 500 2 10 10                              ||00176 R| 10 10 10 10 10 10
00123  |AxisPosition                                      ||00177  |AxisPosition
00124  | 0 0 0 0 0 0                                     ||00178  | 0 0 0 0 0 0
00125  |AxisAutoScale                                     ||00179  |AxisAutoScale
00126 L| 0 0 0 0 0 0                                     ||00180 R| 1 1 1 1 0 0
00127  |                                                  ||00181  |
00128  |                                                  ||00182  |
00129  |[Axis Scales]                                     ||00183  |[Axis Scales]
00130  |                                                  ||00184  |
00131  |AxisTick                                          ||00185  |AxisTick
00132 L| 0.0001 0 0 0 0 1 0 0 0 0 100 0 0 0 0 1 0 0 0 0 0 ||00186 R| 1 0 0 0 0 1 0 0 0 0 1 0 0 0 0 1 0 0 0 0 0 0 0 0 0
00133  |AxisSubTick                                       ||00187  |AxisSubTick
00134 L| 0 0 0 0 0 0 0 0 0 0 100 0 0 0 0 0 0 0 0 0 0 0 0||00188 R| 0 0 0 0 0 0 0 0 0 0 0 0 0 0 0 0 0 0 0 0 0 0 0 0 0
00135  |AxisNum                                           ||00189  |AxisNum
00136 L| 0.0005 0 0 0 0 2 0 0 0 0 0 100 1000 0 0 2 0 0 0 0||00190 R| 2 0 0 0 0 2 0 0 0 0 2 0 0 0 0 2 0 0 0 0 0 0 0 0 0
00137  |AxisGrid                                          ||00191  |AxisGrid
00138  | 0 0 0 0 0 0 0 0 0 0 0 0 0 0 0 0 0 0 0 0 0 0 0 0 0||00192  | 0 0 0 0 0 0 0 0 0 0 0 0 0 0 0 0 0 0 0 0 0 0 0 0 0
00139  |AxisMin                                           ||00193  |AxisMin
00140 L| -3.4E+38 0 0 0 0 -3.4E+38 0 0 0 0 100 100 1000 0 ||00194 R| 0 0 0 0 0 0 0 0 0 0 0 0 0 0 0 0 0 0 0 0 0 0 0 0 0
00141  |AxisMax                                           ||00195  |AxisMax
00142 L| 3.4E+38 0 0 0 0 3.4E+38 0 0 0 0 5000 800 2000 0 0||00196 R| 10 0 0 0 0 10 0 0 0 0 10 0 0 0 0 10 0 0 0 0 0 0 0 0
00143  |AxisNumStyle                                      ||00197  |AxisNumStyle
00144 L| 2 0 0 0 0 0 0 0 0 0 1 0 0 0 0 0 0 0 0 0 0 0 0 0 0||00198 R| 0 0 0 0 0 0 0 0 0 0 0 0 0 0 0 0 0 0 0 0 0 0 0 0 0
00145  |                                                  ||00199  |
00146  |                                                  ||00200  |
00147  |[Aixs LogScales]                                   ||00201  |[Aixs LogScales]
00148  |                                                  ||00202  |
00149  |LogTick                                           ||00203  |LogTick
00150 L| 1 5 1 1 1 1                                     ||00204 R| 1 1 1 1 1 1
00151  |LogNum                                            ||00205  |LogNum
00152 L| 2 10000 2 100 2 2                               ||00206 R| 2 2 2 2 2 2
00153  |LogGrid                                           ||00207  |LogGrid
00154  | 0 0 0 0 0 0                                     ||00208  | 0 0 0 0 0 0
00155  |AxisNumStyle                                      ||00209  |AxisNumStyle
00156 L| 0 2 0 2 0 0                                     ||00210 R| 0 0 0 0 0 0
00157  |                                                  ||00211  |
00158  |                                                  ||00212  |
00159  |[Axis Options]                                    ||00213  |[Axis Options]
00160  |                                                  ||00214  |
00161  |FrameColor                                        ||00215  |FrameColor
00162  | 1 1 1 1 1 1                                     ||00216  | 1 1 1 1 1 1
00163  |TickColor                                         ||00217  |TickColor
00164  | 1 1 1 1 1 1                                     ||00218  | 1 1 1 1 1 1
00165  |TickLength                                        ||00219  |TickLength
00166  | 2 2 2 2 2 2                                     ||00220  | 2 2 2 2 2 2
00167  |TickAngle                                         ||00221  |TickAngle
00168  | 90 -90 -90 90 90 -90                             ||00222  | 90 -90 -90 90 90 -90
00169  |NumColor                                          ||00223  |NumColor
00170  | 1 1 1 1 1 1                                     ||00224  | 1 1 1 1 1 1
00171  |NumFont                                           ||00225  |NumFont
00172  | 0 0 0 0 0 0                                     ||00226  | 0 0 0 0 0 0
00173  |NumDistance                                       ||00227  |NumDistance
00174  | 0 0 0 0 0 0                                     ||00228  | 0 0 0 0 0 0
00175  |NumAngle                                          ||00229  |NumAngle
00176  | -90 90 90 -90 -90 90                             ||00230  | -90 90 90 -90 -90 90
00177  |NumWidth                                          ||00231  |NumWidth
00178  | 1 1 1 1 1 1                                     ||00232  | 1 1 1 1 1 1
00179  |NumHeight                                         ||00233  |NumHeight
00180  | 4 4 4 4 4 4                                     ||00234  | 4 4 4 4 4 4
00181  |NumSpace                                          ||00235  |NumSpace
00182  | 0 0 0 0 0 0                                     ||00236  | 0 0 0 0 0 0
00183  |LabelColor                                        ||00237  |LabelColor
00184  | 1 1 1 1 1 1                                     ||00238  | 1 1 1 1 1 1
00185  |LabelFont                                         ||00239  |LabelFont
00186  | 0 0 0 0 0 0                                     ||00240  | 0 0 0 0 0 0
00187  |LabelDistance                                     ||00241  |LabelDistance
00188  | 0 0 0 0 0 0                                     ||00242  | 0 0 0 0 0 0
00189  |LabelAngle                                        ||00243  |LabelAngle
00190  | -90 90 90 -90 -90 90                             ||00244  | -90 90 90 -90 -90 90
00191  |LabelWidth                                        ||00245  |LabelWidth
00192  | 1 1 1 1 1 1                                     ||00246  | 1 1 1 1 1 1
00193  |LabelHeight                                       ||00247  |LabelHeight
00194  | 5 5 5 5 5 5                                     ||00248  | 5 5 5 5 5 5
00195  |LabelSpace                                        ||00249  |LabelSpace
00196  | 0 0 0 0 0 0                                     ||00250  | 0 0 0 0 0 0
```

```
00197  |LabelPosition                                        ||00251  |LabelPosition
00198  | 50 50 50 50 100 100                                 ||00252  | 50 50 50 50 100 100
00199  |GridColor                                            ||00253  |GridColor
00200  | 1 1 1 1 1 1                                         ||00254  | 1 1 1 1 1 1
00201  |GridLength                                           ||00255  |GridLength
00202  | 100 140 100 140 0 0                                 ||00256  | 100 140 100 140 0 0
00203  |GridAngle                                            ||00257  |GridAngle
00204  | 90 -90 -90 90 90 -90                                ||00258  | 90 -90 -90 90 90 -90
00205  |GridStyle                                            ||00259  |GridStyle
00206  | 2 2 2 2 2 2                                         ||00260  | 2 2 2 2 2 2
00207  |GridPeriod                                           ||00261  |GridPeriod
00208  | 1 1 1 1 1 1                                         ||00262  | 1 1 1 1 1 1
00209  |                                                     ||00263  |
00210  |                                                     ||00264  |
00211  |[Data Operators]                                     ||00265  |[Data Operators]
00212  |                                                     ||00266  |
00213  |DataZero                                             ||00267  |DataZero
00214  | 0 0 0 0 0 0 0 0 0 0 0 0 0 0 0 0 0 0                 ||00268  | 0 0 0 0 0 0 0 0 0 0 0 0 0 0 0 0 0 0
00215  |DataPower                                            ||00269  |DataPower
00216  | 1 1 1 1 1 1 1 1 1 1 1 1 1 1 1 1 1 1                 ||00270  | 1 1 1 1 1 1 1 1 1 1 1 1 1 1 1 1 1 1
00217  |DataMag                                              ||00271  |DataMag
00218  | 1 1 1 1 1 1 1 1 1 1 1 1 1 1 1 1 1 1                 ||00272  | 1 1 1 1 1 1 1 1 1 1 1 1 1 1 1 1 1 1
00219  |DataShift                                            ||00273  |DataShift
00220  | 0 0 0 0 0 0 0 0 0 0 0 0 0 0 0 0 0 0                 ||00274  | 0 0 0 0 0 0 0 0 0 0 0 0 0 0 0 0 0 0
00221  |DataMin                                              ||00275  |DataMin
00222  | -3.4E+38 -3.4E+38 -3.4E+38 -3.4E+38 -3.4E+38 -3.4||00276  | -3.4E+38 -3.4E+38 -3.4E+38 -3.4E+38 -3.4E+38 -3.4
00223  |DataMax                                              ||00277  |DataMax
00224  | 3.4E+38 3.4E+38 3.4E+38 3.4E+38 3.4E+38 3.4E+38 3||00278  | 3.4E+38 3.4E+38 3.4E+38 3.4E+38 3.4E+38 3.4E+38 3
00225  |DataSmooth                                           ||00279  |DataSmooth
00226  | 0 0 0 0 0 0 0 0 0 0 0 0 0 0 0 0 0 0                 ||00280  | 0 0 0 0 0 0 0 0 0 0 0 0 0 0 0 0 0 0
00227  |DataAxis                                             ||00281  |DataAxis
00228  | 0 0 0 0 0 0 0 0 0 0 0 0 0 0 0 0 0 0                 ||00282  | 0 0 0 0 0 0 0 0 0 0 0 0 0 0 0 0 0 0
00229  |DataClip                                             ||00283  |DataClip
00230  | 3 3 3 3 3 3 3 3 3 3 3 3 3 3 3 3 3 3                 ||00284  | 3 3 3 3 3 3 3 3 3 3 3 3 3 3 3 3 3 3
00231  |                                                     ||00285  |
00232  |                                                     ||00286  |
00233  |[File Operators]                                     ||00287  |[File Operators]
00234  |                                                     ||00288  |
00235  |FileOprator                                          ||00289  |FileOprator
00236  | 0 0 0 0 0 0 0 0 0                                   ||00290  | 0 0 0 0 0 0 0 0 0
00237  |FileObject                                           ||00291  |FileObject
00238  | 0 0 0 0 0 0 0 0 0                                   ||00292  | 0 0 0 0 0 0 0 0 0
00239  |                                                     ||00293  |
00240  |                                                     ||00294  |
00241  |[MathFunction Operators]                             ||00295  |[MathFunction Operators]
00242  |                                                     ||00296  |
00243  |CalcSwitch                                           ||00297  |CalcSwitch
00244 L| 1 0 0 0 0 0 0 0 0                                   ||00298 R| 0 0 0 0 0 0 0 0 0
00245  |AutoRange                                            ||00299  |AutoRange
00246  | 1 1 1 1 1 1 1 1 1                                   ||00300  | 1 1 1 1 1 1 1 1 1
00247  |ParamRange                                           ||00301  |ParamRange
00248 L| 0.002 0.0001 -0.0000095 0 10 1 0 10 1 0 10 1 0 10||00302 R| 0 10 1 0 10 1 0 10 1 0 10 1 0 10 1 0 10 1 0 10 1
00249  |Xfunction                                            ||00303  |Xfunction
00250 L|1/x                                                  ||00304 R|x
00251  |x                                                    ||00305  |x
00252  |x                                                    ||00306  |x
00253  |x                                                    ||00307  |x
00254  |x                                                    ||00308  |x
00255  |x                                                    ||00309  |x
00256  |x                                                    ||00310  |x
00257  |x                                                    ||00311  |x
00258  |x                                                    ||00312  |x
00259  |                                                     ||00313  |
00260  |Yfunction                                            ||00314  |Yfunction
00261  |y                                                    ||00315  |y
00262  |y                                                    ||00316  |y
00263  |y                                                    ||00317  |y
00264  |y                                                    ||00318  |y
00265  |y                                                    ||00319  |y
00266  |y                                                    ||00320  |y
00267  |y                                                    ||00321  |y
00268  |y                                                    ||00322  |y
00269  |y                                                    ||00323  |y
00270  |                                                     ||00324  |
00271  |                                                     ||00325  |
00272  |                                                     ||00326  |
00273  |[Math Variables]                                     ||00327  |[Math Variables]
00274 L|X =1163                                              ||       |
00275 L|X =1163                                              ||       |
00276 L|X =1163                                              ||       |
00277 L|X =1163                                              ||       |
00278 L|X =1163                                              ||       |
00279 L|X =1163                                              ||       |
00280 L|X =1163                                              ||       |
00281 L|X =1163                                              ||       |
00282 L|X =1163                                              ||       |
00283 L|X =1168                                              ||       |
```

```
00284 L|X =1528                                     ||       |
00285 L|Y =N                                        ||       |
00286  |                                            ||00328  |
00287  |                                            ||00329  |
00288  |                                            ||00330  |
00289  |[Fitting Parameters]                        ||00331  |[Fitting Parameters]
00290  |                                            ||00332  |
00291  |FitFunctions                                ||00333  |FitFunctions
00292  |                                            ||00334  |
00293  |                                            ||00335  |
00294  |                                            ||00336  |
00295  |                                            ||00337  |
00296  |                                            ||00338  |
00297  |                                            ||00339  |
00298  |                                            ||00340  |
00299  |                                            ||00341  |
00300  |                                            ||00342  |
00301  |                                            ||00343  |
00302  |InitialParams                               ||00344  |InitialParams
00303  | 0                                          ||00345  | 0
00304  |                                            ||00346  |
00305  |Range                                       ||00347  |Range
00306  | -3.4E+38 3.4E+38 -3.4E+38 3.4E+38          ||00348  | -3.4E+38 3.4E+38 -3.4E+38 3.4E+38
00307  |                                            ||00349  |
00308  |                                            ||00350  |
00309  |[Differentiation Parameters]                ||00351  |[Differentiation Parameters]
00310  |                                            ||00352  |
00311  |Orders                                      ||00353  |Orders
00312  | 0 0 0 0 0 0 0 0 0                          ||00354  | 0 0 0 0 0 0 0 0 0
00313  |                                            ||00355  |
00314  |                                            ||00356  |
00315  |[Integration Parameters]                    ||00357  |[Integration Parameters]
00316  |                                            ||00358  |
00317  |FileNumber                                  ||00359  |FileNumber
00318  | 0                                          ||00360  | 0
00319  |Range                                       ||00361  |Range
00320  | -3.4E+38 3.4E+38                           ||00362  | -3.4E+38 3.4E+38
00321  |                                            ||00363  |
00322  |                                            ||00364  |
00323  |[Sort Parameters]                           ||00365  |[Sort Parameters]
00324  |                                            ||00366  |
00325  |Orders                                      ||00367  |Orders
00326  | 0 0 0 0 0 0 0 0 0                          ||00368  | 0 0 0 0 0 0 0 0 0
00327  |                                            ||00369  |
00328  |                                            ||00370  |
00329  |[Mark Styles]                               ||00371  |[Mark Styles]
00330  |                                            ||00372  |
00331  |MarkColor                                   ||00373  |MarkColor
00332  | 1 2 3 4 5 6 7 8 0                          ||00374  | 1 2 3 4 5 6 7 8 0
00333  |MarkType                                    ||00375  |MarkType
00334  | 0 0 0 0 0 0 0 0 0                          ||00376  | 0 0 0 0 0 0 0 0 0
00335  |MarkSize                                    ||00377  |MarkSize
00336  | 2 2 2 2 2 2 2 2 2                          ||00378  | 2 2 2 2 2 2 2 2 2
00337  |                                            ||00379  |
00338  |                                            ||00380  |
00339  |[Error-bar Styles]                          ||00381  |[Error-bar Styles]
00340  |                                            ||00382  |
00341  |ErrorBarColor                               ||00383  |ErrorBarColor
00342  | 1 2 3 4 5 6 7 8 0                          ||00384  | 1 2 3 4 5 6 7 8 0
00343  |ErrorBarType                                ||00385  |ErrorBarType
00344  | 1 1 1 1 1 1 1 1 1                          ||00386  | 1 1 1 1 1 1 1 1 1
00345  |ErrorBarSize                                ||00387  |ErrorBarSize
00346  | 1 1 1 1 1 1 1 1 1                          ||00388  | 1 1 1 1 1 1 1 1 1
00347  |                                            ||00389  |
00348  |                                            ||00390  |
00349  |[Bar Styles]                                ||00391  |[Bar Styles]
00350  |                                            ||00392  |
00351  |BarColor                                    ||00393  |BarColor
00352  | 1 2 3 4 5 6 7 8 0                          ||00394  | 1 2 3 4 5 6 7 8 0
00353  |BarType                                     ||00395  |BarType
00354  | 0 0 0 0 0 0 0 0 0                          ||00396  | 0 0 0 0 0 0 0 0 0
00355  |BarSize                                     ||00397  |BarSize
00356  | 0 0 0 0 0 0 0 0 0                          ||00398  | 0 0 0 0 0 0 0 0 0
00357  |                                            ||00399  |
00358  |                                            ||00400  |
00359  |[Line Styles]                               ||00401  |[Line Styles]
00360  |                                            ||00402  |
00361  |LineColor                                   ||00403  |LineColor
00362  | 1 2 3 4 5 6 7 8 0                          ||00404  | 1 2 3 4 5 6 7 8 0
00363  |LineType                                    ||00405  |LineType
00364  | 1 2 3 4 5 6 7 1 1                          ||00406  | 1 2 3 4 5 6 7 1 1
00365  |LineSize                                    ||00407  |LineSize
00366  | 0 1 2 3 10 15 15 0 0                       ||00408  | 0 1 2 3 10 15 15 0 0
00367  |Spline                                      ||00409  |Spline
00368  | 0 0 0 0 0 0 0 0 0                          ||00410  | 0 0 0 0 0 0 0 0 0
00369  |                                            ||00411  |
00370  |                                            ||00412  |
```

```
00371 |[Arrow Styles]                                    ||00413 |[Arrow Styles]
00372 |                                                  ||00414 |
00373 |ArrowColor                                        ||00415 |ArrowColor
00374 | 1 1 1 1 1 1 1 1 1 1                              ||00416 | 1 1 1 1 1 1 1 1 1 1
00375 |ArrowType                                         ||00417 |ArrowType
00376 | 0 0 0 0 0 0 0 0 0 0                              ||00418 | 0 0 0 0 0 0 0 0 0 0
00377 |ArrowSize                                         ||00419 |ArrowSize
00378 | 2 2 2 2 2 2 2 2 2 2                              ||00420 | 2 2 2 2 2 2 2 2 2 2
00379 |ArrowX1                                           ||00421 |ArrowX1
00380 | 0 0 0 0 0 0 0 0 0 0                              ||00422 | 0 0 0 0 0 0 0 0 0 0
00381 |ArrowY1                                           ||00423 |ArrowY1
00382 | 90 80 70 60 50 40 30 20 10 0                     ||00424 | 90 80 70 60 50 40 30 20 10 0
00383 |ArrowX2                                           ||00425 |ArrowX2
00384 | 10 10 10 10 10 10 10 10 10 10                    ||00426 | 10 10 10 10 10 10 10 10 10 10
00385 |ArrowY2                                           ||00427 |ArrowY2
00386 | 90 80 70 60 50 40 30 20 10 0                     ||00428 | 90 80 70 60 50 40 30 20 10 0
00387 |LineType                                          ||00429 |LineType
00388 | 1 1 1 1 1 1 1 1 1 1                              ||00430 | 1 1 1 1 1 1 1 1 1 1
00389 |LineSize                                          ||00431 |LineSize
00390 | 0 0 0 0 0 0 0 0 0 0                              ||00432 | 0 0 0 0 0 0 0 0 0 0
00391 |                                                  ||00433 |
00392 |                                                  ||00434 |
00393 |[Note Styles]                                     ||00435 |[Note Styles]
00394 |                                                  ||00436 |
00395 |NoteColor                                         ||00437 |NoteColor
00396 | 1 1 1 1 1 1 1 1 1 1                              ||00438 | 1 1 1 1 1 1 1 1 1 1
00397 |NoteFont                                          ||00439 |NoteFont
00398 | 0 0 0 0 0 0 0 0 0 0                              ||00440 | 0 0 0 0 0 0 0 0 0 0
00399 |NoteRefPoint                                      ||00441 |NoteRefPoint
00400 | 0 0 0 0 0 0 0 0 0 0                              ||00442 | 0 0 0 0 0 0 0 0 0 0
00401 |NoteX                                             ||00443 |NoteX
00402 | 10 10 10 10 10 10 10 10 10 10                    ||00444 | 10 10 10 10 10 10 10 10 10 10
00403 |NoteY                                             ||00445 |NoteY
00404 | 90 80 70 60 50 40 30 20 10 0                     ||00446 | 90 80 70 60 50 40 30 20 10 0
00405 |NoteWidth                                         ||00447 |NoteWidth
00406 | 1 1 1 1 1 1 1 1 1 1                              ||00448 | 1 1 1 1 1 1 1 1 1 1
00407 |NoteHeight                                        ||00449 |NoteHeight
00408 | 4 4 4 4 4 4 4 4 4 4                              ||00450 | 4 4 4 4 4 4 4 4 4 4
00409 |NoteSpace                                         ||00451 |NoteSpace
00410 | 0 0 0 0 0 0 0 0 0 0                              ||00452 | 0 0 0 0 0 0 0 0 0 0
00411 |NoteDirection                                     ||00453 |NoteDirection
00412 | 0 0 0 0 0 0 0 0 0 0                              ||00454 | 0 0 0 0 0 0 0 0 0 0
00413 |NoteStrings                                       ||00455 |NoteStrings
00414 |                                                  ||00456 |
00415 |                                                  ||00457 |
00416 |                                                  ||00458 |
00417 |                                                  ||00459 |
00418 |                                                  ||00460 |
00419 |                                                  ||00461 |
00420 |                                                  ||00462 |
00421 |                                                  ||00463 |
00422 |                                                  ||00464 |
00423 |                                                  ||00465 |
00424 |                                                  ||00466 |
00425 |[Caption Styles]                                  ||00467 |[Caption Styles]
00426 |                                                  ||00468 |
00427 |CaptionColor                                      ||00469 |CaptionColor
00428 | 0 0 0 0 0 0 0 0 0                                ||00470 | 0 0 0 0 0 0 0 0 0
00429 |CaptionFont                                       ||00471 |CaptionFont
00430 | 0 0 0 0 0 0 0 0 0                                ||00472 | 0 0 0 0 0 0 0 0 0
00431 |CaptionRefPoint                                   ||00473 |CaptionRefPoint
00432 | 0 0 0 0 0 0 0 0 0                                ||00474 | 0 0 0 0 0 0 0 0 0
00433 |CaptionX                                          ||00475 |CaptionX
00434 | 0 0 0 0 0 0 0 0 0                                ||00476 | 0 0 0 0 0 0 0 0 0
00435 |CaptionY                                          ||00477 |CaptionY
00436 | 150 145 140 135 130 125 120 115 110              ||00478 | 150 145 140 135 130 125 120 115 110
00437 |CaptionWidth                                      ||00479 |CaptionWidth
00438 | 1 1 1 1 1 1 1 1 1                                ||00480 | 1 1 1 1 1 1 1 1 1
00439 |CaptionHeight                                     ||00481 |CaptionHeight
00440 | 4 4 4 4 4 4 4 4 4                                ||00482 | 4 4 4 4 4 4 4 4 4
00441 |CaptionSpace                                      ||00483 |CaptionSpace
00442 | 0 0 0 0 0 0 0 0 0                                ||00484 | 0 0 0 0 0 0 0 0 0
00443 |CaptionDirection                                  ||00485 |CaptionDirection
00444 | 0 0 0 0 0 0 0 0 0                                ||00486 | 0 0 0 0 0 0 0 0 0
00445 |CaptionStrings                                    ||00487 |CaptionStrings
00446 |                                                  ||00488 |
00447 |                                                  ||00489 |
00448 |                                                  ||00490 |
00449 |                                                  ||00491 |
00450 |                                                  ||00492 |
00451 |                                                  ||00493 |
00452 |                                                  ||00494 |
00453 |                                                  ||00495 |
00454 L|@1 DATA01.TXT   A main title of presented Graph   ||00496 R|
00455 |                                                  ||00497 |
00456 |[End of file]                                     ||00498 |[End of file]
```

Figure 9. In using df.exe tool extraction, schematic illustration of differences between Arrhen.GPR and INIT.GPR for executable parameters on graph plot tool GP.exe is shown. INIT.GPR is completely similar with the list in Figure 8. On the other hand, Arrhen.GPR has a data of data01.txt that has four groups of data as shown in Table 1 and four groups of linear line in Arrhenius relationship plotting on temperature inverse and legalism D value as shown in Figure 10

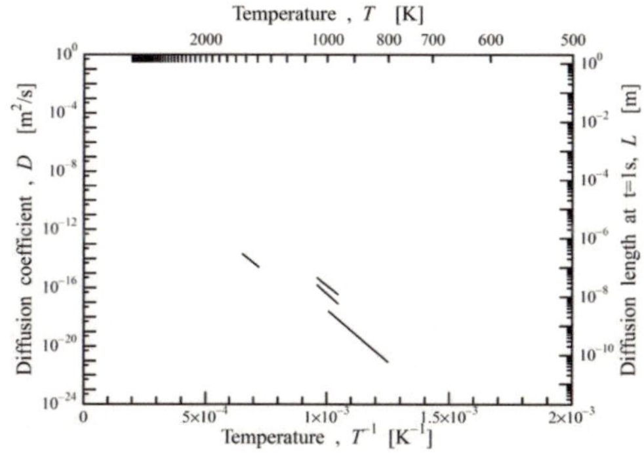

Figure 10. Schematic illustration plots, e.g., for 4 groups of linear line in Arrhenius relationship plotting on temperature inverse and legalism D value; in Table 1, there are four lines of Arrhenius plots, respectively

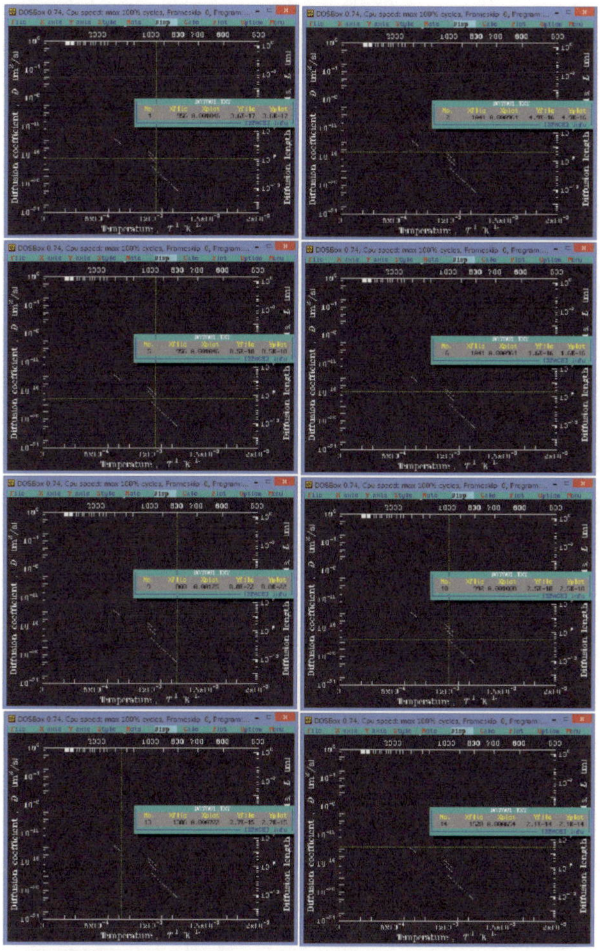

Figure 11. GP.exe schematic illustrations for searching the plots and their points, which, e.g., plots for four groups of linear line in Arrhenius relationship plotting on temperature inverse and legalism D value. There is, e.g., eight point edges of the right and left on the four linear lines

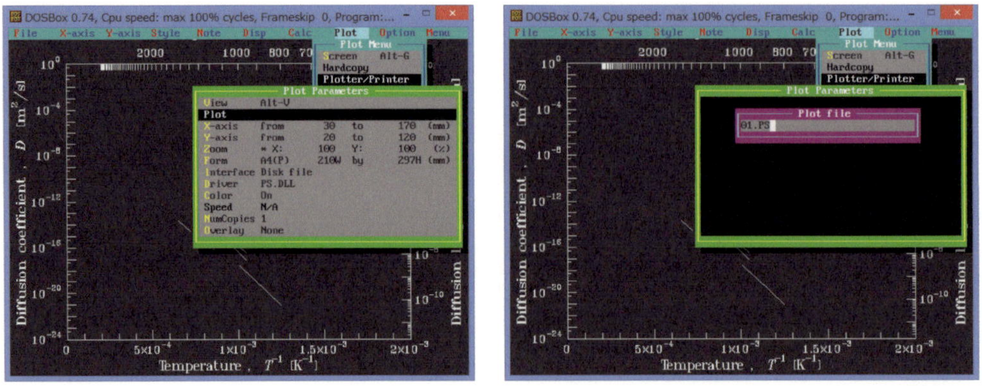

Figure 12. GP.exe schematic illustrations for creating the high-resolution PostScript picture as shown in Figure 11, which the file of 01.ps for common forms of Arrhenius plots using GP.exe. User can reproduce the similar frame Arrhenius plots using another diffusion data, to replace the data (filename from the data01.TXT to another filename, e.g., data02.TXT) into the *.GPR graph parameter file. Meanwhile, the user can transform precisely from 01.ps to 01gw.pdf (PDF: Portable Document Format) using freeware tool kit Ghostscript or gswin32c.exe with a textline command as follows: " C:/Program Files (x86)/gs/gs9.04/bin/gswin32c.exe" -dNOPAUSE -dBATCH -sDEVICE=pdfwrite -r600 -sOutputFile=01gw.pdf -c 300000 setvmthreshold save pop -f 01.ps "

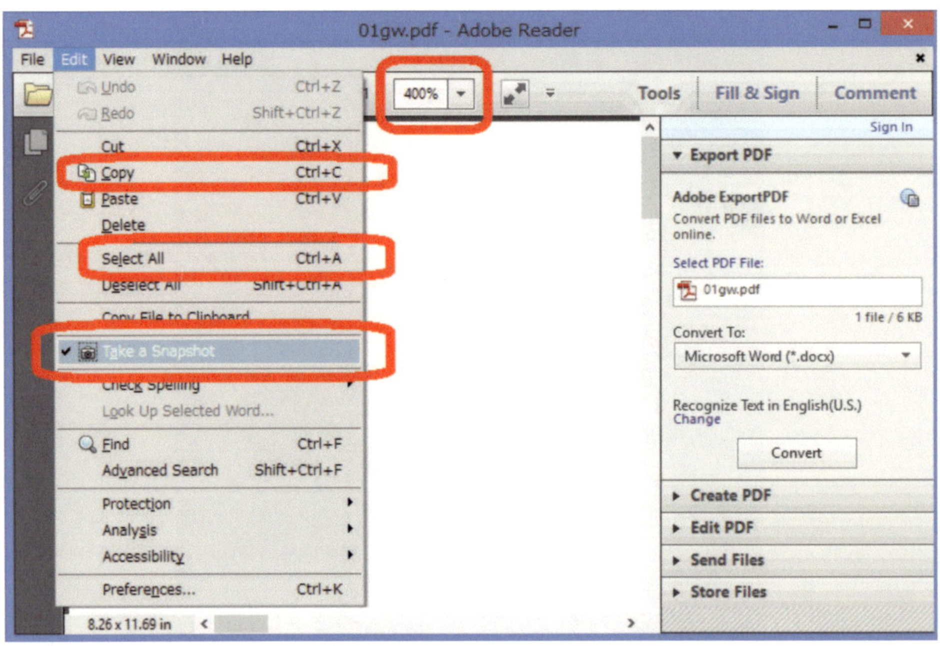

Figure 13. Adobe reader schematic illustrations for creating the high-resolution GIF (Graphics Interchange Format) picture as all pictures shown are presented in this paper. Using freeware of Adobe Reader or Adobe Acrobat Reader, the user opens the PDF of 01gw.pdf, sets the magnitude to "400 %," activates "Take a Snapshot," chooses "Select All," and finally chooses "Copy." All through the process, the user can copy the graphic data onto the Windows OS, Mac OS, and Linux OS clipboard

http://dx.doi.org/10.5772/60993

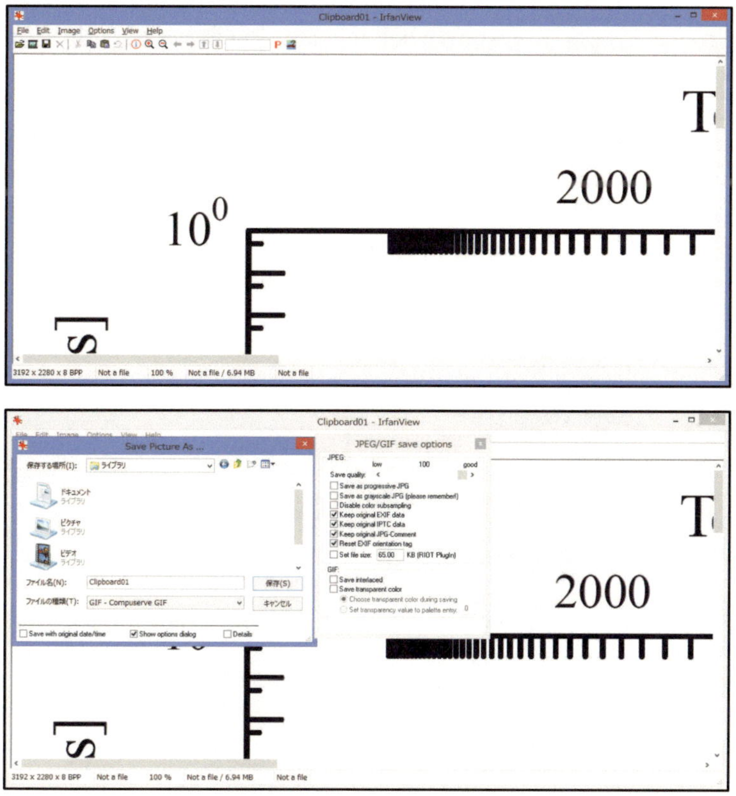

Figure 14. Image processing software schematic illustrations for creating the high-resolution GIF (Graphics Interchange Format) picture as all pictures shown are presented in this paper. For example, using the freeware "IrfanView," the user opens the menu "Save Picture As" of clipboard picture data and pastes it by "Paste Ctrl+V," and the user can copy and paste through the graphic Windows OS clipboard examples. Finally, almost 94 kByte of compact-size and high-resolution GIF file was created via AWK, DOSBox, GP.exe, Gswin32c.exe, Adobe-Reader, and IrfanView. This high-resolution GIF file of around 94 kByte would be user friendly for making documentation with graphic pictures. The Other standard graphic file formats of JPEG, PNG, BMP, TIF, etc., are able to apply in a similar procedure with high-performance IrfanView

3. Technique for information processing

3.1. Command line technique for PDF from PS

How to create a worksite shortcut for the elevated command prompt in Windows 8 and 10:

1. Open search by typing Ctrl+S and enter "cmd" in the search box, as shown below on the upper left. 2. Right click the Command Prompt results and choose Open file location, as shown below on the upper right. 3. Copy the shortcut Command Prompt to the worksite, e.g., C:/prog/gp/worksite/. It is necessary for instructions to include the file and directory name within length of 8 and 3, because of the software of legacy-type DOS.

4. In the worksite, e.g., C:/prog/gp/worksite/, right click the shortcut Command Prompt to open the Command Prompt Property, as shown below on the lower left. 5. In the Command Prompt Property, delete the description in the "Start in," as shown below on the lower right. 6. In the worksite, e.g., C:/prog/gp/worksite/, execute the GP.exe and save a 01.PS in the worksite. 6. In the worksite, e.g., C:/prog/gp/worksite/, click the shortcut Command Prompt; then paste the textline from a described text file to the Command Prompt window, where the textline would be " C:/Program Files (x86)/ gs/gs9.04 /bin/ gswin32c.exe" -dNOPAUSE -dBATCH -sDEVICE=pdfwrite -r600 –sOutputFile=01gw.pdf -c 300000 setvmthreshold save pop -f 01.ps " (e.g., there was already a need for the environment of installed gswin32 toolkit).

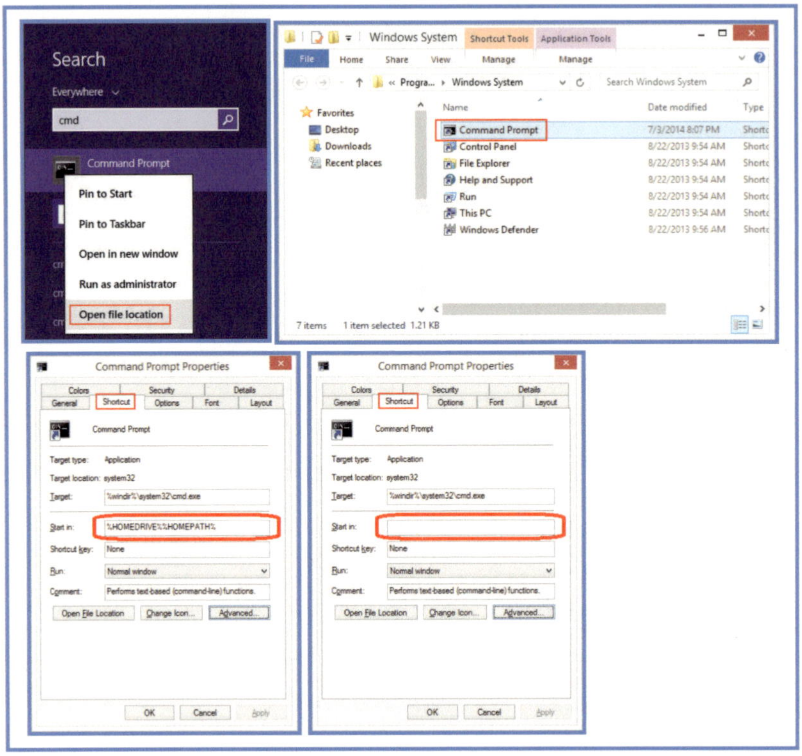

Figure 15. Schematic illustrations of command line technique for PDF from PS

3.2. Method to narrow down the diffusion database

How to narrow down the overflowed results of diffusion database from the over-100 score:

1. Open Web-based diffusion coefficient database presented NIMS, National Institute for Materials Science, Japan.
2. Select "Advanced Search."
3. Select "Diffusant." And input in "Included matched form" as shown below (e.g., if Fe would be entered, the search result includes Fe, 57Fe, Fe57, Fe59, Fe55, etc.).

http://dx.doi.org/10.5772/60993

4. If the result score would be over-100 data, you should better narrow down the over-100 scored data using Q of activation energy.

5. If from 0 to blank would be entered, selected data would be only Arrhenius relations paired data without single temperature diffusion data, as shown below.

6. Input an integer both of the "form" normally; e.g., it should be from 0 to blank, from 0 to 100, from 101 to 200, from 251 to blank, etc.

3.3. Method to change the character strings into numbers at one dash

How to change the character strings of the spreadsheet into number lines at one dash:

1. Open a fresh spreadsheet that would be, e.g., MS Excel.

2. Open Web-based diffusion coefficient database presented NIMS, National Institute for Materials Science, Japan.

3. Use a method to narrow down the diffusion database. Then on the narrowed-down database "frameset" list, which would be selected all of "frameset" on it, and then "copy" to store into a clipboard data zone, then paste on to the spreadsheet from the clipboard.

4. Now at the area of a spreadsheet, e.g., MS Excel, character strings, e.g., D_0 2.00E-05, Q 264, T_{min} 1123, and T_{max} 1699, still should be the character strings in Figure 2.

5. At one dash, the area of spreadsheet character strings, e.g., 2.00E-05, 264, 1123, and 1699, are selected and "Copy" is selected to store into a clipboard data zone and then it is pasted on to the high-end text editor, e.g., terapad.exe (Japanese only), etc., from the clipboard.

6. On the high-end text editor, these character strings would be, e.g., 2.00E-05?, 264?, 1123?, and 1699?; then they should be changed into 2.00E-05, 264, 1123, and 1699 by using "displacement" function effect.

7. Finally, at one dash, they should be copied – all of the 2.00E-05, 264, 1123, and 1699 numbers – into a clipboard data zone and then pasted to override onto the similar area of a spreadsheet character strings.

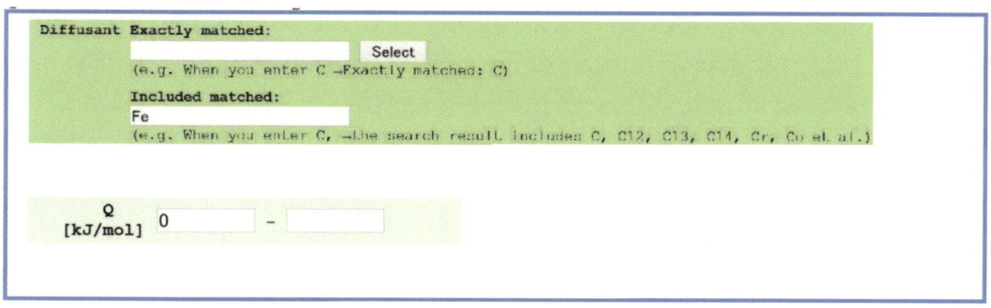

Figure 16. Schematic illustrations on how to narrow down the overflowed results of diffusion database from the over-100 score

4. Procedures and results of metallic systems

In the presented work by use of the AWK-GP-PDF, just suggested system procedure, the so-called big data via NIMS, National Institute for Materials Science, Japan, database was able to discuss it at once on one figure. First of all, in this research, Fe, Co, and Ni of metallic magnetic material were chosen and discussed through the system of AWK-GP-PDF.

4.1. Metallic magnetic material (Fe system)

In the metallic magnetic material of Fe system, in the presented work by use of the AWK-GP-PDF system procedure, Arrhenius plot of 725 line data which has activation energy Q (kJ/mol) of 0 to 150 has 97 lines, Q of 151 to 200 has 99 lines, Q of 201 to 230 has 87 lines, Q of 231 to 250 has 82 lines, Q of 251 to 260 has 62 lines, Q of 261 to 275 has 68, Q of 276 to 285 has 82, Q of 286 to 300 has 68, and Q of 301 and over has 80 lines. In the presented search, diffusant included matches of Fe, 57Fe, Fe57, Fe59, and Fe55. Self-diffusion and other diffusion mechanism are mixture and bridged diffuse, but it would be observed mainstream in the figure. The so-called big data via NIMS, National Institute for Materials Science, Japan, database was able to discuss it at once on one figure. In this research, Fe, Co, and Ni of metallic magnetic material were chosen and discussed through the system of AWK-GP-PDF.

4.2. Metallic magnetic materials (Co system)

In the metallic magnetic material of Co system, in the presented work by use of the AWK-GP-PDF system procedure, Arrhenius plot of 220 line data which has activation energy Q (kJ/mol) of 0 to 220 has 82 lines, Q of 221 to 300 has 98 lines, and Q of 301 and over has 80 lines. In the presented search, diffusant included matches of Co, Co60, and Co57. Self-diffusion and other diffusion mechanisms are mixture and bridged diffuse, but it would be observed mainstream in Figure 17, in the middle position of the graph.

4.3. Metallic magnetic materials (Ni system)

In the metallic magnetic material of Ni system, in the presented work by use of the AWK-GP-PDF system procedure, Arrhenius plot of 582 line data which has activation energy Q (kJ/mol) of 0 to 120 has 67 lines, Q of 121 to 180 has 93 lines, Q of 181 to 220 has 78 lines, Q of 221 to 260 has 99 lines, Q of 261 to 280 has 75 lines, Q of 281 to 295 has 90 lines, and Q of 296 and over has 80 lines. In the presented search, diffusant included matches of Ni, Ni63, Ni66, and Ni59. Self-diffusion and other diffusion mechanisms are mixture and bridged diffuse, but it would be observed mainstream in Figure 17, in the lower position of the graph.

4.4. Metallic magnetic materials

In Figure 17, there are shown Arrhenius relationships with horizontal axis of temperature T inverse and vertical axis of logalism diffusion coefficient D via self-diffusion and other diffusion mixed mechanisms with diffusant Fe, Co, and Ni materials, respectively. Plots for

http://dx.doi.org/10.5772/60993

complex phenomena are shown; T of infinity D_0 should be D_0(Fe) > D_0 (Co) > D_0 (Ni), for it seems like it would be a relation of their atomic radii. Additionally, it seems like activation energies of Co are smaller than those of Ni as shown in Figure 17.

4.5. Cu, Zn, Al, Ga, Cr, and Mn systems

Cu, Zn, Al, Ga, Cr, and Mn systems of metal and alloy are useful and attractive materials in several industrial-purpose products; so in the following sections, the Arrhenius plots of self-diffusions and other diffusion mechanisms have been exemplified.

4.5.1. Cu system

In the metallic Cu system, in the presented work by use of the AWK-GP-PDF system procedure, Arrhenius plot of 153 line data which has activation energy Q (kJ/mol) of 0 to 200 has 94 lines and Q of 201 and over has 59 lines. In the presented search, diffusant included matches of Cu, Cu64, and Cu67. Self-diffusion and other diffusion mechanisms are mixture and bridged diffuse, but it would be observed mainstream in Figure 18, in the upper graph.

4.5.2. Zn system

In the metallic Zn system, in the presented work by use of the AWK-GP-PDF system procedure, Arrhenius plot of 175 line data which has activation energy Q (kJ/mol) of 0 to 120 has 80 lines, Q of 121 to 200 has 72 lines, and Q of 251 and over has 23 lines. In the presented search, diffusant included matches of Zn, Zn65, Zn95, and Zn69. Self-diffusion and other diffusion mechanisms are mixture and bridged diffuse, but it would be observed mainstream in Figure 18, in the lower position of the graph.

4.5.3. Summary of the metallic Cu or Zn system

In Figure 18, Arrhenius relationships with horizontal axis of temperature T inverse and vertical axis of logalism diffusion coefficient D via self-diffusion and other diffusion mixed mechanisms with diffusant Cu and Zn materials, respectively, are shown. Plots for complex phenomena are shown; T of infinity D_0 should be D_0(Cu) > D_0(Zn), for it seems like it would be a relation of their atomic radii. Additionally, in relation with Figure 17, D_0 should be D_0(Fe) > D_0(Co) > D_0(Ni) > D_0(Cu) > D_0(Zn). Moreover, it seems that Q of activation energies Q(Fe) > Q(Co) > Q(Ni) > Q(Cu) > Q(Zn).

4.5.4. Al system

In the metallic Al system, in the presented work by use of the AWK-GP-PDF system procedure, Arrhenius plot of 111 line data which has activation energy Q (kJ/mol) of 0 to 300 has 80 lines and Q of 301 and over has 31 lines. In the presented search, diffusant included matches of Al, Al26, and Al27. Self-diffusion and other diffusion mechanisms are mixture and bridged diffuse, but it would be observed mainstream in Figure 19, in the upper part of graph.

4.5.5. Ga system

In the metallic Ga system, in the presented work by use of the AWK-GP-PDF system procedure, Arrhenius plot of 63 line data which has activation energy Q (kJ/mol) of 0 and over has 63 lines. In the presented search, diffusant included matches of Ga, Ga67, Ga72, Ga69, and Ga71. Self-diffusion and other diffusion mechanisms are mixture and bridged diffuse, but it would be observed mainstream in Figure 19, at the lower position of the graph.

4.5.6. Summary of the metallic Al or Ga system

In Figure 19, Arrhenius relationships with horizontal axis of temperature T inverse and vertical axis of logalism diffusion coefficient D via self-diffusion and other diffusion mixed mechanisms with diffusant Al and Ga of the similar 3 valence bonding numbers, respectively, are shown. Plots for complex phenomena are shown; T of infinity D_0 should be $D_0(\mathrm{Al}) > D_0(\mathrm{Ga})$, for it seems like it would be a relation of their atomic radii.

Additionally, in relation with Figures 17 and 18, D_0 should almost be $D_0(\mathrm{Al}) > D_0(\mathrm{Ga}) > D_0(\mathrm{Fe}) > D_0(\mathrm{Co}) > D_0(\mathrm{Ni}) > D_0(\mathrm{Cu}) > D_0(\mathrm{Zn})$. Moreover, it seems like it would be that Q of activation energies is $Q(\mathrm{Al}) > Q(\mathrm{Ga}) > Q(\mathrm{Fe}) > Q(\mathrm{Co}) > Q(\mathrm{Ni}) > Q(\mathrm{Cu}) > Q(\mathrm{Zn})$.

4.5.7. Cr system

In the metallic Cr system, in the presented work by use of the AWK-GP-PDF system procedure, Arrhenius plot of 205 line data which has activation energy Q (kJ/mol) of 0 to 230 has 79 lines, Q of 231 to 300 has 90 lines, and Q of 300 and over has 36 lines. In the presented search, diffusant included matches of Cr, Cr51, and Cr48. Self-diffusion and other diffusion mechanisms are mixture and bridged diffuse, but it would be observed mainstream in Figure 20, in the upper part of the graph.

4.5.8. Mn system

In the metallic Mn system, in the presented work by use of the AWK-GP-PDF system procedure, Arrhenius plot of 111 line data which has activation energy Q (kJ/mol) of 0 to 250 has 83 lines and Q of 251 and over has 28 lines. In the presented search, diffusant included matches of Mn, Mn54, and Mn55. Self-diffusion and other diffusion mechanisms are mixture and bridged diffuse, but it would be observed mainstream in Figure 20, at the lower position of the graph.

4.5.9. Summary of the metallic Cr and Mn system

In Figure 20, Arrhenius relationships with horizontal axis of temperature T inverse and vertical axis of logalism diffusion coefficient D via self-diffusion and other diffusion mixed mechanisms with diffusant Cr and Mn materials, respectively, are shown. Plots for complex phenomena are shown; T of infinity D_0 should be $D_0(\mathrm{Cr}) > D_0(\mathrm{Mn})$, for it seems like it would be a relation of their atomic radii. Additionally, in relation with Figure 17, D_0 should almost be

D_0(Cr) = D_0(Fe) > D_0(Co) > D_0(Ni) > D_0(Mn). Also it seems like it would be that Q of activation energies is Q(Cr) = Q(Fe) > Q(Co) > Q(Ni) > Q(Mn).

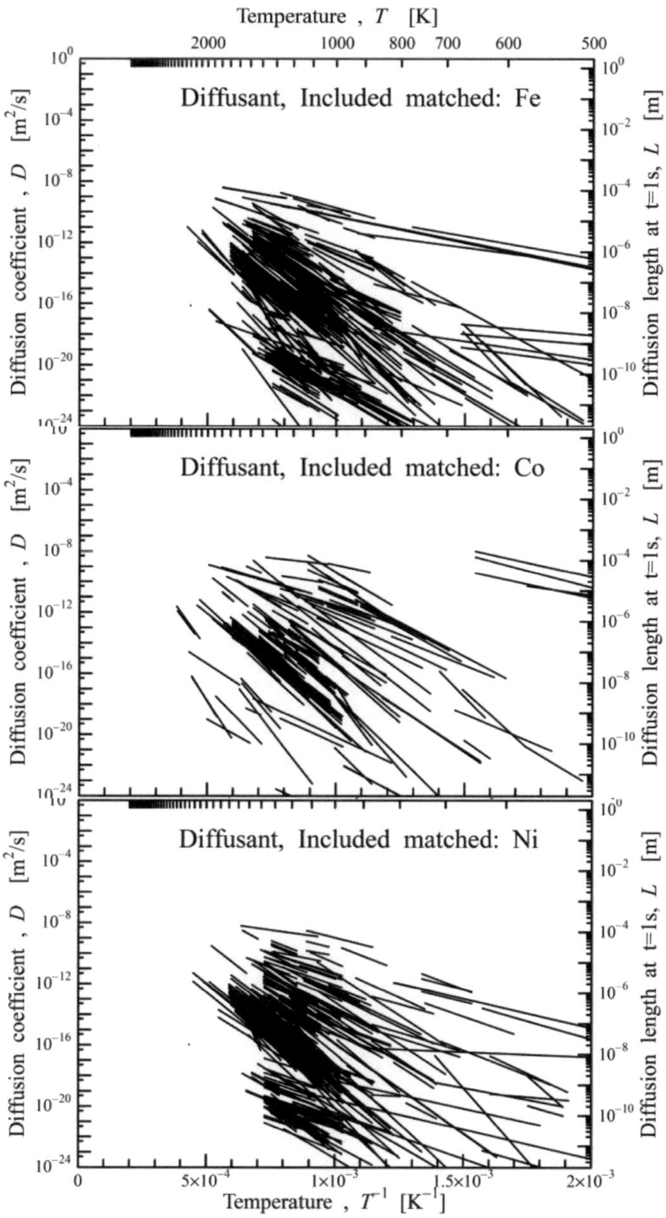

Figure 17. Arrhenius relationship with horizontal axis of temperature T inverse and vertical axis of logalism diffusion coefficient D via self-diffusion and other diffusion mixed mechanisms with diffusant Fe, Co, and Ni materials, respectively. Plots for complex phenomena are shown; T of infinity D_0 should be D_0(Fe) > D_0(Co) > D_0(Ni), for a relation of their atomic radii

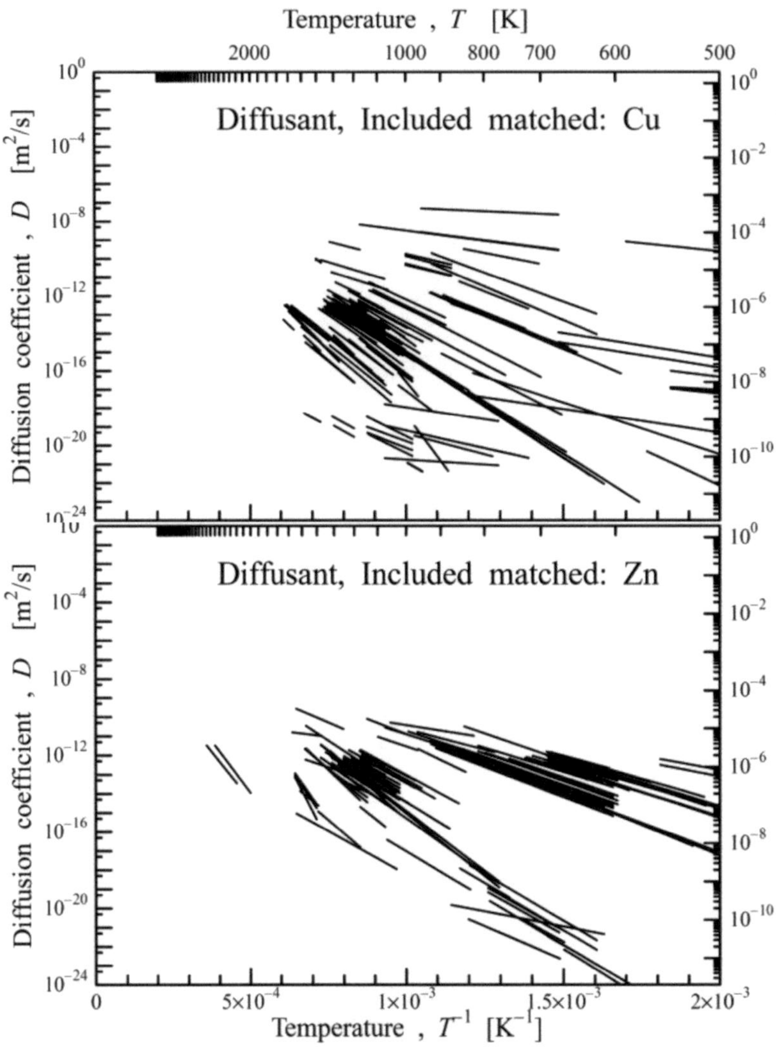

Figure 18. Arrhenius relationship with horizontal axis of temperature T inverse and vertical axis of logalism diffusion coefficient D via self-diffusion and other diffusion mixed mechanisms with diffusant Cu and Zn materials, respectively. Plots for complex phenomena are shown; T of infinity D_0 should be D_0(Cu) > D_0(Zn), for a relation of their atomic radii. Additionally, in relation with Figure 17, D_0 should be D_0(Fe) > D_0(Co) > D_0(Ni) > D_0(Cu) > D_0(Zn). Also it looks like Q of activation energy is Q(Fe) > Q(Co) > Q(Ni) > Q(Cu) > Q(Zn)

http://dx.doi.org/10.5772/60993

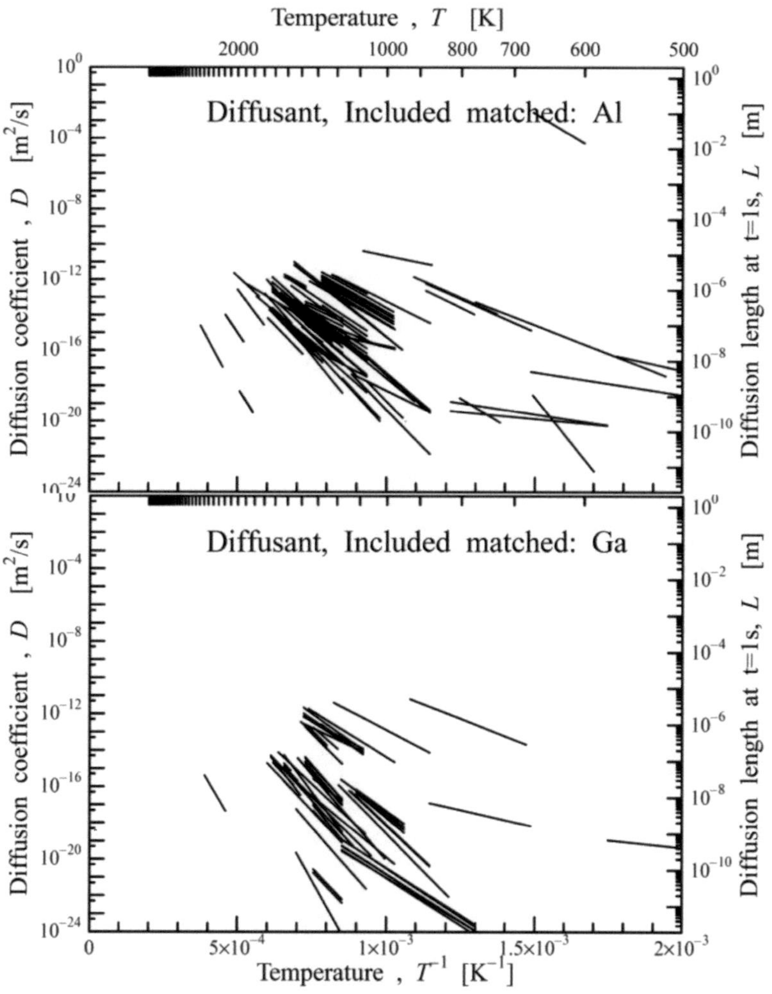

Figure 19. Arrhenius relationship with horizontal axis of temperature T inverse and vertical axis of logalism diffusion coefficient D via self-diffusion and other diffusion mixed mechanisms with diffusant Al and Ga of the similar 3 valence bonding numbers, respectively. Plots for complex phenomena are shown; T of infinity D_0 should be D_0(Al) > D_0(Ga), for a relation of their atomic radii. Additionally, in relation with Figures 17 and 18, D_0 should almost be D_0(Al) > D_0(Ga) > D_0(Fe) > D_0(Co) > D_0(Ni) > D_0(Cu) > D_0(Zn). Also it looks like Q of activation energy is Q(Al) > Q(Ga) > Q(Fe) > Q(Co) > Q(Ni) > Q(Cu) > Q(Zn)

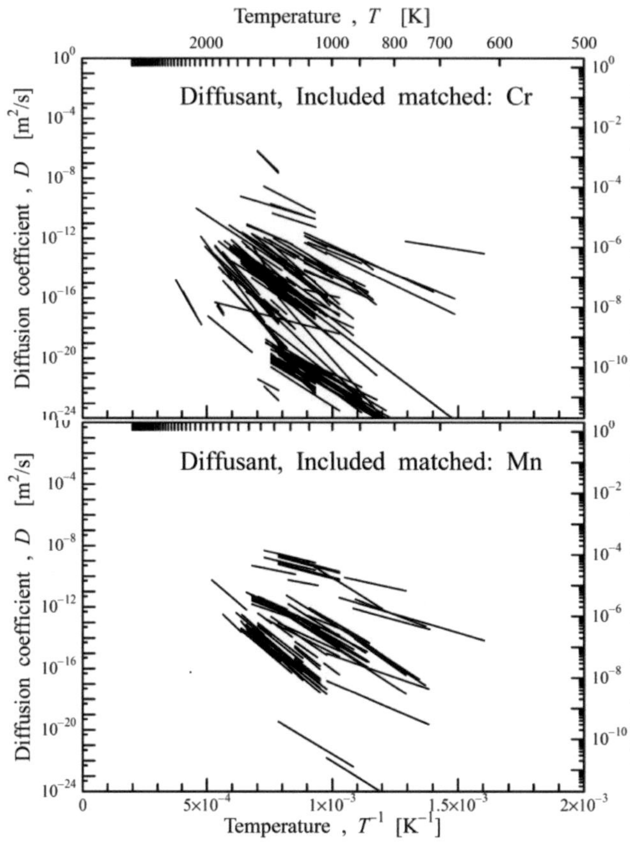

Figure 20. Arrhenius relationship with horizontal axis of temperature T inverse and vertical axis of logalism diffusion coefficient D via self-diffusion and other diffusion mixed mechanisms with diffusant Cr and Mn materials, respectively. Plots for complex phenomena are shown; T of infinity D_0 should be D_0(Cr) > D_0(Mn), for a relation of their atomic radii. Additionally, in relation with Figure 17, D_0 should almost be D_0(Cr) = D_0(Fe) > D_0(Co) > D_0(Ni) > D_0(Mn). Also it looks like Q of activation energy is Q(Cr) = Q(Fe) > Q(Co) > Q(Ni) > Q(Mn)

5. Conclusion

In Fe, Co, Ni, Cu, Zn, Al, Ga, Cr, and Mn alloys, considerable work has been done to obtain reliable value of diffusion coefficient, particularly because of the importance of physical constant values in specified materials. Meanwhile, free-of-charge Web-based diffusion coefficient database presents NIMS with over 8,000 diffusion data. It has been successfully provided.

In the present work, firstly, instructions to narrow down the diffusion database, to calculate using a specific spreadsheet for minimum temperature T_{min} vs diffusion coefficient $D(T_{min})$ and maximum temperature T_{max} vs diffusion coefficient $D(T_{max})$, to reform text file format using

http://dx.doi.org/10.5772/60993

AWK language, and to use computer drawing programs GP.exe to make an Arrhenius plot picture have been constructed through the process of Web-connected and numerical-based technique. Addition secondary to plot 9 kinds of Arrhenius relations Fe, Co, Ni, Cu, Zn, Al, Ga, Cr, and Mn to be comparison among the relations has been drawing.

Mainly, the tendency of the plots for complex phenomena, T of infinity D_0, regarding the relation of their atomic radii has been shown. Meanwhile, also the tendency Q of activation energy was discussed.

It was the tutorial on high-resolution PDF builder using the freeware in GP.EXE that was designed until 1999 to make smart graphs for publication with powerful data analysis ability such as Arrhenius relations T inverse and legalism plot. And now it is shown that the GP.EXE has been useful for genuine data processing even in 2015.

Finally, it is concluded that graph plot tool GP.exe and its extracted high-resolution PostScript and PDF with common forms of optimized Arrhenius plots using Arrhen.GPR showed good performance, because it produced a similar frame with the Arrhenius plots using diffusion data to replace into the GPR graph parameter file.

Acknowledgements

The research work presented was partly supported by JSPS KAKENHI Grant-in-Aid for Challenging Exploratory, Research Number 25560092. The author was favored to have the assistance of Dr. I. A. Figueroa in Universidad Nacional Autonoma de Mexico who contributed to the experimental work in accomplishing sample preparations at the University of Sheffield, UK (in 2016), for discussion of diffusion and thermal property in metal and alloy. The author also would like to express appreciation to Dr. Sergio Gonzalez Sanchez (Universitat Autonoma de Barcelona) and Mr. P. J. J. Hawksworth in the University of Sheffield (in 2006, the first author staying as a visiting researcher). The author is indebted to Professor H. A. Davies at University of Sheffield, UK (in 2016), Professor Yoshiaki Iijima, Kazuaki Fukamichi, and Ken-ichi Hirano in Tohoku University for drawing their attention to presented researches.

Author details

Kazu-masa Yamada[1*] and Nobuaki Matsuhashi[2]

*Address all correspondence to: yama@js8.so-net.ne.jp, yama@hakodate-ct.ac.jp

1 Department of Electrical and Electronic Engineering, (Department of Production Systems Engineering), National Institute of Technology, Hakodate College, Japan

2 Electrical and Computer Engineering Course, (Department of Industrial Systems Engineering), National Institute of Technology, Hachinohe College, Japan

References

[1] P. Heitjans, J. Karger, eds., Diffusion in Condensed Matter: Methods, Materials, Models (second edition), Birkhauser, ISBN 3-540-20043-6 (2005).

[2] A. Einstein, "Uber die von der molekularkinetischen Theorie der Warme geforderte Bewegung von in ruhenden Flussigkeiten suspendierten Teilchen", Ann. Phys. 17(8), pp. 549–560 (1905).

[3] Kazu-masa Yamada, "Numerical Solutions for Structural Relaxation of Amorphous Alloys Studied by Activation-Energy-Spectrum Model", Applications of Calorimetry in a Wide Context – Differential Scanning Calorimetry, Isothermal Titration Calorimetry and Microcalorimetry, InTech, Editor: Amal Ali Elkordy, ISBN 978-953-51-0947-1, pp. 343–364 (2012).

[4] Diffusion Database (Kakusan), Copyright National Institute for Materials Science (NIMS), http://diffusion.nims.go.jp/index_en.html (accessed 12 Jan 2015).

[5] Kazu-masa Yamada and Nobuaki Matsuhashi, "OpenGL Assisted Dynamically 4-Dimensional Animation E-Textbook Associated with Perovskit Di-electrics", Proceedings of IEEE Technically Co-sponsored Science and Information Conference 2014 (SAI2014/August), pp. 912–920 (2014/August).

[6] Kazu-masa Yamada and Nobuaki Matsuhashi, "Java Based Numerical Extractable Molecular Dynamics Calculation", Proceedings of 2014 International Conference on Mechanical Design, Manufacture and Automation Engineering (MDMAE 2014/January), Published by DEStech Publications, Inc., pp. 502–507 (2014/March).

[7] Kazu-masa Yamada, Nobuaki Matsuhashi, "Extended 4-Dimensional OpenGL E-Book Associated with Electric Material", Article Published in International Journal of Advanced Computer Science and Applications(IJACSA), Special Issue on Extended Papers from Science and Information Conference 2014, pp. 114–123(2014/November).

[8] A.V. Aho, B.W. Kernighan, P.J. Weinberger, "Awk – A Pattern Scanning and Processing Language" (second edition) (1978) http://inst.eecs.berkeley.edu/~selfpace/sp.unix.stuff/awk.pdf (accessed 12 Jan 2015).

Section 3

Development of Commercial Light Magnesium Alloys

Chapter 4

Native Oxide Films on AZ31 and AZ61 Commercial Magnesium Alloys – Corrosion Behaviour, Effect on Isothermal Oxidation and Sol–gel Thin Film Formation

Sebastián Feliu (Jr), Amir A. El Hadad,
Violeta Barranco, Irene Llorente,
Federico R. García-Galván,
Antonia Jiménez-Morales and Juan Carlos Galván

Additional information is available at the end of the chapter

http://dx.doi.org/10.5772/60721

Abstract

The authors present a review of their recent research work in an endeavour to interpret the influence of native oxide films on the corrosion behaviour of commercial AZ31 and AZ61 magnesium alloys or on the oxidation kinetics in air at 200°C. The tendency of some of these thin films to be sufficiently protective in mild or weak corrosive environments is examined. For obtaining oxide films with different protective properties, some of the specimens are tested with the surface in the as-received condition, while others are tested immediately after mechanical polishing. The technique applied to characterise thin (thickness of just a few nanometres) oxide films present on the surface of alloys has basically been XPS (X-ray photoelectron spectroscopy) in combination with ion sputtering. Oxidation resistance of the alloys is quantified by thermo gravimetric (TG) curves and their corrosion rate is evaluated by Electrochemical Impedance Spectroscopy (EIS) and hydrogen evolution measurement in chloride solutions with different aggressivity. Emphasis is placed on the possible effects of: (a) the different thickness of the native oxide films formed on the polished surfaces on the corrosion behaviour of the alloys; and (b) the different film homogeneity and uniformity on the oxidation results. Finally, an attempt will be made to learn more about the influence of the native oxide films that cover the substrate on the subsequent growth and protective behaviour of the sol–gel coatings.

Keywords: native oxide film, corrosion, magnesium alloys, surface chemistry, chloride, electrochemical impedance spectroscopy, x-ray photoelectron spectroscopy

1. Introduction

This chapter focuses on Mg–Al alloys, a material which has been transformed from a topic of scientific interest to a material of broad technological and commercial significance in the last two decades. From a practical point of view, Magnesium (Mg) is a promising structural material for transportation industries due to its low density (1.74 g/cm^3), high strength-to-weight ratio, good castability and mechanical properties, making it attractive for use in the automotive, aerospace and electronics industries, where light weighting is increasingly important. Especially for automotive industry, using lightweight materials like Mg can improve the fuel economy and have a positive impact on the emission of greenhouse gases and environment [1]. Unfortunately, magnesium and its alloys intrinsically have high chemical reactivity, which makes them susceptible to oxidation in various corrosive environments. The poor corrosion resistance of the magnesium alloy restricts its widespread application [2]. A better understanding of their corrosion mechanism, and of the factors influencing their corrosion behaviour, is expected to lead to more corrosion-resistant Mg alloys [3].

There are many scientific works that have studied the relationship between the microstructure of the alloy (quantity and distribution of the β-phase precipitates) or between Al content in the bulk of the alloys and their corrosion resistance [4-10]. Far fewer researchers have studied the chemical composition of the thin passivating layer on the outermost surface (thickness of about 3nm) which spontaneously grows in contact with the atmosphere or after short heat treatments at low temperature [11-15]. As it has been recently noted by Song [16], the rupture of an initially formed film can to a great extent influence the film growth behaviour in a later stage; the pitting distribution on a metal surface in an early stage will affect the final corrosion morphology; the initial surface cleanness of steel may sometimes determine the steel passivity. It is of great interest to learn how, when and where the transition suddenly starts. Predicting the occurrence of the transition on a metal surface is difficult, but it is a challenging goal that must be attained in corrosion research.

Like other metals, magnesium oxidises rapidly when exposed to the air or in an aqueous solution, giving rise to the formation of passivating oxide films of just a few nanometres thickness. According to Mott and Cabrera's theory on the growth of very thin oxide films [17], the electric field associated with the presence of oxygen ions adsorbed on the metal surface is enormous at first, but quickly weakens as the film thickness grows, and after a few minutes, the reaction virtually ceases. In dry air, at room temperature, the oxide reaches a limited thickness of about 4 nm, but can grow in thickness owing to the formation of magnesium hydroxide in a humid atmosphere and/or at a higher temperature.

From a scientific point of view, the literature contains controversial views relating to the effect of thin native oxide films characteristics on the corrosion performance of magnesium alloys. Chen et al. [18] reported that unlike other active light metals, such as Al and Ti, Mg alloys do not form a naturally passivating oxide film. In contrast, Nordlien et al. [19] and [20] and Lunder et al. [21] suggested that the oxide formed naturally on Mg–Al alloy surfaces is stable and protective and responsible for the passivity of the surface in the presence of an aqueous environment. Santamaria et al. [15] report that the films formed soon after mechanical treatment and immersion in aqueous electrolyte have a bilayer structure, consisting of an ultra-thin MgO inner layer (similar to 2.5 nm) and a $Mg(OH)_2$ external layer. Recently, Atrens et al. [22] proposed the partially protective surface film (PPSF) mechanism for Mg corrosion that assumes that the Mg surface is covered by a partially protective film, even in a concentrated chloride solution. However, the chemical composition and microstructure of the oxide films and their protection properties are still not clear [23].

According to Jeurgens et al. [24] and Saleh et al [25], the thermal oxidation of metallic alloys at low temperatures (e.g. at $T < 600$ K) and for short times has only scarcely been investigated. The detailed chemical composition and constitution of the oxide films formed on such alloy surfaces at low temperatures for short heating times are unknown. Furthermore, there is no comprehensive knowledge of the effect of the concurrent processes of chemical segregation and preferential oxidation on both the developing oxide-film microstructure and the induced compositional changes in the alloy subsurface.

Czerwinski [26] studied the oxidation behaviour of AZ91D Mg alloy at different temperatures. His results showed that AZ91D exhibited protective oxidation only at a temperature of 197°C, while at higher temperatures, the behaviour was non-protective and associated with the formation of oxide nodules and their coalescence into a loose fine-grained structure.

One of the main obstacles to characterise the thermally oxidised film formed on the surface of magnesium alloys after heating at low temperatures (200°C) is its small thickness. As Czerwinski [26] reported, the maximum oxide thickness achieved during heating (as converted from weight gain data) was equal to 64 nanometres for 197°C. This film thickness is too small to produce a sufficient signal for conventional materials characterisation techniques (Optical microscopy, SEM/EDX, XRD or TEM) [27]. The use of the XPS surface analysis technique makes it possible to reduce the analysed thickness to only 3 nm and supplies information on the oxidation state of the detected element.

One of the most effective ways to prevent corrosion of metals is to separate the metallic surface from the corrosive medium by mean of coating films. In this context, the sol-gel coatings have received considerable attention over the years, but recently there has been an upsurge of interest in these technologies due to the urgent necessity of searching an alternative way to the use of conventional primer coatings and pretreatments based in the use of Cr(VI) compounds. The great growth of the attention devoted to the sol-gel coatings is in part due to the new environmental protection legislation which restricts the use of chromium Cr(VI) salts because of their being highly toxic and carcinogenic. In the scientific literature, one can find valuable review papers that can be a good approach guide to understand the problems associated with the sol-gel science and technology [28-31].

The development of hybrid organic–inorganic sol-gel coating for protecting aircraft light alloys and improving environmentally benign surface treatments for aerospace alloys are topics of high priority around the world. In this way, Vreugdenhil et al. since 2001 described a process to develop Self-assembled NAnophase Particle (SNAP), which was based on the use of glycidylpropyltrimethoxysilane and tetramethylorthosilicate [32]. Afterwards, in 2003, these authors described a new route to improve the SNAP coating process. This new process was based on the use of amino-silanes as crosslinking agents 3-aminopropyltrimethoxysilane, 3-(2-aminoethyl) aminopropyltrimethoxysilane and 3-(trimethoxysilyl) propyldiethylenetriamine [33]. In a recent work, Li et al. have prepared a series of hybrid organic–inorganic coatings based on silica–epoxy composite coatings with the sol-gel method by using the amino-silane 3-aminopropyl triethoxysilane as a coupling agent and tetraethoxysilane to afford chemical bonding to form silica networks organic–inorganic hybrid coatings [34].

The production of hybrid sol-gel coatings is based in the co-hydrolysis and polycondensation of two organopolysiloxane precursors. One of them is the precursor of inorganic chain and another one is the former of the organic chain of the sol-gel network. Tetramethylorthosilicate and tetraethylorthosilicate are typically used as precursors of silica network [28-31]. On the other hand, aminopropyltrimethoxysilane, glycidylpropyltrimethoxysilane, methacryloyloxypropyltrimethoxysilane are commonly used as precursors of the organic chain. These molecules contain active groups that can react chemically with both the inorganic and organic substances, which can be used to design tailored multifunctional coatings with multiple properties (scratch, wear and corrosion resistance, superhydrophobicity, bioactivity and biocompatibility, environmentally benign antifouling, etc.) [28-31]. The use of mercapto silanes as the 3-mercaptopropyltrimethoxysilane can also be a good choice for preparing sol–gel coatings for light alloys [35].

Garcia Heras et al. have also prepared sol–gel coatings based on siloxane bonded units by starting from an organic–inorganic hybrid coatings by using as precursors γ-methacryloxypropyltrimethoxysilane and tetramethylorthosilicate for corrosion protection of zinc substrates Cerium nitrate hexahydrate in three different concentrations was added. The anticorrosive performance of the Ce^{3+} ions entrapped in the resulting hybrid silica sol–gel network occurs by means of the inhibitor effect and the self-healing mechanism (probably due to the $Ce(OH)_3$ precipitation) [36]. Recently, Harb et al. have reported a similar study where they also correlated the structural properties of cerium doped organic–inorganic hybrid coatings with their efficiency for the corrosion protection of carbon steel. The films were prepared via the sol–gel route from radical polymerisation of methyl methacrylate with 3-methacryloxy propyltrimethoxysilane followed by acidic hydrolysis and condensation of tetraethoxysilane [37]. A promising research line consists in introducing phosphorus precursors into organic–inorganic hybrid sol–gel coatings. In this way, El Hadad et al. have introduced triethylphosphite in sol–gel coatings based on mixtures of γ-methacryloxypropyltrimethoxysilane and tetramethylorthosilicate, not just for improving the corrosion protection properties of the coatings but also their biocompatibility [38].

Before, Khramov et al. had also used hybrid organic–inorganic coatings with phosphonate functionalities for evaluating their corrosion protection performance when they are used as surface treatments for magnesium materials. These coatings have been processed via a sol–gel route by hydrolysis and condensation of a mixture of diethylphosphonatoethyltriethoxy-silane and tetraethoxy-silane obtaining very promising results [39]. In line with this idea, Lamaka et al. have synthesised hybrid organic–inorganic sols by copolymerisation of epoxy-siloxane and titanium or zirconium alkoxides by also using tris(trimethylsilyl) phosphate as additive to confer additional corrosion protection to magnesium-based alloy [40]. Barranco et al. have also evaluated the behaviour against corrosion of four optimised sol–gel coating systems (inorganic, hybrid organic–inorganic, containing zirconium and containing cerium ions) on magnesium alloys. These coatings were obtained by the sol–gel process from tetra-methoxysilane and diethoxydimethylsilane as organopolysiloxane precursors and were doped with Ce^{3+} to make evident their usefulness as autonomous protective coatings as well as a pretreatment prior to acrylic topcoat [41].To finish this section it might be interesting to mention a recent paper published by the group of Murillo-Gutierrez. These authors proposed the use of bilayer architectures as a strategy to improve the protection properties of sol–gel coatings on magnesium alloys. Because of the high reactivity of the magnesium alloys, a sol–gel coating based on a single layer presents defects consisting of macro-pores and protuber-ances, which opens access for corrosive species to reach the metallic substrate. This undesired effect can be overcome by using a sol–gel coating based on a bilayer architecture. For such a purpose, they use sol–gel formulations based on 3-glycidyloxypropyl-trimethoxysilane and aluminium-tri-sec-butoxide [42].

In this chapter, the authors present a review of their research work over the last three years on the correspondence between the chemical composition and physical characteristics of mag-nesium oxide layers present on the surface of commercial Mg–Al alloys and a series of metal properties including its microstructure, corrosion resistance, oxidation performance and sol–gel film formation. Most of the experimental information and results have been taken from published works by the authors [27, 43-47].

In the subsequent paragraphs, focused on the native oxide film, information is presented in relation with the following points:

1. Correspondence between thickness of the naturally formed oxide films on the surface of AZ31 and AZ61 alloys and a series of properties connected with the type of magnesium alloy, presence of second-phase particles in the magnesium matrix and the manufacturing conditions.

2. The corrosion resistance of the native oxide film in saline solutions of different aggres-siveness.

3. The homogeneity and uniformity of the surface layer, which is closely related to the resistance by the magnesium alloys to the thermal oxidation.

4. The chemical changes in the native oxide surface film of the AZ61 alloy induced by the thermal oxidation at low temperatures and for short times, reinforcing its anticorrosive properties.

5. The behaviour of a silane hybrid sol–gel coating designed to be applied onto magnesium alloy surfaces for their corrosion protection.

2. Experimental section

Samples of wrought magnesium aluminium alloy (AZ31 or AZ61) in plates of 3 mm thickness were supplied by Magnesium Elektron Ltd, Manchester, UK. The chemical compositions of the tested magnesium alloys, AZ31 and AZ61, are listed in Table 1.

Alloy	Chemical Composition (wt%)						
	Al	Zn	Mn	Si	Fe	Ca	Mg
AZ31	3.1	0.73	0.25	0.02	0.005	0.0014	Balance
AZ61	6.2	0.74	0.23	0.04	0.004	0.0013	Balance

Table 1. Chemical composition of AZ31 and AZ61 alloys (wt%).

The following nomenclature is used in the remainder of the paper to designate the four dual combinations tested: AZ31-O, AZ31-P, AZ61-O and AZ61-P, where the letters O and P, which accompany the alloy type, denote: O= original surface condition (e.g. as-received condition); P= polished surface condition.

This research compares the behaviour of specimens of the above alloys in the following surface conditions:

Specimens in the as-received condition, where the untreated surfaces were only cleaned with distilled water and dried with hot air.

Freshly polished specimens were dry ground through successive grades of silicon carbide abrasive paper, from P600 to P2000, followed by finishing with 3 and 1 μm diamond paste, cleaned in distilled water and dried with hot air.

Two etching reagents were used: (a) nital 2%, 2 ml HNO_3 + 98 ml H_2O, to reveal the constituents and general microstructure of alloy AZ61 and (b) 4.6 g picric acid + 10 ml acetic acid + 70 ml ethanol + 10 ml H_2O to reveal the grain boundaries of alloy AZ31.

The thermal treatment was very simple, consisting of the horizontal exposure of 2 cm × 2 cm square specimens of the AZ31 and AZ61 alloys in a convective stove at 200 °C in air for 5, 20 and 60 min.

For preparation of sol–gel coatings, γ-methacryloxypropyltrimethoxysilane (MAPTMS) (98% from Aldrich) and tetramethoxysilane (TMOS) (98% from Fluka) have been used as received. Sols were prepared starting from a mixture of 4 mol of MAPTMS and 1 mol of TMOS. Ethanol and water were added with the molar ratio (TMOS + MAP)/water/ethanol of 1/7/8 [47]. The organic–inorganic hybrid coatings were deposited on the AZ31 and AZ61 by using a dip coating technique with a withdrawal speed of 9 cm/min and holding time of 60 s. The coated MAPTMS/TMOS-AZ31 alloy and MAPTMS/TMOS-AZ61 alloy systems were then placed in a furnace for 2 h at 120°C for curing. The rest of MAPTMS/TMOS sol was placed 12 h at 120 °C and crushed to convert it into a powder form [47].

AZ31 and AZ61 alloys were oxidised in identical conditions in a thermogravimetric analyser (TGA) (TA instruments Q600 SDT) using cylindrical specimens of 4 mm in diameter by 2 mm in height (weight approximately 100 mg). The apparatus was capable of accommodating a

http://dx.doi.org/10.5772/60721

specimen with a maximum weight of 0.5 g and had a measurement accuracy of 0.1 μg. The reaction temperature was monitored by a Pt/Pt–Rh thermocouple.

Weight change kinetics were measured in air under isothermal conditions at a temperature of 200 °C. The heating rate before reaching the isothermal condition was 50 °C/min.

The microstructure and chemical composition of the magnesium alloys were examined by the scanning electron microscopy (SEM, Hitachi S-4800) and by energy dispersive X-ray spectroscopy (EDS, Oxford energy dispersive X-ray microanalysis system) attached to SEM, respectively.

The surface topography of the specimens was measured ex situ by atomic force microscope (model 5100 AFM/SPM from Agilent Technologies) in taping mode with a resolution of 512 × 512 points.

Photoelectron spectra were recorded using a Fisons MT500 spectrometer equipped with a hemispherical electron analyser (CLAM 2) and an Mg Kα X-ray source operated at 300 W. The energies of all spectra were referenced to the C 1s peak at 285.0 eV. For acquiring the concentration profiles (distribution of elements as a function of specimen thickness), the surface was sputtered by argon ion bombardment (AIB) with a sputtering rate of 5 nm/min based on a SiO_2 reference. The data were analysed using a Xpspeak 4.1 software.

The morphology of the attack on the corroded surface was examined at low magnification and a camera was used to take the photographic images.

To measure the corrosion rate of magnesium alloys during immersion, hydrogen evolution detection tests were conducted at ambient temperature. The specimens were put in the beaker and then a funnel was located over the samples to collect the volume of evolved hydrogen during the corrosion process in a burette above the funnel.

Electrochemical impedance measurements were conducted in 0.006 M NaCl and 0.6 M NaCl at room temperature (25 °C). An AUTOLAB potentiostat, model PGSTAT30, with frequency response analyser (FRA) software was used. A three-electrode cell was used employing the magnesium alloy as working electrode with an exposed area of 1.0 cm^2, a graphite electrode as counter electrode and a saturated Ag/AgCl (saturated KCl) electrode as reference electrode, respectively. The impedance data were collected at the open circuit potential with a 10 mV sinusoidal AC perturbation over a frequency range of 100 kHz to 1 mHz. The Nyquist shapes of the completed experiments are fitted using the Zview program [48], and the impedance data are acquired from the fitting results using the equivalent electrical circuit.

3. Results and discussion

3.1. Characterisation of magnesium alloys

3.1.1. Morphology and structure of the outer oxide layer formed on AZ31 and AZ61 magnesium alloys with different surface conditions

Figure 1 compares the surface morphologies on the as-received (O) and polished (P) surface conditions for: (a, c) AZ31 alloy and (b, d) AZ61 alloy. As can be seen, the metallic surface of the AZ31-O specimen appears to be covered by a large number of white precipitated particles (Figure 1a), whereas in the surface of the AZ61-O, these particles are not apparently visible (Figure 1b). In a previous study [45], we obtained EDX analyses on the white particles and an increase in the Al, Zn and Mn content and the decrease in the Mg content compared to the darker regions was observed. SEM examinations of the specimens in as-received condition (Figures 1a and 1b) revealed a rough surface covered with non-uniform features, very different to that of the polished surface (Figures 1c and 1d).

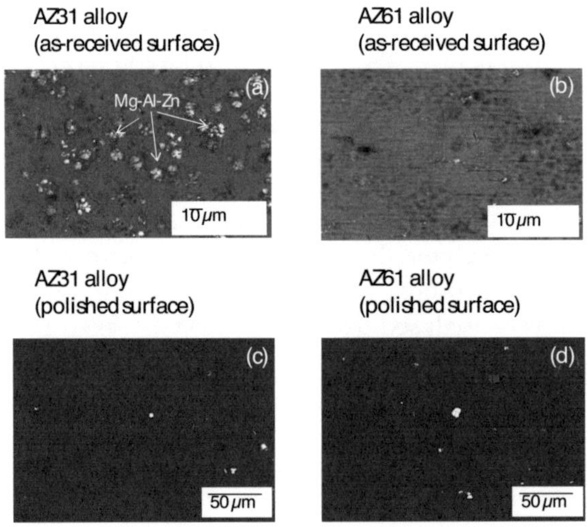

Figure 1. (a, b) SEM morphologies of the surfaces in the AZ31-O and AZ61-O and (c, d) AZ31-P and AZ61-P specimens (figure adapted from [45] and [46]).

Table 2 compares the surface roughness values on the AZ31-O and AZ61-O specimens with those corresponding to the AZ31-P and AZ61-P ones. Nanometric scale details of the typical surface roughness exhibited by the tested specimens are given in Figure 2.

SPECIMENS	RMS (nm)
AZ31-O	150.0
AZ61-O	138.6
AZ31-P	7.3
AZ61-P	5.6

Table 2. Roughness values obtained with atomic force microscope. The values are average of four determinations.

Figure 3 shows the SEM microstructure of the tested materials. Both alloys have an equiaxed grain structure typical of wrought magnesium alloys. The average grain sizes are approximately 5-100 μm (Figure 3a) and 40-100 μm (Figure 3b) in AZ31 and AZ61, respectively, but in AZ61 irregular-shaped divorced eutectic β-$Mg_{17}Al_{12}$ phases were seen at grain boundaries

http://dx.doi.org/10.5772/60721

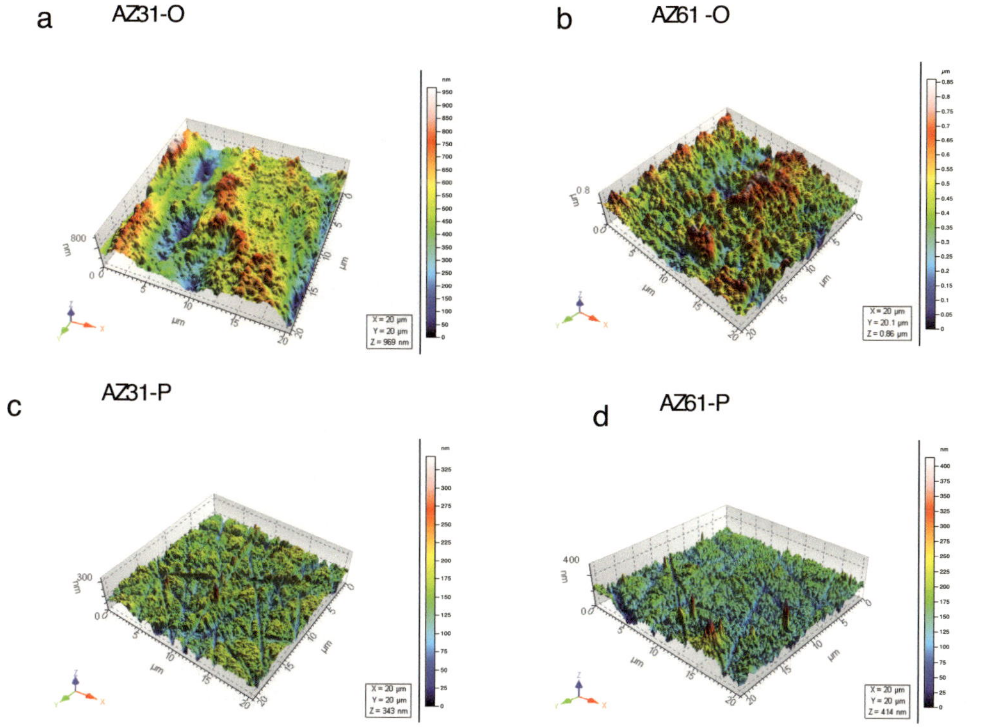

Figure 2. (a, b) Atomic-force Microscope (AFM) images of the surfaces in the AZ31-O and AZ61-O and (c, d) AZ31-P and AZ61-P specimens.

(Figure 3b). Microstructural features often promote the formation of points of attack on the surface of the specimens exposed to the test solution due to the discontinuities that are caused in the oxide film. In the case of AZ61, the small amounts of β-phase ($Mg_{17}Al_{12}$), which is formed at the grain boundaries (Figure 3b), can act as a galvanic cathode accelerating the corrosion of the α-Mg grain [49] and [50]. In contrast, it can be found that the alloying element Al in AZ31 alloy is completely dissolved into Mg matrix and no precipitation of β-phase is observed (Figure 3a). As will be seen later, the rate of corrosion is strongly influenced by the microstructure of the tested specimens.

3.1.2. Surface chemistry of the AZ31 and AZ61 magnesium alloys with different surface conditions

Figure 4 compares the C1s (a-d), O1s (e-h), Mg 2p (i-l) and Al 2s (m-p) XPS high-resolution spectra of the oxide films formed spontaneously on the surface of the AZ31 and AZ61 alloys after 5 minutes of sputtering. Attention is drawn to the absence of carbon on the surface of the specimens (Figures 4a-4d). The O 1s spectra (Figures 4e-4h) can be fitted to two components, which include the high binding energy component at 533.2 eV corresponding to the Al_2O_3 for the case of AZ31-P and AZ61-P specimens (Figures 4e and 4f), or $MgAl_2O_4$ for AZ31-O and AZ61-O specimens (Figures 4g and 4h) [51], and the low binding energy, most intense component at 531.2 eV relating the MgO [52].There are two components of similar intensity in

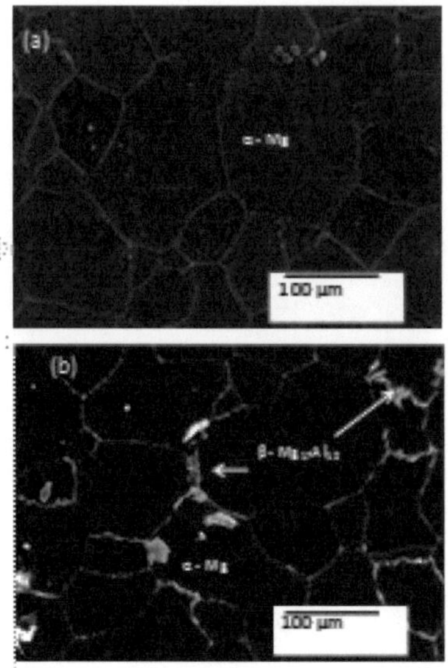

Figure 3. (a) SEM micrographs of AZ31 and (b) AZ61 magnesium alloys.

the Mg2p peak observed on the AZ31-P specimen (Figure 4i), in which the low binding energy component at 48.6 eV is attributed to metallic Mg, and the broader high binding energy peak centred at 50.8 eV corresponds to magnesium oxide (MgO). It can be seen that the ratio of the intensity of magnesium oxide component to the intensity of metallic magnesium component on the surface of the AZ61-P specimen (Figure 4j) is higher than on the AZ31-P specimen (Figure 4i). This is explained by accepting that the oxide film thickness on the AZ61-P specimen is greater. For AZ31-O and AZ61-O specimens (Figures 4k and 4l), the spectra are fairly similar, containing one single component at a binding energy of 50.8 eV associated with the presence of magnesium in the form of $MgAl_2O_4$ [51]. The Al2s spectra obtained on the AZ31-P, AZ31-O and AZ61-O specimens (Figures 4m, 4o and 4p) show a broad peak centred at 119.3–119.7 eV which indicates the presence of Al oxide and/or Al hydroxide. Attention is drawn to the presence of a significant metallic Al signal, which appears at approximately 117.3 eV, in the spectrum obtained on AZ61-P specimen (Figure 4n). The absence of a significant metallic aluminium signal on the AZ31-P, AZ31-O and AZ61-O specimens (Figures 4m, 4o and 4p) suggests that no appreciable quantities of Al^0 are located in the outer region of the material. In contrast, this aluminium is in metallic or free form on the surface layers of the AZ61-P specimen (Figure 4n).

The thickness of the native oxide film (d_{o}) on the surface of the magnesium specimens was calculated using the expression given by Strohmeier [53]

$$d_0(nm) = \lambda_{oxide} \sin\theta ln \frac{(I_{oxide}\ x\ \lambda_{metal}\ x\ N_{metal})}{(I_{metal}\ x\ \lambda_{oxide} x\ N_{oxide})} + 1 \quad (1)$$

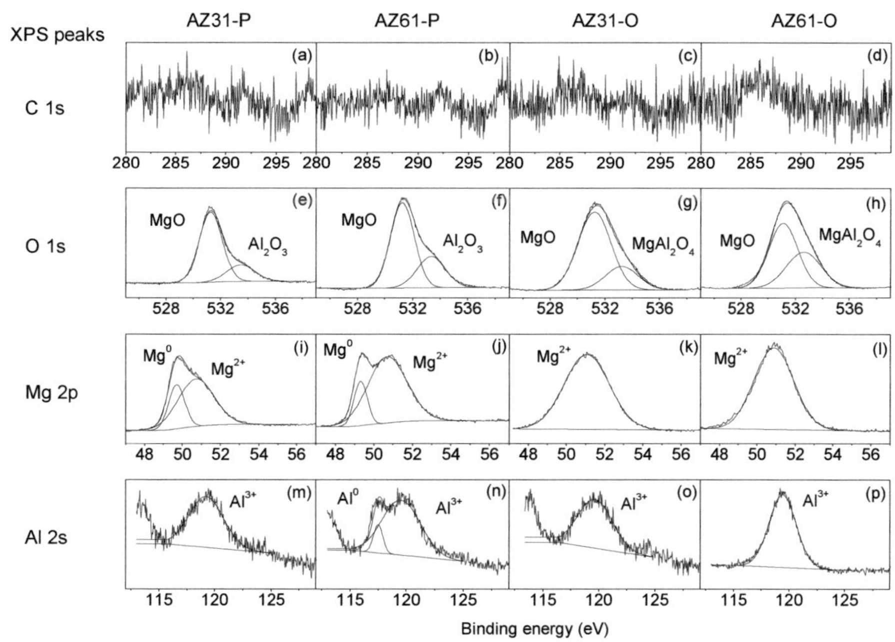

Figure 4. (a-d) Variation in the C1s, (e-h) O1s, (i-l) Mg2p and (m-p) Al2s high-resolution peaks obtained by XPS on the surface of the specimens after 5 min sputtering as a function of surface conditions and alloy type (figure adapted from [27]).

where I_{oxide} and I_{metal} denotes the component intensities, λ_{metal} and λ_{oxide} the mean free paths of photoelectrons of the metal atoms and the metal oxide, θ is the take-off angle for the analysed electrons, $\theta = 45°$ in this work; and N_{metal} and N_{oxide} correspond to the number density of the magnesium in metallic Mg and the surface layer. For the current investigation, an N_{metal}/N_{oxide} ratio of 1.24 was used [54]. The λ values for magnesium metal and magnesium oxide were kept as 3.0 and 2.6 nm [55-57], respectively. Figure 5 shows the oxide film thickness as calculated from Eq. (1) for the Mg2p peak after 3 min and 4.25 min (not shown), 5 min (Figures 4i and 4l) and 6 min of sputtering (not shown) in AZ31-P and AZ61-P specimens; and after 17 min, 20 min, 25 min and 30 min of sputtering (not shown) in AZ31-O and AZ61-O specimens. The thickness values have been determined by extrapolation to zero sputtering time.

From XPS analysis of oxide film thickness, it is observed that this value in the AZ61-P specimen is some 2 nm thicker than on the AZ31-P specimen (Figure 5a). The differences in film thickness among the two alloys are probably related with the degree of perfection of the oxide films and differences in diffusion rate of the reaction products [13], [26] and [58]. The higher film thickness values obtained with the AZ61 alloys, which contain more Al than AZ31 alloy, may be related to the heterogeneous structure associated with the β-precipitation phase located in the boundaries (Figure 3b). On the other hand, the presence of a coarse discontinuous outer layer of magnesium and aluminium oxides on the original surface of the AZ31-O and AZ61-O specimens (Figures 1a, 1b, 2a and 2b and Table 2) impedes the measurement of native oxide film thickness.

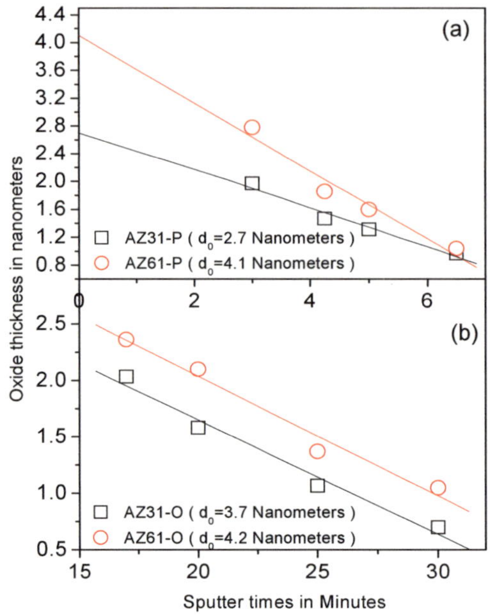

Figure 5. (a) Oxide thickness obtained by XPS on AZ31-P and AZ61-P specimens alloy and (b) in AZ31-O and AZ61-O specimens after sputtering (figure adapted from [27]).

3.2. Corrosion behaviour of magnesium alloys

The effect of native oxide films in chloride solutions at different times is described below.

3.2.1. Immersion tests

3.2.1.1. Hydrogen evolution measurements

Hydrogen evolution is dependent on the nature and composition of oxide films. Figure 6 shows the hydrogen evolution versus time curves (direct measure of the corrosion rate) during immersion in 0.006 M NaCl (a, d), 0.06 M NaCl (b, e) and 0.6 M NaCl (c, f) for 700 hours. No significant differences were observed in these curves for the AZ31-P specimen compared to the AZ31-O specimen, regardless the chloride concentrations of the solution (Figures 6a-6c). In contrast with the AZ31 alloy, significantly lower hydrogen evolution data were observed in the AZ61-P specimens during immersion in 0.006M (Figure 6d) and 0.06M NaCl (Figure 6e) than those corresponding to the AZ61-O substrate. Similar values of hydrogen volume evolved were measured in the AZ61-P specimen compared to those of the AZ61-O specimen during immersion in 0.6M NaCl (Figure 6f).

3.2.1.2. Electrochemical impedance measurements

The evolution of the corrosion process on the AZ31 and AZ61 alloys as a function of the surface conditions has been monitored by means of impedance measurements. It will be observed

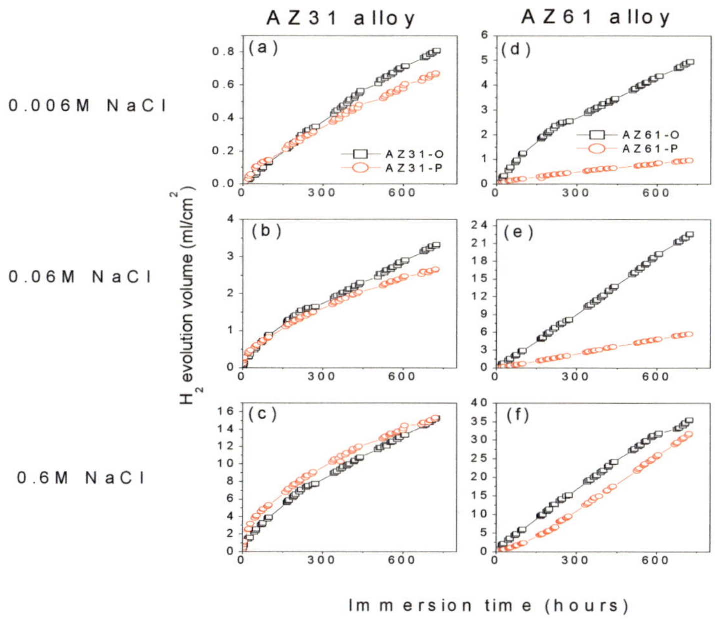

Figure 6. Comparison of the hydrogen evolution as a function of surface conditions and alloy type during 700 hours of immersion in saline solutions with different chloride ion concentrations.

from Figure 7 that all Nyquist diagrams exhibit a capacitive arc at high frequencies and intermediate frequencies (HF and MF) followed by an inductive arc in the low-frequency region (LF).

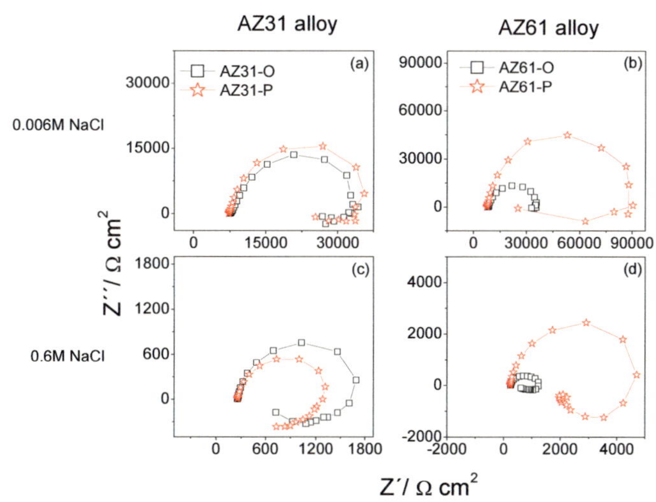

Figure 7. Comparison of the Nyquist plot as a function of substrate surface conditions and alloy type after 1 day of immersion in saline solutions with two different chloride ion concentrations.

In the literature about the corrosion of magnesium alloys, it is normal to associate the diameter of the capacitive loop in the HF/MF region with the charge transfer resistance (R_{CT}) of the corrosion process [9, 59, 60] value which is inversely related to the corrosion current (i_{corr}) through the well-known Stern-Geary equation [61]:

$$i_{corr} = \frac{B}{R_{CT}} \tag{2}$$

B being a proportionality constant which depends on the Tafel slopes, βa and βc. In this study, due to the uncertainty in the βa values from the polarisation curves [45], it was decided to use the constant B obtained empirically from the correlation between R_{CT} measurements and weight loss data [45]. Previous investigation [27] has justified the use of B values of 65 mV for the AZ31 alloy and 120 mV for the AZ61 alloy.

Table 3 compares the charge transfer resistance (R_{CT}) values for the AZ31 and AZ61 alloy specimens as a function of surface conditions and chloride ion concentration after one day of immersion. The corrosion current densities (mA/cm^2) obtained from Eq. (2) were converted to corrosion rates (mm/y) in Table 3 based on Faraday's law and the resulting equation:

$$corrosion\ rate = 22.85 i_{corr} \tag{3}$$

Specimens	Chloride ion concentration (M)	R_{CT} (Ω.cm^2)	Corrosion rate (mm/y)
AZ31-O	0.006	25000	0.06
AZ31-P		29000	0.05
AZ61-O		28000	0.10
AZ61-P		82000	0.03
AZ31-O	0.6	1500	1.0
AZ31-P		1100	1.3
AZ61-O		1000	2.7
AZ61-P		4000	0.7

Table 3. Variation in R_{CT} and corrosion rate values as a function of alloy-type surface condition and chloride ion concentration, after one day of immersion.

No significant differences in these values were observed in the AZ31-P specimen compared with those corresponding to the AZ31-O specimen (Table 3). In contrast with the AZ31 alloy, significantly lower corrosion rate values than the AZ61-O specimens were observed in those corresponding to the AZ61-P specimens (Table 3). It is interesting to note that similar trends regarding the corrosion behaviour are deduced from these electrochemical values as from the hydrogen evolution ones in 0.006 M NaCl (Figures 6a, d) and 0.06 M NaCl ones (Figures 6b, e).

The series of photographs in Figure 8 show that on the alloy AZ31 filiform corrosion is initiated almost immediately after immersion, especially on the AZ31-O specimen. The population of filaments expands in less than one day across the entire exposed surface. On the AZ61 alloy, the initiation and evolution of filiform corrosion is significantly slower than on AZ31, and clusters of propagating filaments do not appear until after 19 h of immersion on the as-received surface or 70 h on the polished surface. The small points in Figure 8 for AZ6-P specimen after 1 hour of immersion correspond to hydrogen bubbles growing at and detaching from the polished surface. In general, there is a qualitative agreement between the largest fraction of the corrosion area of the samples (Figure 8) and the electrochemical measurements (Table 3).

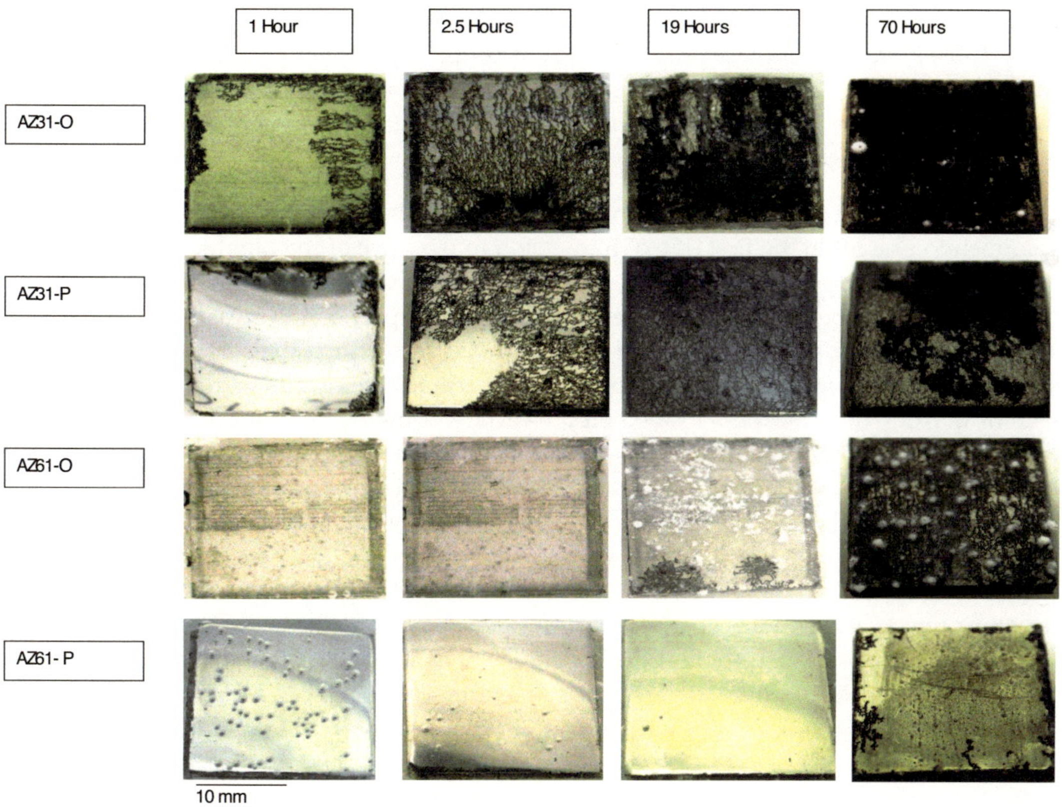

Figure 8. Photographic images of the evolution of corrosion morphology with immersion time for AZ31-O, AZ31-P, AZ61-O and AZ61-P specimens immersed horizontally about 50 mm below the surface of test solution (figure adapted from [44]).

As the results of the immersion tests, the native oxide film formed on the AZ31-P specimen does not reveal any particularly significant difference in corrosion resistance compared to those formed on the AZ31-O specimen (Figures 6a-c and Table 3).

Aluminium-rich particles containing high manganese, which are present in the alloy surface, can be detrimental to the corrosion resistance of Mg–Al alloys [62] The attack can be initiated at points close to these particles which, acting as a galvanic cathode, accelerates alpha phase corrosion. In the present work, the presence of abundant white particles has been observed on the surface of the AZ31-O specimen (Figure 1a), while, these are practically irrelevant on the surface of the AZ31-P ones (Figure 1b). Curiously, the corrosion results in Figures 6a-c, Figures 7a, c and Table 3 reveal no special effect of the initial presence or absence of these particles on the surface of the AZ31 specimens.

In the immersion test in the 0.006 and 0.06 M NaCl, attention is drawn to the notably lower corrosion rate on the AZ61-P specimens compared to the AZ61-O specimens over 30 days testing (Figures 6d, e). This difference may be due to the strong protective effect of the oxide film that forms spontaneously on the polished AZ61 alloy surface, much more uniform, homogeneous and continuous than that on the as-received surfaces (Figures 1b, d, Figures 2b, d and Table 2). As long as these properties do not noticeably deteriorate with testing time, said film will prevent or slow the enhanced micro-galvanic action due to the abundant presence of β-phase in this alloy (Figure 3b) in posterior immersion in saline solutions of weak (0.006 M NaCl) or medium (0.06 M NaCl) aggressiveness.

One point emerging from the corrosion rate values in Table 3 is the clear trend of the AZ61-P specimen to present in the first day of immersion in presence of the highly aggressive 0.6 M NaCl solution lower values than the AZ31-P specimen. Also, this effect seems to be reflected in the series of photographs in Figure 8, where a large part of the surface of the AZ61-P specimen retains for a quite some time the initial shin, while soon the surface of the AZ31-P specimen appears affected by the corrosion. Comparing these results with the XPS determinations of oxide film thickness formed on the surface of the AZ61-P and AZ31-P specimens, one clearly see a tendency towards a decrease in corrosion rate values as the thickness of the spontaneously formed MgO film increases (Figure 5a). This correspondence between electrochemical values and the native oxide film thickness suggests that said film somehow controls the corrosive attack in the early stages of the immersion tests in the 0.6 M NaCl solution. It is logical to relate a decrease in the corrosion rate with the strengthening of the oxide film that usually protects the surface of magnesium alloys. Any increase in the native oxide film thickness decreases the percentage of active points or oxide-free areas on the metallic surface, inhibiting the tendency (stimulated by the presence of Cl- ions) for metallic ions to pass from the reactive bare surface into the aqueous solution. It is important to mention that, after an initial period which does not last for more than about 7 days, no effect attributable to the initial surface condition of the AZ61-P specimen is generally seen (Figure 6f).

3.3. Isothermal oxidation

There is the possibility that some of the more significant changes observed in the composition and specific characteristics of the metallic surface for the AZ31 and AZ61 alloys commented on previously, may exert an appreciable effect on the alloy oxidation kinetics at low temperatures. In this respect, attention is drawn to: (a) very thin MgO layer, only few nanometers thick, grown on the surface of AZ31-P and AZ61-P specimens; (b) the lack of continuity of the oxide film on the surface of AZ31-O and AZ61-O specimens resulting from the fabrication of the wrought alloy.

http://dx.doi.org/10.5772/60721

Figure 9 shows that the AZ31-P and AZ61-P specimens present much lower increases in weight gain during isothermal oxidation than the corresponding AZ31-O and AZ61-O specimens. As can be seen in Figure 9, 35–60 min of heating at 200 °C produces a very small weight gain, around only 0.6 μg/cm^2, which is similar for the two alloys. It is generally accepted that the growth of compact MgO films is controlled by solid-state diffusion through adherent oxide areas followed by the reaction with oxygen at the oxide/gas interface; hence, a lack of easy-paths for fast Mg transport could be a possible explanation for a highly protective behaviour [26]. From Eq. (4) [63]:

$$D_L = 1.0\text{x}10^{-6}\exp\left(\frac{-150000}{\text{RT}}\right)\text{m2/s} \tag{4}$$

Diffusivity (*DL*) of Mg within the MgO lattice at 473 K is as low as 2.67 × 10^{-23} m^2/s, justifying negligible weight gain.

This behaviour is not detected in the case of the AZ31-O and AZ61-O specimens (Figure 9), which tends to suggest that the increase in weight gain with the oxidation time is dependent on the initial surface condition of the studied alloys. As Table 2 shows, roughness values of the AZ31-O and AZ61-O specimens are more than ten times greater than for the AZ31-P and AZ61-P specimens. The difference in weight gain with heating time between the as-received and polished surfaces (Figure 9) may be in agreement with the very heterogeneous and likely defective surface layer present on the as-received surfaces (Figure 2) compared to the continuity of the oxide films formed on the polished specimens.

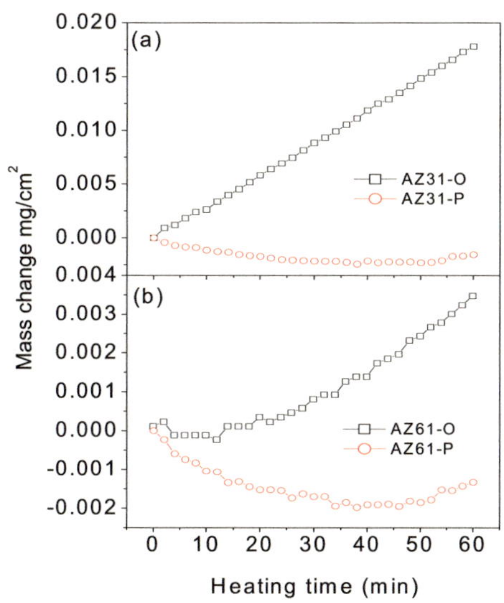

Figure 9. (a) Evolution of weight gain values obtained in the AZ31-O and (b) AZ61-O specimens as a function of the time of oxidation at 200°C in air compared with those of the same alloys in polished condition (figure adapted from [45]).

The evolution with sputtering time of the Al2s high-resolution XPS spectrum obtained on the oxide film formed on the surface of the AZ61-O and AZ61-P oxidised for 60 minutes in air at 200°C is shown in Figure 10. Attention is drawn to the high intensity of the Al^0 signal on the Al2s peak obtained on the AZ61-P specimen surface after 5 (Figure 10d) or 10 minutes of sputtering (Figure 10f) compared to its absence on the AZ61-O specimen surface (Figures 10c and 10e). This difference in trend suggests that a large part of the Al is in metallic state in the subsurface layers of the oxidised AZ61-P specimens, and as aluminium oxide or $MgAl_2O_4$ type species in the layers close to the outermost surface in the case of the heat-treated AZ61-O specimens. Jeurgens et al. [24] have investigated the growth kinetics and the evolution of the chemical composition and constitution of the initial oxide film grown on Mg-based MgAl surfaces by dry thermal oxidation, and observed that within the grown oxide films adjacent to the alloy/oxide interface, the interstitial sites were preferentially occupied by Al cations. In the present study, it is speculated that the homogeneous, uniform and compact native oxide layer present on the surface of the polished specimens (Figures 2c, d) inhibits significantly the oxidation or solid-state diffusion of the Al alloying element within the exposure times and temperature chosen in our study, inhibiting the growth of Al_2O_3 or $MgAl_2O_4$ type species [27].

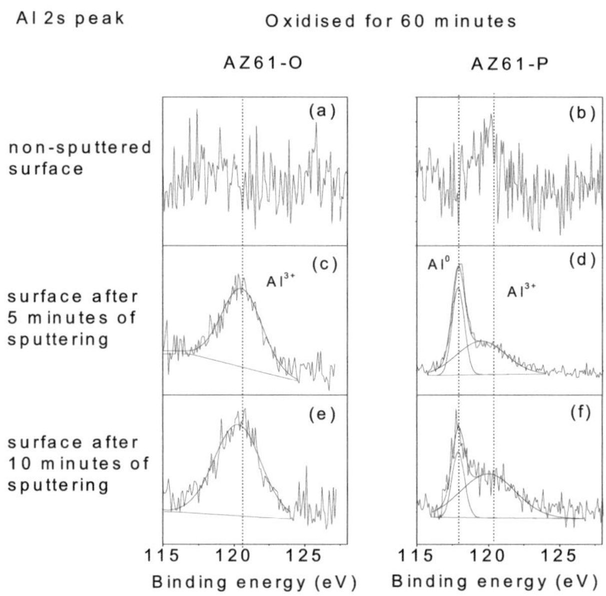

Figure 10. Variation of the Al2s high-resolution peak obtained by XPS on the surface of the AZ61-O and AZ61-P oxidised for 60 minutes as a function of the time of sputtering (figure adapted from [45] and [46]).

3.4. Native oxide film and sol–gel coating

In the XPS analysis of the external surface of the sol–gel coatings formed on the AZ31 substrate after the curing process, attention is drawn to the detection of significant amounts of magnesium (Table 4), which indicated the significant presence of discontinuities or pores in the coating that leave the substrate surface exposed. In contrast, the absence of magnesium indicates that the surface of the AZ61 alloy substrate is completely covered by the sol–gel coating.

Alloy	%C	%O	%Si	%Mg
AZ31	16	53	29	2
AZ61	14	56	30	0

Table 4. Atomic percentage observed by XPS of the external surface of AZ31 and AZ61 coated with sol–gel.

Figures 11 and 12 display optical images and AFM topography images which represent the visual appearance and topography of the sol–gel coatings formed on the AZ31 substrate. It can be seen that the surface is not uniform, presenting areas without apparent defects coexisting with other areas with visible pores or defects. In contrast, the sol–gel coating formed on the AZ61 alloy substrate looks far more perfect and uniform (Figures 11 and 12). The formation of porous and non-uniform coatings during their deposition on magnesium alloys has been reported by Song and Liu [64]. This behaviour appears to be a result of the hydrogen evolution closely associated with the magnesium dissolution from the substrate.

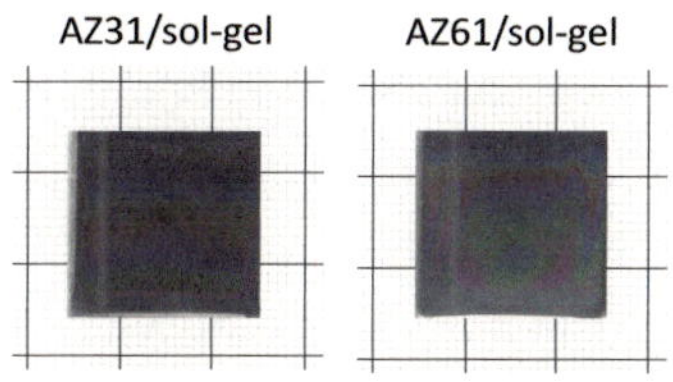

Figure 11. Optical pictures showing sol–gel coatings deposited on AZ31 and AZ61 magnesium alloy substrates.

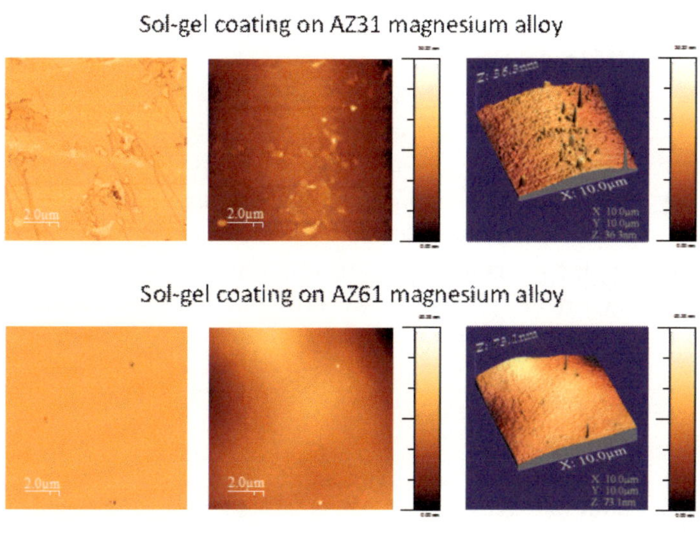

Figure 12. AFM topography images of the sol–gel coating formed on the AZ31 and AZ61 alloys substrates.

In previous studies [27], [43] and [44] with the same alloys immersed in a 0.6 M NaCl saline solution, we observed that the native oxide layer formed on the polished AZ61 surface was more protective than that on the AZ31 surface in the early stages of testing. The difference seemed to be due to the strong protective effect of the uniform, passive film of magnesium oxide that is formed spontaneously on the as-polished surface of the AZ61 magnesium alloy. At the beginning of the sol–gel coating formation, it is likely that this protective and homogeneous surface layer initially present on the polished AZ61 alloy substrate had a greater capability to isolate this alloy from the aqueous electrolyte environment and the effects of hydrogen evolution compared to the AZ31 alloy, resulting in the growth of a more perfect sol–gel coating than that of the AZ31 alloy.

It seems likely that some of the differences that have been revealed in the surface morphology of the sol–gel coatings formed on the AZ31 and AZ61 alloys have an impact on the corrosion behaviour. As commented earlier, the more significant changes that have been observed on the surface of the sol–gel coating, is the presence of defects or pores as a result of the interactions between coating and substrate during curing.

After less than 1 day of exposure, impedance measurements have revealed a significant reduction in the R_{HF} values of the coated AZ31 alloy substrate compared with those observed at the start of exposure (Figures 13 and 14). The presence of microscopic defects or pores in the coating formed on the AZ31 alloy (Figure 12) had the effect of decreasing its barrier properties, with any relevant influence on corrosion protection being lost.

In contrast to the findings with the AZ31 alloy, the sol–gel coating significantly improves the corrosion resistance of the AZ61 alloy as observed in the R_{HF} values obtained in the coated AZ61 samples, clearly exceeding the bare substrate values during the first seven days of corrosion testing in the presence of the highly aggressive 0.6 M NaCl solution (Figures 13 and 14). It seems likely that the uniformity and the absence of visible defects on the sol–gel coating formed on the AZ61 slow down the corrosion process by physically blocking the active sites on the metal surface (i.e. the presence of β-phase), diminishing the rate at which Cl^- ions are transported through the sol–gel coating.

The results obtained in this study show that the native oxide films spontaneously developed on AZ31 and AZ61 magnesium alloys prior coating exert a strong influence not just on the sol-gel thin film formation but also on the posterior corrosion protection behaviour of these sol–gel films when they are exposed to immersion tests to saline aqueous solutions. The chemical composition of the Mg–Al alloy is also a determinant factor to explain the subsequent growth of corrosion products on the metal substrates during immersion tests and the interactions of the corrosion products with the sol–gel films, exerting a remarkable effect on their posterior protection performance.

4. Conclusions

1. This chapter has summarised some of our work in an attempt to learn more about the influence of the native oxide films on the surface of AZ31 and AZ61 commercial magnesium alloys on their corrosion resistance in saline solutions of different aggressiveness.

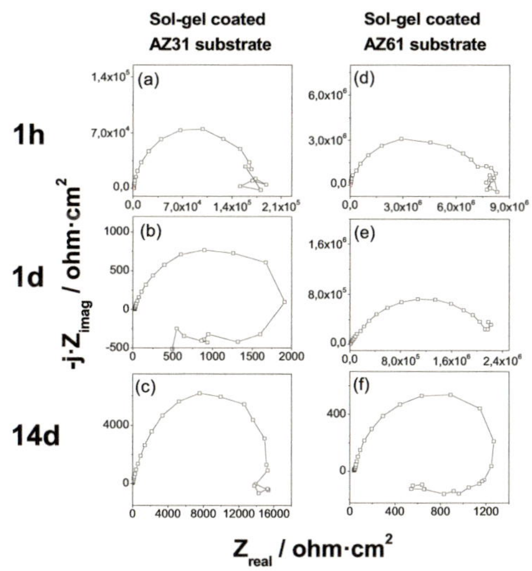

Figure 13. Variation in Nyquist plots as a function of the coated alloy-type substrate during immersion tests in 0.6 M NaCl aqueous solution.

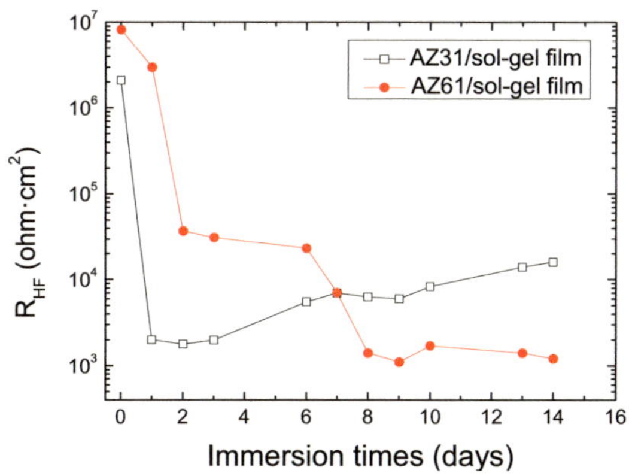

Figure 14. Variation in R_{HF} values as a function of the coated alloy-type substrate during immersion tests in 0.6 M NaCl aqueous solution.

2. The strong protective effect of the native oxide film that forms spontaneously on the polished AZ61 specimen, much more perfect and protective than the film on the as-received surface, prevents the fast micro-galvanic corrosion process present on the two-phase AZ61 alloy in weak or mild corrosive environments (0.006 M and 0.06 M NaCl solutions). In presence of the highly aggressive 0.6 M NaCl solution, the expected beneficial effect of the native oxide film has been shown in this investigation during the first hours or days of corrosion testing.

3. In contrast to the AZ61 alloy, it is important to mention that no effect attributable to the native oxide film is generally seen on the corrosion properties of the AZ31 alloy.

4. The results suggest some kind of link between the thickness of the native oxide film formed on the polished surfaces and the increase in the protective properties. From XPS determinations, it is found that the oxide film formed on the polished AZ61 alloy is some 1.4 nm thicker than on AZ31 alloy.
5. The difference in the isothermal oxidation behaviour at 200°C between the as-received and polished surfaces may be in agreement with the very heterogeneous and likely defective surface layer present on the as-received surfaces compared to the more uniform, homogeneous and continuous native oxide film formed on the polished surfaces. The native oxide layer present on the surface of the alloys inhibits significantly the oxidation or solid-state diffusion of the Al alloying element within the exposure times and temperature chosen in our study, reducing the adverse effects on the protective properties associated to the growth of Al_2O_3 or $MgAl_2O_4$ type species.
6. AFM, SEM and XPS analyses revealed that the sol–gel coatings formed on the surface of AZ61 alloy are far more perfect, uniform and free from microscopic defects than those formed on the AZ31 alloy. This behaviour is attributed to the effect of the native oxide film initially present on the surface of the AZ61 alloy, which inhibits the attack of magnesium. However, with the AZ31 alloy, the native oxide film is not sufficiently protective to prevent magnesium dissolution and associated hydrogen evolution, causing microscopic pores during the curing process of the sol–gel film formation.

Acknowledgements

The authors express their gratitude to Prof. S. Feliu for several clarifying and stimulating discussions during the course of this work. They also gratefully acknowledge financial support for this work from the Ministry of Economy and Competitivity of Spain (MAT 2009-13530 and MAT2012-30854).

Author details

Sebastián Feliu (Jr)[1*], Amir A. El Hadad[2], Violeta Barranco[3], Irene Llorente[1], Federico R. García-Galván[1], Antonia Jiménez-Morales[4] and Juan Carlos Galván[1]

*Address all correspondence to: sfeliu@cenim.csic.es

1 Centro Nacional de Investigaciones Metalúrgicas (CENIM), CSIC, Madrid, Spain

2 Physics Department, Faculty of Science, Al-Azhar University, Cairo, Egypt

3 Instituto de Ciencia de Materiales de Madrid (ICMM), CSIC, Madrid, Spain

4 Universidad Carlos III de Madrid, Departamento de Ciencia e Ingeniería de Materiales e Ingeniería Química, Leganés, Spain

http://dx.doi.org/10.5772/60721

References

[1] Wang C, Jiang B, Liu M, Ge Y. Corrosion characterization of micro-arc oxidization composite electrophoretic coating on AZ31B magnesium alloy. Journal of Alloys and Compounds 2015; 621(5) 53-61.

[2] Seifzadeh D, Bezaatpour A, Joghani RA. Corrosion inhibition effect of N, N'-bis (2-pyridylmethylidene)-1, 2-diiminoethane on AZ91D magnesium alloy in acidic media. Transactions of Nonferrous Metals Society of China 2014; 24(11) 3441-3451.

[3] Cao F, Shi Z, Song GL, Liu M, Dargusch M. Atrens A. Influence of casting porosity on the corrosion behaviour of Mg0.1Si. Corrosion Science. DOI:10.1016/j.corsci.2015.02.002.

[4] Ballerini G, Bardi U, Bignucolo R, Ceraolo G. About some corrosion mechanisms of AZ91D magnesium alloy. Corrosion Science 2005; 47(9) 2173-2184.

[5] Song GL, Atrens A, Wu XL, Zhang B. Corrosion behaviour of AZ21, AZ501 and AZ91 in sodium chloride. Corrosion Science 1998; 40(10) 1769-1791.

[6] Song GL, Atrens A, Dargusch M. Influence of microstructure on the corrosion of diecast AZ91D. Corrosion Science 1999; 41(2) 249-273.

[7] Ambat R, Aung NN, Zhou W. Evaluation of microstructural effects on corrosion behaviour of AZ91D magnesium alloy. Corrosion Science 2000; 42(8) 1433-1455.

[8] Song GL, Atrens A. Corrosion mechanisms of magnesium alloys. Advanced Engineering Materials 1999; 1(1) 11–33.

[9] Mathieu S, Rapin C, Hazan J, Steinmetz P. Corrosion behaviour of high pressure diecast and semi-solid cast AZ91D alloys. Corrosion Science 2002; 44(12) 2737–2756.

[10] Galicia G, Pebere N, Tribollet B, Vivier V. Local and global electrochemical impedances applied to the corrosion behaviour of an AZ91 magnesium alloy. Corrosion Science 2009; 51(8) 1789-1794.

[11] Fournier V, Marcus P, Olefjord I. Oxidation of magnesium. Surface and Interface Analysis 2002; 34(1) 494-497.

[12] Fotea C, Callaway J, Alexander MR. Characterisation of the surface chemistry of magnesium exposed to the ambient atmosphere. Surface and Interface Analysis 2006; 38(10) 1363-1371.

[13] McIntyre NS, Chen C, Role of impurities on Mg surfaces under ambient exposure conditions. Corrosion Science 1998; 40(10) 1697-1709.

[14] Chen C, Splinter SJ, Do T, McIntyre NS. Measurement of oxide film growth on Mg and Al surfaces over extended periods using XPS. Surface Science 1997; 382 (1-3) L652-L657.

[15] Santamaria M, Di Quarto F, Zanna S, Marcus P. Initial surface film on magnesium metal: A characterization by X-ray photoelectron spectroscopy (XPS) and photocurrent spectroscopy (PCS). Electrochimica. Acta 2007; 53(3) 1314-1324.

[16] Song GL. The grand challenges in electrochemical corrosion research. Frontiers in Materials 2014; DOI: 10.3389/fmats.2014.00002.

[17] Cabrera N, Mott NF. Theory of the oxidation of metals. Reports on Progress in Physics 1948; 12 163-184.

[18] Chen XB, Birbilis N, Abbott TB. Review of corrosion-resistant conversion coatings for magnesium and its alloys. Corrosion 2011; 67(3) Article Number: 035005.

[19] Nordlien JH, Nisancioglu K, Ono S, Masuko N. Morphology and structure of oxide films formed on MgAl alloys by exposure to air and water. Journal of the Electrochemical Society 1996; 143(8) 2564-2572.

[20] Nordlien JH, Nisancioglu K, Ono S, Masuko N. Morphology and structure of water-formed oxides on ternary MgAl alloys. Journal of the Electrochemical Society 1997; 144(2) 461-466.

[21] Lunder O, Lein JE, Aune TK, Nisancioglu K. The role of Mg17Al12 phase in the corrosion of Mg alloy AZ91. Corrosion 1989; 45(9) 741-748.

[22] Atrens A, Song GL, Cao F, Shi Z, Bowen PK. Advances in Mg corrosion and research suggestions. Journal of Magnesium and Alloys 2013; 1(3) 177-200.

[23] Song YW, Han EH, Dong KH, Shan DY, Yim CD, You BS. Study of the corrosion product films formed on the surface of Mg–xZn alloys in NaCl solution. Corrosion Science 2014; 88 215-225.

[24] Jeurgens LPH, Vinodh MS, Mittemeijer EJ. Initial oxide-film growth on Mg-based MgAl alloys at room temperature. Acta Materialia 2008; 56(17) 4621-4634.

[25] Saleh H, Weling T, Seidel J, Schmidtchen M, Kawalla R, Mertens FORL, Vogt HP. An XPS study of native oxide and isothermal oxidation kinetics at 300°C of AZ31 twin roll cast magnesium alloy. Oxidation of Metals 2014; 81(5-6) 529-548.

[26] Czerwinski F. The oxidation behaviour of an AZ91D magnesium alloy at high temperatures. Acta Materialia 2002; 50(10) 2639-2654.

[27] Feliu Jr. S, Maffiotte C, Samaniego A, Galván JC, Barranco V. Effect of the chemistry and structure of the native oxide surface film on the corrosion properties of commercial AZ31 and AZ61 alloys. Applied Surface Science 2011; 257(20) 8558-8568.

[28] Zheludkevich ML, Salvado IM, Ferreira MGS. Sol-gel coatings for corrosion protection of metals. Journal of Materials Chemistry 2005; 15(48) 5099-5111.

[29] Wang D, Bierwagen G R. Sol-gel coatings on metals for corrosion protection. Progress in Organic Coatings 2009; 64 (4) 327-338.

[30] Zheng SX, Li JH. Inorganic-organic sol gel hybrid coatings for corrosion protection of metals. Journal of Sol-Gel Science and Technology 2010; 54(2)174-187.

[31] Figueira RB, Silva CJR, Pereira EV. Organic–inorganic hybrid sol–gel coatings for metal corrosion protection: a review of recent progress. Journal of Coatings Technology Research 2015; 12(1) 1-35.

[32] Vreugdenhil AJ, Balbyshev VN, Donley MS. Nanostructured silicon sol-gel surface treatments for Al 2024-T3 protection. Journal of Coatings Technology 2001; 73(915) 35-43.

[33] Khramov AN, Balbyshev VN, Voevodin NN, Donley MS. Nanostructured sol-gel derived conversion coatings based on epoxy- and amino-silanes. Progress in Organic Coatings 2003; 47(3–4) 207–213.

[34] Li WL, Huang D, Xing XY, Tang JJ, Xing YJ, Li XJ, Zhang JD. Study the factors affecting the performance of organic-inorganic hybrid coatings. Journal of Applied Polymer Science 2014; 131(21) 41010.

[35] Vignesh RB, Edison TNJI, Sethuraman MG. Sol–gel coating with 3-mercaptopropyltrimethoxysilane as precursor for corrosion protection of aluminium metal. Journal of Materials Science & Technology 2014; 30 (8) 814–820.

[36] Garcia-Heras M, Jimenez-Morales A, Casal B, Galvan JC, Radzki S, Villegas MA. Preparation and electrochemical study of cerium-silica sol-gel thin films. Journal of Alloys and Compounds 2004; 380 (1-2) 219-224.

[37] Harb SV, Santos FCD, Caetano BL, Pulcinelli SH, Santilli CV, Hammer P. Structural properties of cerium doped siloxane-PMMA hybrid coatings with high anticorrosive performance. RSC Advances 2015; 5 (20) 15414-15424.

[38] El Hadad AA, Barranco V, Jiménez-Morales A, Hickman GJ, Galván JC, Perry, CC. Triethylphosphite as a network forming agent enhances in vitro biocompatibility and corrosion protection of hybrid organic-inorganic sol-gel coatings for Ti6Al4V alloys. Journal of Materials Chemistry B 2014; 2 (45) 7955-7963.

[39] Khramov AN, Balbyshev VN, Kasten LS, Mantz RA. Sol-gel coatings with phosphonate functionalities for surface modification of magnesium alloys. Thin Solid Films 2006; 514 (1-2) 174-181.

[40] Lamaka SV, Montemor MF, Galio AF, Zheludkevich ML, Trindade C, Dick LF, Ferreira MGS. Novel hybrid sol-gel coatings for corrosion protection of AZ31B magnesium alloy. Electrochimica Acta 2008; 53 (14) 4773-4783.

[41] Barranco V, Carmona N, Galván JC, Grobelny M, Kwiatowski L. Electrochemical study of tailored sol-gel thin films as pre-treatment prior to organic coating for AZ91 magnesium alloy. Progress in Organic Coatings 2010; 68(4) 347-355.

[42] Murillo-Gutiérrez NV, Ansart F, Bonino JP, Kunst SR, Malfatti CF. Architectural optimization of an epoxy-based hybrid sol-gel coating for the corrosion protection of a cast Elektron21 magnesium alloy. Applied Surface Science 2014; 309 62-73.

[43] Feliu Jr. S, Maffiotte C, Samaniego A, Galvan JC, Barranco V. Effect of naturally formed oxide films and other variables in the early stages of Mg-alloy corrosion in NaCl solution. Electrochimica Acta 2011; 56(12) 4454-4565.

[44] Samaniego A, Llorente I, Feliu Jr. S. Combined effect of composition and surface condition on corrosion behaviour of magnesium alloys AZ31 and AZ61. Corrosion Science 2013; 68 66-71.

[45] Feliu Jr. S, Samaniego A, Barranco V, El-Hadad AA, Llorente I, Adeva P. The effect of low temperature heat treatment on surface chemistry and corrosion resistance of commercial magnesium alloys AZ31 and AZ61 in 0.6 M NaCl solution. Corrosion Science 2014; 80 461-472.

[46] Feliu Jr. S, Samaniego A, Barranco V, El-Hadad AA, Llorente I, Serra C, Galvan JC. A study on the relationships between corrosion properties and chemistry of thermally oxidised surface films formed on polished commercial magnesium alloys AZ31 and AZ61. Applied Surface Science 2014; 295 219-230.

[47] El-Hadad AA, Barranco V, Samaniego A, Llorente I, García-Galván FR, Jímenez-Morales A, Galván JC, Feliu Jr. S. Influence of substrate composition on corrosion protection of sol–gel thin films on magnesium alloys in 0.6 M NaCl aqueous solution. Progress in Organic Coatings 2014; 77(11) 1642-1652.

[48] ZView Software, Version 3.1c; Scribner Associates Inc.: Southern Pines, NC, USA, 2007.

[49] Song GL, Atrens A. Understanding magnesium corrosion – a framework for improved alloy performance. Advanced Engineering Materials 2003; 5(12) 837–858.

[50] Liu M, Schmutz P, Uggowitzer PJ, Song GL, Atrens A. The influence of yttrium (Y) on the corrosion of Mg–Y binary alloys. Corrosion Science 2010; 52(11) 3687–3701.

[51] Feliu Jr. S, Maffiotte C, Galván JC, Pardo A, Merino MC, Arrabal R. The application of X-ray photoelectron spectroscopy in understanding corrosion mechanisms of magnesium and Mg-Al Alloys. The Open Surface Science Journal 2011; 3 1-14.

[52] Kim J, Wong KC, Wong PC, Kulinich SA, Metson JB, Mitchell KAR. Characterization of AZ91 magnesium alloy and organosilane adsorption on its surface. Applied Surface Science 2007; 253(9) 4197-4207.

[53] Strohmeier BR. An ESCA method for determining the oxide thickness on aluminium-alloys. Surface and Interface Analysis 1990; 15(1) 51-56.

[54] Liu M, Zanna S, Ardelean H, Frateur I, Schmutz P, Song GL, Atrens A, Marcus P. A first quantitative XPS study of the surface films formed, by exposure to water, on Mg

and on the Mg–Al intermetallics: Al3Mg2 and Mg17Al12. Corrosion Science 2009; 51(5) 1115-1127.

[55] Tanuma S, Powell CJ, Penn DR. Calculations of electron inelastic mean free paths. 2. Data for 27 elements over the 50–2000 eV range. Surface and Interface Analysis 1991; 17(13) 911-926.

[56] Akkerman A, Boutboul T, Breskin A, Chechik R, Gibrekhterman A, Lifshitz Y. Inelastic electron interactions in the energy range 50 eV to 10 keV in insulators: alkali halides and metal oxides. Physica Status Solidi (b) 1996; 198(2) 769-784.

[57] Powell CJ, Jablonski A. NIST electron inelastic-mean-free-path database, SRD 71, US Department of Commerce National Institute of Standards and Technology, Gaithersburg, MD, 2000.

[58] Feliu Jr. S, Bartolomé MJ. Influence of alloying elements and etching treatment on the passivating films formed on aluminium alloys. Surface and Interface Analysis 2007; 39(4) 304-316.

[59] Pebere N, Riera C, Dabosi F. Investigation of magnesium corrosion in aerated sodium-sulfate solution by electrochemical impedance spectroscopy. Electrochimica Acta 1990; 35(2) 555-561.

[60] Makar L, Kruger J. Corrosion studies of rapidly solidified magnesium alloys. Journal of the Electrochemical Society 1990; 137(2) 414-421.

[61] Stern M, Geary AL. Electrochemical polarization 1. A theoretical analysis of the shape of polarization curves. Journal of the Electrochemical Society 1957; 104(1) 56-63.

[62] Zeng R, Zhang J, Huang W, Dietzel W, Kainer KU, Blawert C, Ke W. Review of studies on corrosion of magnesium alloys. Transactions of Nonferrous Metals Society of China. 2006;16(1) s763-s771.

[63] Lea C, Molinari C. Magnesium diffusion, surface segregation and oxidation in Al–Mg alloys. Journal of Materials Science 1984; 19(7) 2336-2352.

[64] Song GL, Liu MH. The effect of Mg alloy substrate on "electroless" E-coating performance. Corrosion Science 2011; 53(11) 3500-3508.

Section 4

Ni-P Alloys for Imprinting on Pyrex Glass

Chapter 5

Application of Ni-P Alloys to A Mold Material for Thermal Imprinting on Pyrex Glass

Harutaka Mekaru

Additional information is available at the end of the chapter

http://dx.doi.org/10.5772/60408

Abstract

The author developed a low-cost mold for imprinting on Pyrex glass. The mold was fabricated by the processing of micro-/nano-patterns made by FIB etching on an amorphous Ni-P alloy layer electroless-plated on Cu/Cr/Si, Ni/Ti/Si, and Inconel-600 substrates, followed by a thermal treatment, since a 400 °C for 3 h thermal treatment was known to increase the hardness of a Ni-P layer. The wt.% of P was changed to 4, 8, and 16; and scratch test and imprint experiments on the Pyrex glass showed Ni-16P/polished-Inconel-600 mold giving a best result. We also successfully demonstrated the making of line/space patterns, microlenses, and AIST logos on Pyrex glass. It was also experimentally proved that a Ni-P alloy can be used as a mold material to imprint on Pyrex glass at high temperature around 640 °C under vacuum without employing any release agent.

Keywords: Ni alloy, Thermal imprint, Hot embossing, Pyres glass, FIB, Microlens

1. Introduction

Nanoimprint lithography [1] has received considerable attention from the industry because it is capable of fabricating nanostructures at relatively low costs. Currently, UV-nanoimprint [2], [3] is beginning to make valuable contributions to the advancement of next-generation semiconductor manufacturing technology. In thermal nanoimprint [1], patterning on molds has been relatively difficult, especially when the application is related to optical components. In recent years, many applications of thermal nanoimprint technology to plastic and glass materials have been developed [4]. In chemical and biochip devices, expensive disposable glass chips have largely been replaced by cheaper plastic chips. Plastic chips can be produced rapidly by mass production with molding techniques. However, glass chips are still used when the chips must be heated and cooled, or optical transparency demanded by high-resolution evaluation is required. Therefore, we are developing micro- and nanoimprint lithography technology for glass. A glass-nanoimprint lithography experiment was carried out, and the author succeeded in molding line and space patterns with linewidths of 300 nm on Pyrex glass (glass transition temperature T_g=560 °C) and quartz (T_g=1300 °C) [5], [6]; and an original thermal nanoimprint system was also developed [7], [8]. Moreover, a diamond mold for specializing in the imprinting on Pyrex glass was also developed [9]. However, the manufacturing application of this technique might not go far because the materials used here, such as glass-like carbon (GC) mold, and the diamond mold are expensive. In these glass-nanoimprint lithography experiments, GC that was selected as the mold material is still stable at a temperature of 1650 °C. However, there remain some problems when GC is used industrially as a mold material. GC is an expensive material, and is difficult to machine because of its fragility. Moreover, GC is brittle and it cannot withstand pressures of over 5.23 MPa [10]. As a result, it is quite difficult to make molds with high-aspect-ratio patterns.

2. Ni-P/Cu/Cr/Si mold

The fabrication technology of metallic micro - and nanostructures by plating is widely used in the manufacturing of wiring for electronic devices, the processing of optical components, and chemical and biochip devices. An electroformed Ni is often chosen as a material for metallic molds used for thermal nanoimprint lithography of engineering plastics such as polymethyl acrylate and polycarbonates because the extraction process is rapid and its machining is easy. Moreover, electroformed Ni is suitable for molding with high loads because the mechanical strength on Ni significantly improved after alloying. For glass nanoimprint technology, widely employed in the industry, the development of comparatively inexpensive Ni alloy molds like Ni-P [11], [12] and Ni-W [13] has resulted in significant advancements in the area. After searching through many Ni alloys [14], [15], [16], we chose an amorphous Ni-P alloy to serve as mold material for our work. The rationale behind this choice was the ease of working with the amorphous material, and that its mechanical characteristics improve with thermal treatment. In addition, when a nanopattern is formed on a curved surface such as, on a cylinder mold used with roller nanoimprint lithography [17], electroformed Ni–P alloy is selected in a great majority of cases. So we thought of making the Ni alloy as amorphous, and then use it as a mold for Pyrex-glass nanoimprint lithography.

2.1. Fabrication and patterning

In an amorphous material, the atomic arrangement is roughly in a thermodynamically metastable state. A lot of amorphous metals are employed in alloy plating with metals that can usually plate nonmetals or high-melting point metals. In these alloy-plating methods, the size of the crystal becomes smaller as the proportion of the secondary element increases. The entire film adopts an amorphous structure with a sufficient content of a secondary element [18]. As a mold material, we chose to use the amorphous Ni–P alloy in conjunction with an electroless-plating method because it allows easy fabrication. For this experiment, 4-inch Si wafers were used for substrates on which a chrome layer of 0.1 μm thickness was formed by sputtering for adhesion-force improvement. Then a copper layer of 1 μm thickness was deposited on it by sputtering as a seed layer. Next an amorphous Ni–P alloy layer of 10 μm thickness was formed on this substrate. Table 1 shows the optimum conditions for electroless-plating with amorphous Ni–P alloy and for electroplating with pure Ni, which was used for a comparison. The determination that the Ni–P alloy was, in fact, amorphous, was made using the measurement results from X-ray diffraction. Figure 1 shows an X-ray-diffraction spectra of electroplated pure Ni layer and electroless-plated Ni–P alloy. The X-ray-diffraction peaks of fcc Ni(111) and Ni(200) observed in the pure Ni layer is broad, and an amorphous halo peak can be observed in the Ni–P alloy. The intensity of the Ni(111) diffraction line decreases gradually as the content of P increases in the Ni–P alloy. When the content of P becomes 11 wt.% or more, a broad spectrum, which is peculiar to amorphous materials [11], is observed. Therefore, we concluded that the electroless-plated Ni–P alloy (with a content of P=8 wt.%) was sufficiently amorphous. The mold pattern was etched with focused-ion-beam (FIB) lithography. The FIB system used in this experiment was an FB-2000A from the Hitachi High-Technologies Corporation. We drew the pattern by using an aperture of 100 μm diameter, in consideration of a trade-off between the processing speed and the accuracy of the pattern. Figure 2 shows the processing speed for Si, pure Ni, and amorphous Ni-P alloy. The total ion-beam charge irradiated to an area of 1 μm^2 square was calculated according to this procedure. This value was plotted against the horizontal axis. The process depth produced by FIB for each substrate was measured by using a three-dimensional (3D) optical profiler NewView 5000 (Zygo); and was plotted against the vertical axis. In the result, the processing speed using FIB with amorphous Ni-P alloy was found to be almost same as that with the Si. In addition, a remarkable difference appears between the pure Ni and the amorphous Ni-P alloy in relation to the process accuracy. Figure 3(a) is a scanning-electron microscope (SEM) image of line and space patterns with linewidths of 500 nm and 1 μm made on a pure Ni substrate using FIB with an irradiated-ion-beam charge of 80 $\mu C/\mu m^2$. A Hitachi S-3000H SEM was used for the observations. The insertion figure is an expanded SEM image of the line and space patterns observed from a vertical direction at a depression angle of 30°. The processed surface was very rough, and the crystal grain seemed to have worsened the FIB etching considerably. The crystal size of the pure Ni layer was measured by a transmission electron microscope (TEM). The maximum crystal size was 500 nm. On the other hand, in processing the amorphous Ni-P alloy shown in Fig. 3(b), high process accuracy was achieved, and the surface was very smooth after etching under the same condition as in the case of the pure Ni. It would seem that when the pattern size is smaller than the size of the crystal grains used in the mold material, utilizing an amorphous material is effective, and improves the process accuracy.an aperture of 100 μm diameter, in consideration of the balance between the processing

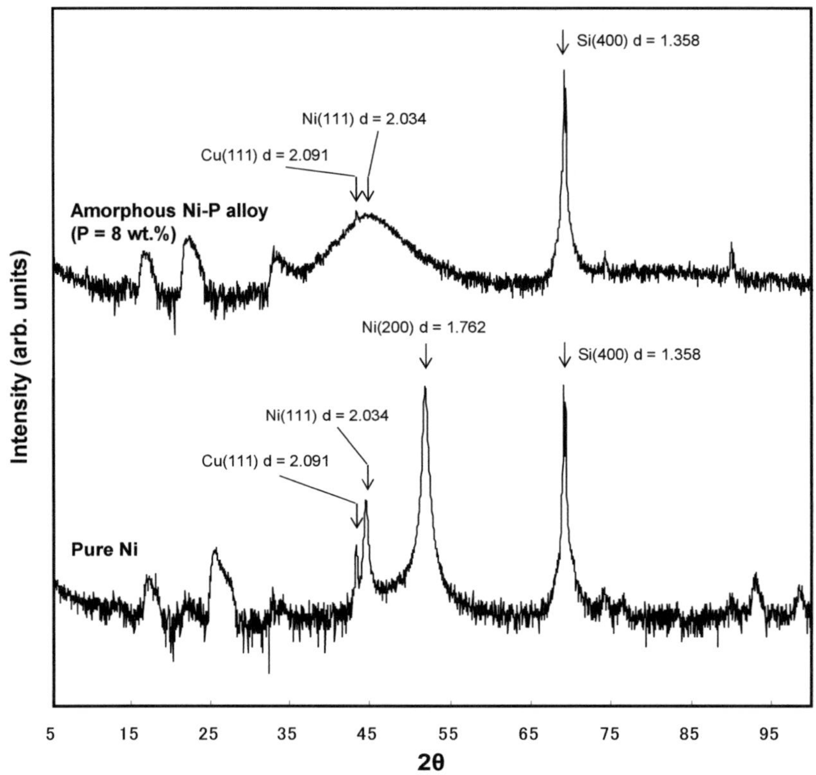

Figure 1. X-ray-diffraction spectra of electroplated pure Ni and electroless-plated Ni-P alloy.

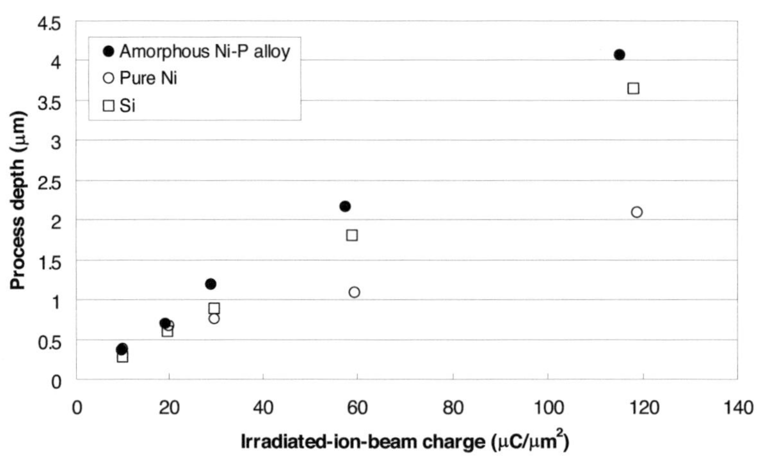

Figure 2. Relation between irradiated-ion-beam charge and process depth of Si, electroplated pure Ni, and electroless-plated Ni-P alloy.

http://dx.doi.org/10.5772/60408

	Ni-P electroless plating	Ni electroplating
Content ratio of P	8 wt.%	0 wt.%
Substrate	Cu/Cr/Si, Ni/Ti/Si,	Ni/Ti/Si
Degreasing	ACE CLEAN A-220*: 30 g/L Temperature: 50° C Time: 1 min	OPC CLEAN 91*: 8 ml /L Temperature: 23° C Time: 1 min
Activation	ICP ACCERA*: 200 ml/L Temperature: 25° C Time: 1 min	
Plating	TOP NICORON NAC-A*: 80 ml/L + TOP NICORON NAC-B*: 200 ml/L Temperature: 90° C Time: 50 min	Sulfamine acid nickel: 600 g/L Temperature: 50° C Time: 200 min Current density: 500 A/m^2
Washing	Distilled water Temperature: 25° C	Distilled water Temperature: 23° C

* Product of Okuno Chemical Industries Co., Ltd.

Table 1. Plating conditions of amorphous Ni-P alloy and pure Ni.

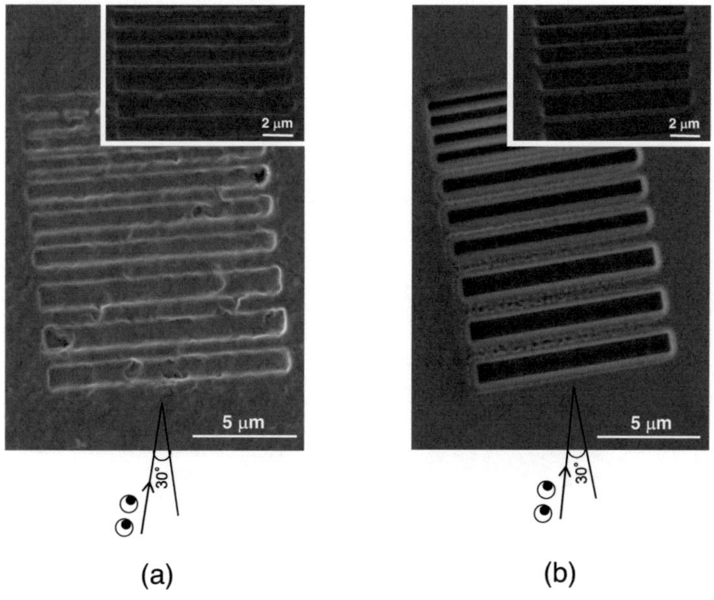

Figure 3. SEM images of patterns etched by FIB on: (a) electroplated pure Ni, and (b) the electroless-plated Ni-P alloy. Expanded SEM images, from a vertical direction, of line and space patterns at a depression angle of 30°.

2.2. Thermal nanoimprint system and experimental procedure

Figure 4 shows a photograph and specifications of a high-temperature thermal nanoimprint system ASHE0201 [7] (Engineering System) used for the glass-nanoimprint lithography experiment. To apply the mold on to a Pyrex glass at high temperature two ceramic loading stages (upper and lower) were employed. To experiment at a high temperature, the upper and bottom stages were made from ϕ90 mm ceramic plates. The backs of the loading stages were fitted with graphite heaters and cooling devices where the later were connected to a chiller system. The graphite heater and a cooling system were built into the back of the ceramic plates. A mold could be heated up to 1400 °C in this system. This system had a vacuum chamber of the bellow's type. When the setup was completed the bellows-type chamber was closed, and then the molecular turbo pump was started following a rotary pump. The loading stages could be decompressed to 0.07 Pa with a rotary pump and a turbo molecular pump. A load of 4.9 kN or less was applied with a servomotor. Figure 5 shows the sample setup and procedure for the thermal imprint experiment. The sequence of the experiment was:

1. The starting point of the upper loading stage drive was detected.

2. A 2-mm-thick GC sheet was put on the bottom loading stage for leveling adjustments.

To fix this sheet, thick GC blocks were arranged.

3. With its pattern side up, an amorphous Ni-P alloy mold was put on the lower stage.

4. A 0.5-mm-thick Pyrex glass was put on top of the amorphous Ni-P mold.

5. Another 2-mm-thick GC sheet was put on top of thnume Pyrex glass.

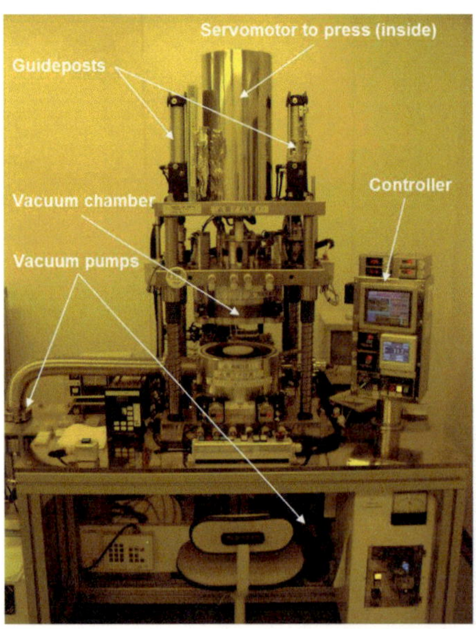

ASHE0201

• Maximum mold size	ϕ 60 mm
• Highest heating temperature	≤ 1400 °C
• Arrival time to the highest heating temperature	15 min
• Highest vacuum level	0.07 Pa
• Maximum contact force	≤ 9.8 kN
• Driving stage method	Servomotor

Figure 4. Photograph and specifications of the high-temperature thermal nanoimprint system used.

http://dx.doi.org/10.5772/60408

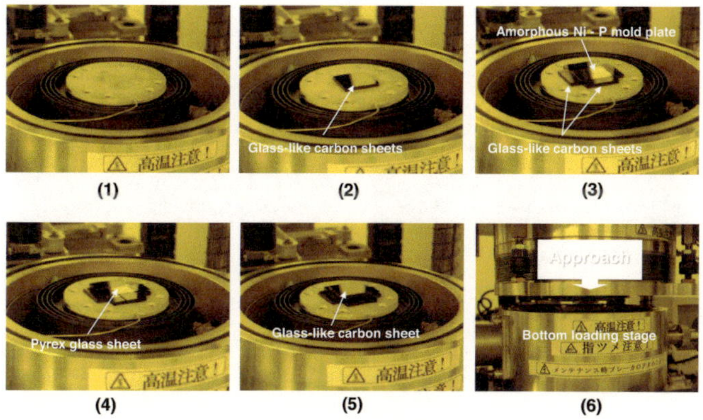

Figure 5. Setup procedure for an amorphous Ni-P alloy mold and Pyrex glass in the high-temperature thermal nano-imprint system.

6. The upper stage was lowered and brought close to the GC sheet until the space between the two was 1 mm or less.

7. The bellow's-type chamber was closed when the setup of the sample was completed; and then the molecular turbo pump was started following the rotary pump.

8. When the pressure in the chamber reached 10 Pa or less, the switches of the graphite heaters mounted on the upper and lower stages were turned on.

9. When the temperature reached a set value, a contact force was applied and maintained for a fixed time.

10. Afterwards, the cooling process was started, and we waited for the amorphous Ni-P mold and Pyrex glass to cool down to 150-200 ^{0}C in the vacuum for about 1 h.

11. The vacuum pump was stopped; the chamber was vented out to the atmosphere, and the upper stage was raised completely.

12. The Ni-P mold and Pyrex glass (adhered to each other) were removed from the chamber and manually pulled apart.

2.3. Experimental results and discussion

Line and space patterns with a depth of 3 μm and linewidths of 500 nm, 750 nm, and 1 μm were produced on an amorphous Ni-P alloy by FIB. After the FIB etching, the amorphous Ni-P alloy was thermally treated for 3 h at 400 °C by using the high-temperature thermal nano-imprint system ASHE0201. The Vickers hardness of the amorphous Ni-P alloy was increased to HV=873 by thermally treating it at 400 °C. In addition, the hardness of the amorphous Ni-P alloy was HV=1048, which is about 4.4 times that of the pure Ni when heated up to 600 °C, which is the molding temperature of the Pyrex glass. Moreover, the transformation of the pattern by thermal treatment was investigated. The SEM images of the line and space patterns before and after the thermal treatment are shown in Fig. 6.

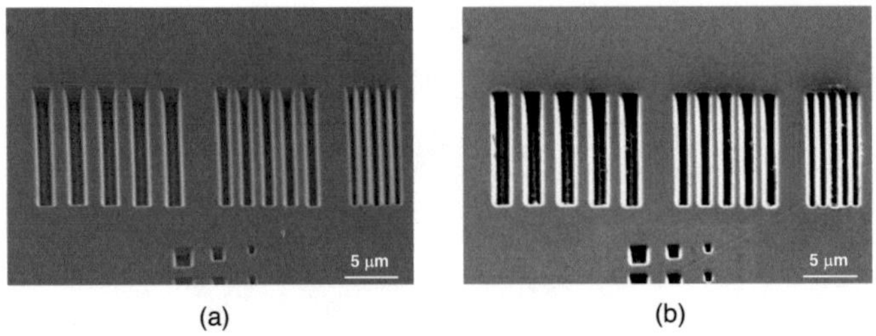

Figure 6. SEM images of line and space patterns on an amorphous Ni-P alloy mold (a) before and (b) after thermal treatment at 400 °C.

Molding conditions in this experiment were: heating temperature of 600 °C, maximum contact force of 5 kN, and contact time of 10 or 20 min. (The influence of the contact time on the molding accuracy was later investigated.) Subsequently, the graphite heaters were turned off and the amorphous Ni-P alloy mold and the Pyrex glass (Iwaki glass, Tg = 560 °C) were let to cool down to 150 °C in the vacuum chamber. For contact times of 10 and 20 min, the SEM images of the line and space patterns on the Pyrex glass are shown in Figs. 7(a) and 7(b), respectively. Moreover, cross-sectional profiles are shown by the white lines in the SEM images, measured with an atomic-force microscope (AFM) SPA-501 (SII NanoTechnology). We achieved the molding line and space patterns with linewidths of 750 nm and 1 μm. Because the heating temperatures were a little low, the Pyrex glass was not completely filled into the mold patterns. Therefore, the surface of the imprinted patterns curved and a clear profile could not be measured. The pattern produced using a 20 min contact time was higher than that for 10 min, although the height of the pattern was less than 3 μm. No variation was observed in the shape of the molding pattern even though the experiment was repeated ten times or more under various molding conditions.

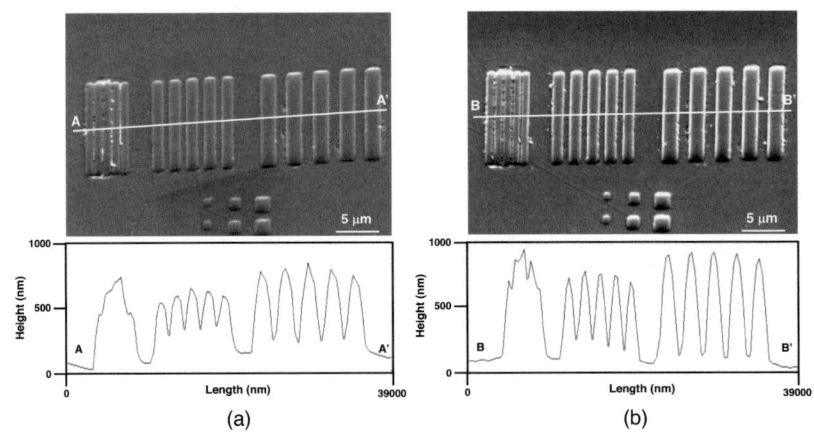

Figure 7. SEM and cross-sectional AFM images of line and space patterns on Pyrex glass after thermal imprint at contact times of: (a) 10 and (b) 20 min.

3. Ni-P/Ni/Ti/Si molds

In the process of fabricating a low-cost Ni-P alloy mold, an amorphous Ni-P plating layer on a silicon substrate was formed employing electroless-plating. However, prior to this plating, a 100-nm-thick Cr film was deposited on a silicon substrate to serve as an interlayer, and the deposited Cr film was then coated with a 1-μm-thick Cu film to serve as a seed layer to initiate the plating process. The electroless-plating resulted in the formation of a10-μm-thick Ni-P alloy layer on top of the Cu layer. Then, by the help of FIB etching, sets of line and space patterns were created on the Ni-P layer. The mold thus made, was used for thermal imprinting on a Pyrex glass [19]. In this thermal naoimprint work, however, the allowable heating temperature was only up to 600 ⁰C. With this restriction the transcription shape of the imprinted patterns on a Pyrex glass turned out to be less than desirable. When the temperature was raised up to of 620 ⁰C, the Ni-P alloy layer began to flake off from the Si substrate. It is believed that this phenomenon was caused by poor adhesive property of the Cr layer. The author then looked into other materials for the job. We used 500-nm-thick nickel layer as a seed layer and a 10-nm-thick Titanium layer as an interlayer that was sputter deposited on a 400-μm-thick Si wafer. Ti was used as an interlayer, because it has been known to increase the adhesive force by its anchoring properties.

3.1. Fabrication and patterning

For the purpose of comparison with an earlier work using the Cu/Cr/Si substrate [19], here the P content was set to 8 wt.%. Moreover, an electroplated layer of pure Ni on a Ni/Ti/Si substrate was also prepared for the purpose of comparison. Table 1 shows the parameters used for the various plating processes involved in these experiments. The mold pattern was created using FIB lithography that employed an EIP-5400 system (Elionix). Here, an ion-beam from a liquid Ga metallic ion source was accelerated by 40 kV to irradiate the surface of the sample. The beam current during the FIB processing was 0.5 nA. The FIB system was equipped with an improved version of the original 3D computer-aided-design (CAD) software. With this software, it was possible to process each layer by resolving 3D shape. The mold pattern in shape of microlenses were created as FIB scanned the concentric circle into which the multi-layer was divided. The etch times for the microlens pattern of curvature radii 12 μm and 20 μm were 113 min and 53 min, respectively. Figures 8(a) and 8(b) show SEM images of the microlens pattern in the curvature radii of 12 μm and 20 μm that were etched into the pure Ni substrate using FIB. Figures 8(c) and 8(d) show SEM images of the microlens pattern in the curvature radii of 12 μm and 20 μm etched into the amorphous Ni-P alloy substrate using FIB.

3.2. Experimental results and discussion

For the thermal imprint on Pyrex glass experiments a high-temperature thermal nanoimprint system ASHE0201 was used. When the temperature of the stage reached 620 ⁰C, a contact force of 1 kN was then applied to the upper stage and was maintained there for 10 min. Afterwards, the mold and the Pyrex glass were cooled down to 200 ⁰C, while still under vacuum. No release agent of any kind was used in the experiment. Figures 9(a) and 9(b) are the SEM images of the microlenses with the curvature radii of 12 μm and 20 μm imprinted on a Pyrex glass Pyrex

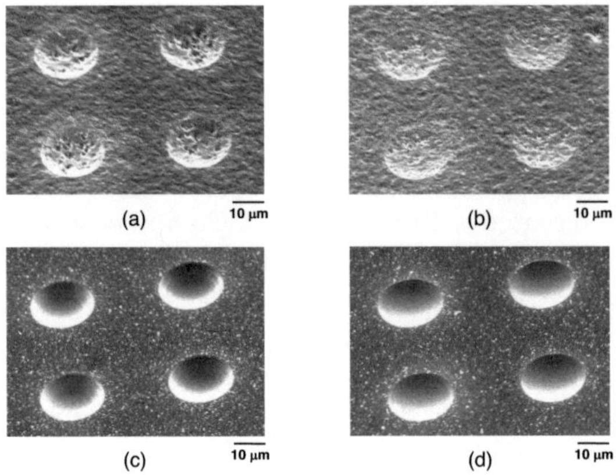

Figure 8. SEM image of a mold pattern of microlenses with the curvature radii of: (a) 12 and (b) 20 μm on the pure Ni mold. SEM image of a mold pattern for microlenses with the curvature radii of: (c) 12 and (d) 20 μm on the amorphous Ni-P alloy mold.

glass by using pure Ni mold. However, in Fig. 9(b) an imprinted microlens could not be observed. This could be related to the microlens pattern becoming indistinguishable in the noisy background of the rough surface. The surface roughness is an important factor in the fabrication of microlenses, especially because processing accuracy on pure Ni has been known to be poor. The irregularity of the surface of the pure Ni mold was transcribed on the Pyrex glass by the influence of the crystal grains on the mold surface. Figures 9(c) and 9(d) are SEM images of microlenses with the curvature radii of 12 μm and 20 μm imprinted on a Pyrex glass by using the amorphous Ni-P alloy mold. The imprinted patterns on a Pyrex glass lens were very smooth as shown in Figs. 9(c) and 9(d), their difference from the pure Ni mold is quite apparent.

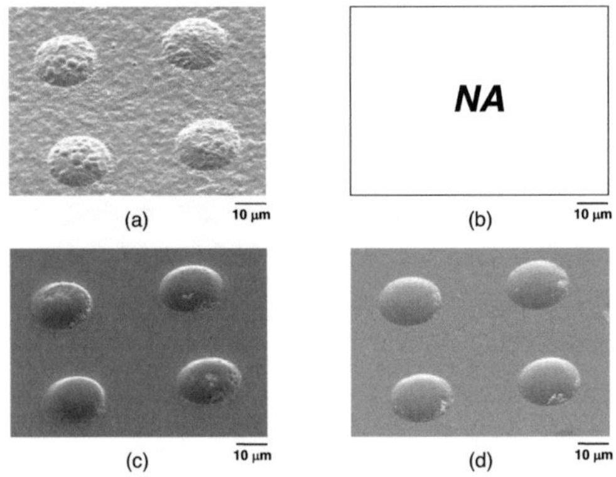

Reprinted from Microelectronic Engineering 2012; 85 873-876. [20]

Figure 9. SEM images of imprinted microlenses with the curvature radii of: (a) 12 and (b) 20 μm on a Pyrex glass using pure Ni mold. SEM images of imprinted microlenses with the curvature radii of: (c) 12 and (d) 20 μm on a Pyrex glass using the amorphous Ni-P alloy mold.

4. Ni-P/Inconel alloy molds

In previous experiments, the amorphous Ni-P alloy layer was electroless-plated on a Si substrate previously sputter-deposited with a 100-nm-thick interlayer of Cr and a 1-μm-thick seed layer of Cu [19]. However, the Ni-P/Cu/Cr/Si mold had weak adhesion forces between its different layers, and their maximum heat-proof temperature was only 600ºC. To address this difficulty, the interlayer was changed to a 10-nm-thick Ti layer, and the seed layer was changed to a 500-nm-thick Ni layer. Thus the mold composed of Ni-P/Ni/Ti/Si layers could then be used to imprint Pyrex glass at a higher temperature of 630 ºC [20]. However, from the imprinting experiments on Pyrex glass it appeared that the heating temperature would need to be 630-640 ºC. However, molds capable of functioning at even higher temperatures up to 650 ºC were considered more desirable. Therefore, we changed the substrate material from Si to Inconel-600.

4.1. Fabrication of molds

4.1.1. Electroless-plating

The author thought that the electroless-plated Ni-P alloy layer flaked off the Si substrate because the adhesion forces with the Cr layer were insufficient. We, therefore, changed the material. A 500-nm-thick Ni layer as a seed layer and a 10-nm-thick Ti interlayer were sputter-deposited on a 400-μm-thick Si. Ti was used because, as an interlayer, it increases the adhesion forces by an anchoring effect. The amorphous Ni-P alloy layer was deposited on the substrate by electroless-plating. For this case, the Ni-P alloy layer was deposited under three different plating conditions such that the content ratios of P could be set at 4, 8, and 16 wt.%. Details of each plating process are summarized in Table 2. The purpose of working with the three kinds of molds was to investigate the influence of the content ratios of P on the characteristics of the mold material. Moreover, an electroplated layer of pure Ni on a Si substrate fabricated under the same conditions as in Table 1 was also prepared to compare the X-ray diffraction data. In this section, the Ni-P alloy layers with content ratios of 4, 8, and 16 wt.% are labeled Ni-4P, Ni-8P, and Ni-16P. After plating, the substrate was diced into small square pieces of sides 30 mm in size; these pieces were thermally treated in a preliminary experiment. Each substrate was heated to 400 ºC for 3 h. Cross-sectional SEM images of the mold substrates of Ni-4P, Ni-8P, and Ni-16P are shown in Figs. 10(a), 10(b), and 10(c). Flaking-off of the plated layer was observed in all types of substrates. The Ni-P plated layer had flaked off from the Ni seed layer and the Ti interlayer in the Ni-4P/Ni/Ti/Si substrate. In the higher P density substrate, the bulk Si of the Si substrate cracked and flaked off while still adhering to the Ni-P plated layer. It appeared that the Ni-P plated layer had flaked off to relieve some internal stress developed during the thermal treatment because of some mismatch in the coefficients of thermal expansion of the two materials (Ni: 17:2 x 10^{-6}/K, Si: 4:4 x 10^{-6}/K, at 1000 K) [21]. We changed the substrate material from Si to Inconel-600 because Inconel-600 is a Ni alloy composed of 72% Ni (+Co), 14–17% Cr, 6–10% Fe [22] and the coefficient of thermal expansion of Inconel-600 (15.8 x 10^{-6}/K, at 973 K [21]) is almost equal to that of the bulk Ni. The size of the Inconel-600 substrate was 30 x 30 x 0.5 mm^3. To investigate the effect of chemical-mechanical polishing

(CMP) on the Inconel-600 substrate before the electroless-plating of the Ni-P layer, two kinds of Inconel-600 substrates were prepared. One was polished to a surface roughness of 10 nm while the other was left unpolished. Then, a 10-μm-thick Ni-P plated layer was deposited on the Si and Inconel-600 substrates under the plating condition described in Table 2. To evaluate the sticking power between the Ni-P layer and the substrate, a scratch test was done with tweezers and repeated ten times; results of this test are shown in Fig. 11.

Content ratio of P	4 wt.%	8 wt.%	16 wt.%
Substrate	Ni/Ti/Si, Inconel 600	Ni/Ti/Si, Inconel 600	Ni/Ti/Si, Inconel 600
Degreasing	Sodium hydroxide 15 g/L Sodium silicate, ortho 30 g/L Sodium alkylbenzene sulfonate 0.1 g/L Temperature: 23 °C Time: 40 s	Dipsol AL-45b) 40 g/L Temperature: 50 °C Time: 3 min	Sodium hydroxide 15 g/L Sodium silicate, ortho 30 g/L Sodium alkylbenzene sulfonate 0.1 g/L Temperature: 23 °C Time: 40 s
Activation	19.93% HCl Temperature: 23 °C Time: 8 min	HCL : 500 mL/L Temperature: RT Time: 0.5 min NiCl strike plating Temperature: 35 °C Time: 4 min	19.93%HCl Temperature: 23 °C Time: 8 min
Electroless plating	Sumer S-795 a) Temperature: 90 °C Time: 60 min	Dipsol NP-1900 b) Temperature: 88 °C Time: 40 min	Sumer S-795 a) Temperature: 90 °C Time: 60 min
Washing	Distilled water Temperature: 23 °C	Distilled water Temperature: RT	Distilled water Temperature: 23 °C

a) Product of JAPAN KANIGEN Co., Ltd.

b) Product of Dipsol Chemicals Co., Ltd.

Table 2. Electroless-plating conditions of amorphous Ni-P alloy.

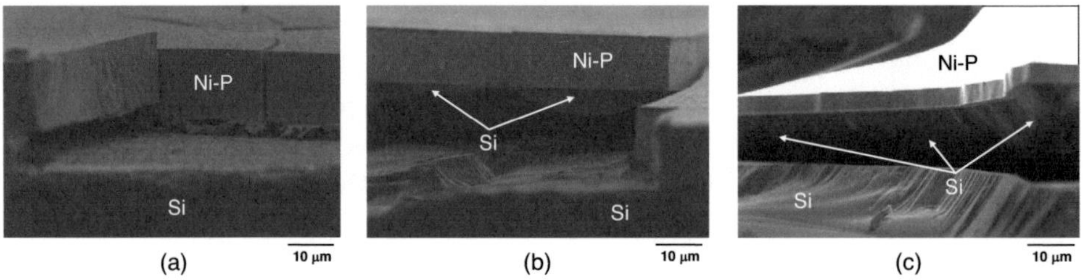

Figure 10. Cross-sectional SEM images of electroless-plated Ni-P alloys that flaked off from the Si substrate after heating at 400 ºC. Content ratio of P: (a) 4, (b) 8, and (c) 16 wt %.

http://dx.doi.org/10.5772/60408

As can be seen from this figure, a Ni-8P/unpolished-Inconel-600 substrate was not fabricated for these tests. The flaking off phenomenon by scratching was not observed in every substrate, although some black spots were observed on the Ni-4P/unpolished-Inconel-600 substrate. In the next step in this experiment, other substrates were thermally treated under the following conditions and sequence: first, the plated substrates were treated thermally at 400 ^{0}C for 3 h. After that, they were heated at 650 ^{0}C for 10 min. The substrates were then left to cool down to room temperature. Then 10 scratch examinations were executed. The Ni-P alloys electroless-plated on a Si substrate flaked off completely as shown in Fig. 12. The Ni-16P layer flaked off from the Si substrate completely even before scratching. This behavior was thought to be caused by the difference in the coefficients of thermal expansion of the Ni-P plated layer and the Si substrate as mentioned earlier. On the other hand, when the Inconel-600 substrate was used, Ni-P plated layers remained firmly adhered to the Inconel-600 substrate.

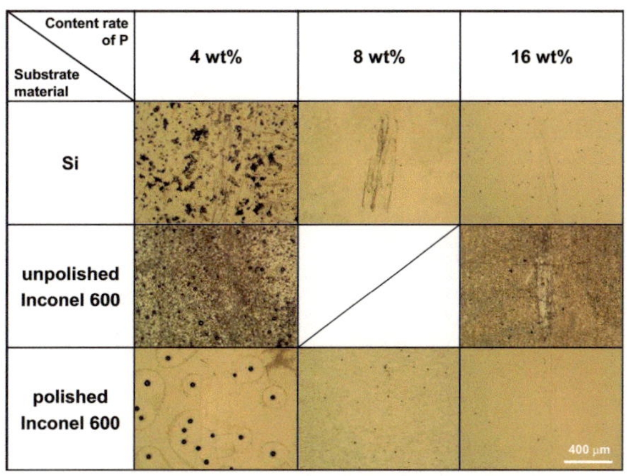

Figure 11. Results of scratch test on Ni-P alloy layers before thermal treatments.

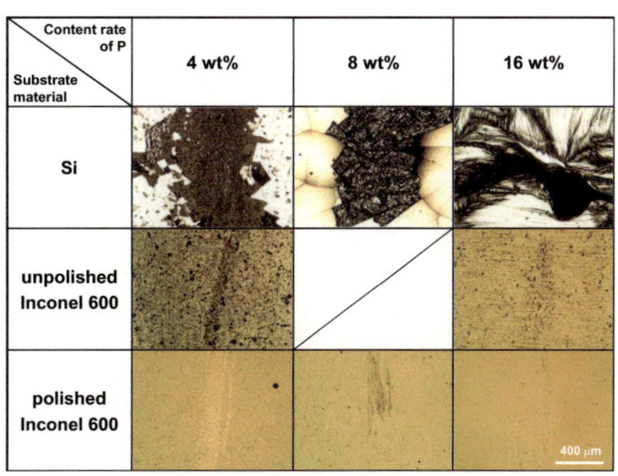

Figure 12. Results of scratch test of Ni-P alloy layers after thermal treatments.

4.1.2. Etching using FIB

A mold pattern was processed with FIB lithography using EIP-5400 system. FIB etching was performed under the same conditions that processed microlenses as shown in Fig. 8. Figure 13 shows the SEM images of the microlens pattern with the curvature radii of 12 and 20 μm etched with FIB on the Ni-P/Inconel-600 substrates before CMP. The image shows that the surface of the Ni-P layer was very uneven, and there were swellings of the same size as of the microlens pattern on the Ni-4P/unpolished-Inconel-600 substrate. This roughness seemed to originate from the surface roughness of the unpolished Inconel-600 substrate. Figure 14 shows the microlens pattern etched by FIB on the Ni-4P/polished Inconel-600 substrate. In this substrate, cracks were seen, and a part of the microlens pattern was affected. On the other hand, on the polished Inconel-600 substrate electroless-plated by Ni-8P and Ni-16P, the microlens pattern had a very smooth surface. From these results, Ni-4P/Inconel-600 seemed to be unsuitable for mold material.

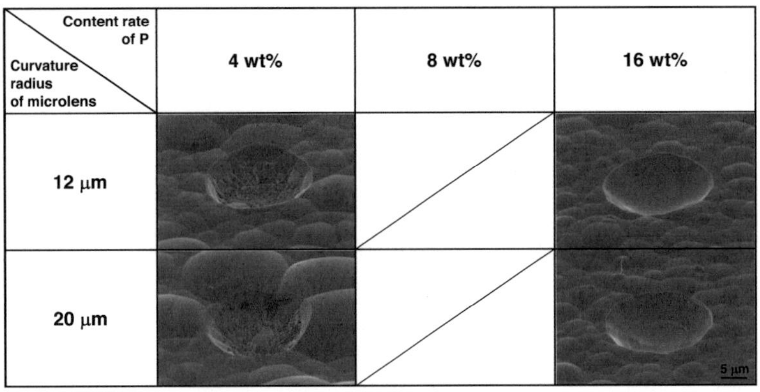

Figure 13. SEM images of microlens patterns on Ni-P alloy mold using unpolished Inconel-600 substrates.

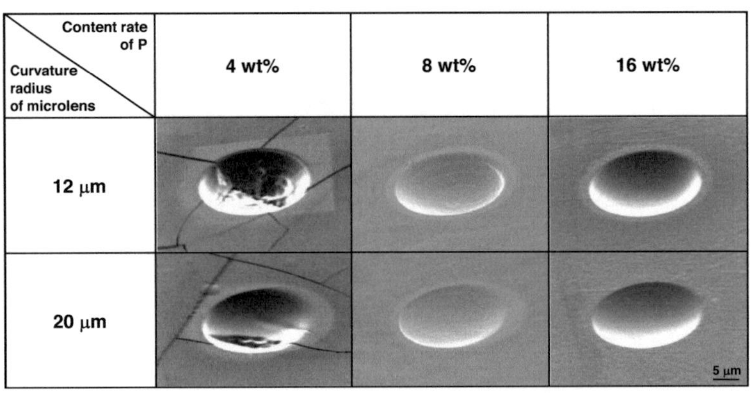

Figure 14. SEM images of microlens patterns on Ni-P alloy mold using polished Inconel-600 substrates.

4.1.3. Thermal treatment

To improve the hardness of the mold, an amorphous Ni-P alloy after FIB etching was thermally treated at 400 °C for 3 h using the same system ASHE0201. Normally, the heating temperature to imprint on Pyrex glass is 600-650 °C. Therefore, after a thermal treatment at 400 °C, the substrate was heated at 600 or 650 °C for 10 min. An Inconel-600 substrate was also thermal-treated under the same conditions for comparison.

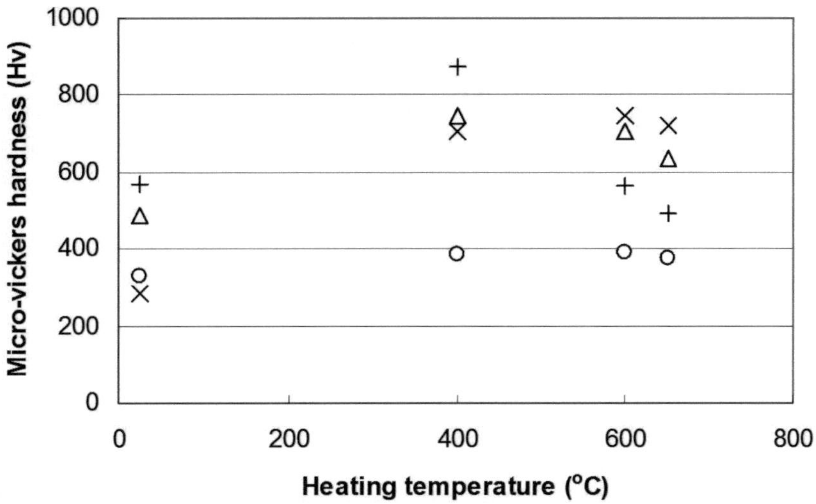

Figure 15. Relation between the heating temperature and the micro-vickers hardness of electroless-plated Ni-P alloys and the Inconel-600 substrate.

Figure 15 shows the results of measuring the micro-vickers hardness of these substrates with a micro-vickers hardness tester Matsuzawa MMT-X3. The micro-vickers hardness of the amorphous Ni-8P alloy increased to Vicker's hardness HV = 850 by thermal treatment at 400 °C. Even if Ni-4P and Ni-16P substrates were heated to 600 °C or more, the rate of the hardness decrease seemed to be gradual (comparatively speaking). In the case of the Inconel-600 substrate on which a Ni-P layer was not plated, the thermal treatment temperature influence on micro-vickers hardness was minor. Stiffening started by a precipitation of Ni_3P when the Ni-P alloy was heated at 400 °C or higher. However, softening of the Ni-P alloy started by heating it at temperatures higher than 450 °C, which is also the temperature at which micro-vickers hardness approaches the value of the Ni_3P crystal [11]. On the other hand, it hardly changes for Inconel-600 and maintains a value lower than HV of 400. This phenomenon can be confirmed from the results of the X-ray diffraction spectrum measured with MXP18AV (Bruker AXS) as shown in Fig. 16. The peaks of Ni(111), Ni(200), Ni(220), and Ni(311) that show crystalline states of Ni can be confirmed in a pure Ni substrate. In the amorphous Ni-P alloy substrate, a clear peak was not seen before thermal treatment at 400 °C; and the entire spectrum appeared as broad as in an amorphous material. In Ni-4P and Ni-16P alloys thermally treated at 400 °C for 3 h, a new peak did not appear. On the other hand, after the Ni-8P alloy was thermally treated for 3 h at 400 ⁰C, new peaks of $Ni_3P(031)$, $Ni_3P(231)$, and $Ni_3P(141)$ that originated in Ni_3P in addition to the peaks of Ni(111), Ni(200), and Ni(220) were seen. In the case of the Ni-8P alloy thermally treated at 400 ⁰C for 3 h, the alloy developed a crystalline form in which its micro-vickers hardness decreased abruptly. A rise in the hardness resulted in a phase change from an amorphous phase into a meta-stable crystal phase when the amorphous Ni-P alloys were thermally treated at 400 ⁰C. In addition, the hardness reaches a maximum when the sample changes into Ni_3P and Ni with a highly crystalline structure. Thus, it is necessary to thermally treat the amorphous Ni-P alloy mold at 400 ⁰C to increase its mechanical strength, and to improve its thermal tolerance at high temperatures. After a thermal treatment at 400 ⁰C for 3 h, Ni-P alloys were heated to 600 or 650 ⁰C for 10 min, and the peaks originating in the crystal of Ni_3P and Ni were observed in all substrates. The crystallization of Ni_3P and Ni progressed when the Ni-P alloys were heated to 450 ⁰C or more. Moreover, when Ni-P alloys were heated to 600 or 650 ⁰C, many fine particles precipitated in the crystal [11]. The Ni-P alloy, in which a crystal grain boundary exists, became fragile. The reason for this is a decrease in the micro-vickers hardness mentioned earlier.

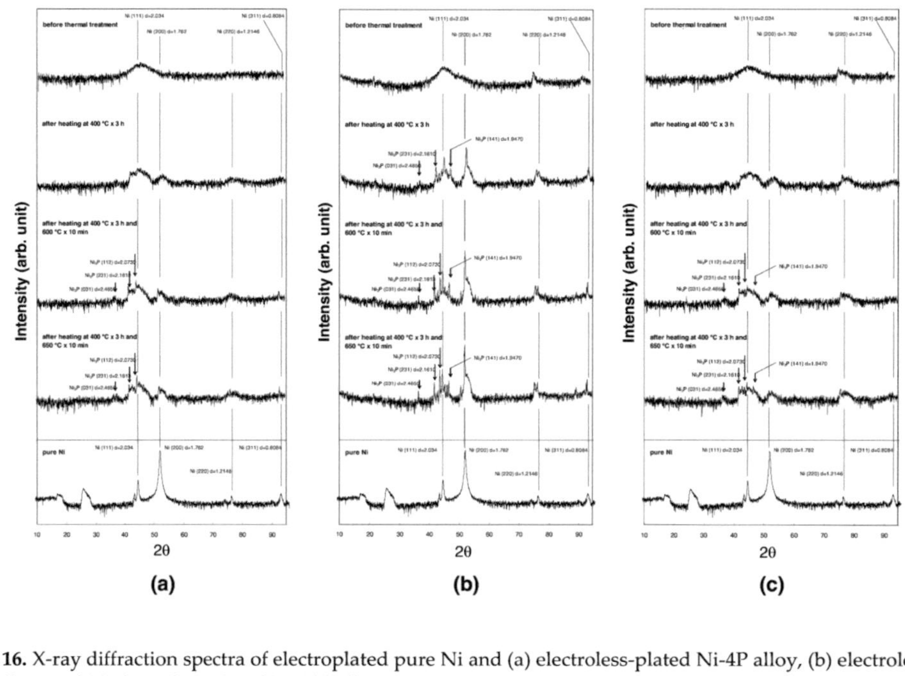

Figure 16. X-ray diffraction spectra of electroplated pure Ni and (a) electroless-plated Ni-4P alloy, (b) electroless-plated Ni-8P alloy, and (c) electroless-plated Ni-16P alloy.

4.2. Thermal imprint on Pyrex glass

The high-temperature thermal nanoimprint system, ASHE0201, was used for the thermal imprint experiment on Pyrex glass. When the temperature reached the set value, a contact force of 400 N was applied and maintained for a fixed time. Afterwards, the cooling process

http://dx.doi.org/10.5772/60408

was started, and we waited for the amorphous Ni-P alloy mold and Pyrex glass to cool down to 200 ⁰C in the vacuum for about 1 h. In the imprinting process, we experimented at heating temperatures of 630 and 640⁰C and changed the contact times to 10 and 20 min. As mentioned previously, a thermal-treatment process was executed without applying a loading force according to a similar procedure with the same system as in the thermal imprint experiment.

4.3. Experimental results and discussion

4.3.1. SEM Observation

Figure 17 shows the SEM images of the microlenses imprinted on Pyrex glass using Ni-4P and Ni-16P/unpolished-Inconel-600 molds. The heating temperature was 640 ⁰C and the contact time was 20 min for these experimental conditions. Surface roughness is a critical factor in the fabrication of microlenses. When unpolished Inconel-600 molds were used, the processing accuracy was low. The irregularity of the surface of Ni-P/unpolished-Inconel-600 molds was transcribed on Pyrex glass because of the influence of the surface roughness. On the other hand, when imprinting with the Ni-P/polished-Inconel-600 molds under the same conditions as with the unpolished Inconel-600 molds, the surface of the mold patterns and imprinted Pyrex glass lenses were very smooth, as shown in Fig. 18; and the difference from Ni-P/unpolished-Inconel-600 mold seemed remarkably clear. However, it was observed that the imprinted patterns were bulldozed sideways. The deformed pattern appeared to be pushed aside considerably when the Ni-8P/polished-Inconel-600 mold was used. The width of the deformation was not large, although a similar phenomenon was seen when the Ni-16P/polished-Inconel-600 mold was used. Imprinting using the Ni-16P/polished-Inconel-600 mold resulted in almost complete microlenses being processed. The heights of the microlenses imprinted under the four molding conditions were measured by a 3D optical profiler, NewView 5000. The heating temperatures were 630 and 640 ⁰C. The contact time was changed to 10 or 20 min. Because the surface was very rough as shown in Fig. 17, the height could not be properly measured when imprinted using Ni-P/unpolished-Inconel-600 molds. Figure 19 shows the measured results of the Pyrex glass microlens with the curvature radius of 12 μm that was imprinted by using the Ni-P/polished-Inconel-600 molds. The maximum depths of the mold pattern on the Ni-4P, Ni-8P, and Ni-16P alloy molds were measured to be 4.4, 4.7, and 4.5 μm, respectively. The maximum heights of the imprinted microlenses were divided by the maximum depths of the corresponding mold pattern, and this value was defined as the filling rate. The filling rate rose when the heating temperature was 640 ⁰C, regardless of which mold was used; and lengthening of the contact time also produced a higher filling rate. It became clear that heating temperature was a more effective parameter to raise the filling rate than the contact time was. The same tendency was observed for microlenses with a curvature radius of 20 μm as shown in Fig. 20. The maximum depths of the mold pattern of Ni-4P, Ni-8P, and Ni-16P alloy molds were measured to be 2.4, 2.3, and 2.2 μm, respectively. The filling rate rose, although the maximum depth of the mold patterns decreased compared with the microlenses with a curvature radius of 12 μm. When the heating temperature and the contact time were 640 ⁰C and 20 min, the filling rate was close to 1 for every mold.

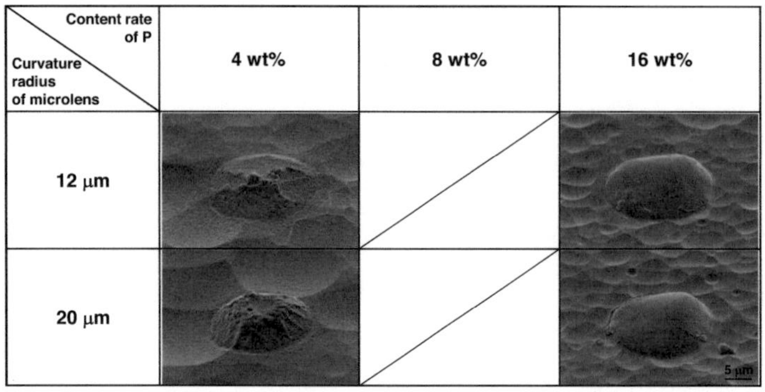

Figure 17. SEM images of microlenses on Pyrex glass using Ni-P/unpolished-Inconel-600 mold.

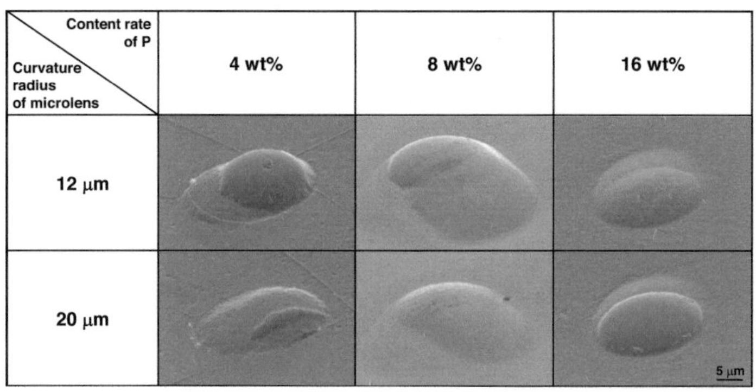

Figure 18. SEM images of microlenses on Pyrex glass using Ni-P/polished-Inconel-600 mold.

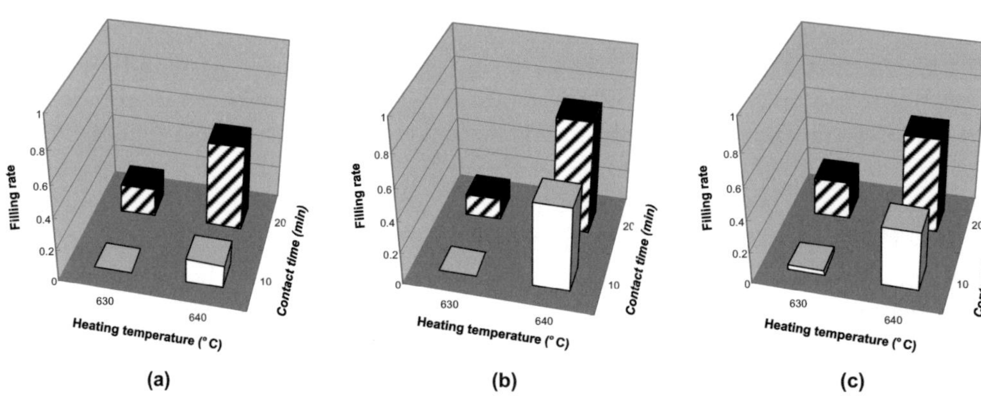

Figure 19. Filling rate of Pyrex glass into microlens patterns with a curvature radius of 12 μm when heating temperature and contact time were changed. The molds used in the imprint process were: (a) Ni-4P/polished-Inconel-600 mold, (b) Ni-8P/polished-Inconel-600 mold, and (c) Ni-16P/polished-Inconel-600 mold.

http://dx.doi.org/10.5772/60408

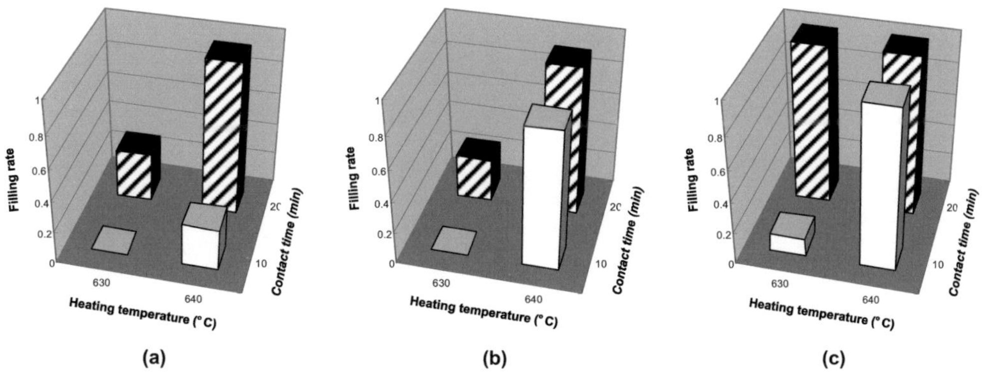

Figure 20. Filling rate of Pyrex glass to microlens patterns with a curvature radius of 20 μm when heating temperature and contact time were changed. The molds used in the imprint process were: (a) Ni–4P/polished-Inconel-600 mold, (b) Ni–8P/polished-Inconel-600 mold, and (c) Ni–16P/polished-Inconel-600 mold.

4.3.2. Profile measurement

The e-beam 3D surface roughness analyzer (ERA-8900) [23] developed by Elionix Inc. was used to evaluate the shape of the microlens. In general, the contrast in the SEM image originates from roughness on the surface of the sample and corresponds to the number of the secondary electrons detected with a second electronic detector. The amount of generation of the secondary electron depends on the angle of the gradient on the surface of the sample and increases linearly between 0 and 75°. Four additional electronic detectors were installed in this device, where a precise roughness and gradual undulation can be observed. It is possible to draw a 3D profile in an arbitrary place based upon the SEM image associated with a high-resolution evaluation [23]. Figure 21 shows cross-sectional profiles of the microlenses measured using ERA-8900. Figures 21(a)–21(c) show the profiles of the microlenses with a curvature radius of 12 μm, and Figs. 21(d)–21(f) show those of microlenses with a curvature radius of 20 μm. In the left, centre, and right figures, the molds used to imprint Pyrex glass were Ni-4P, Ni-8P, and Ni-16P/polished-Inconel-600 molds. Patterns imprinted at a heating temperature of 640 °C for 20 min resulted in the observation of the highest filling rate. The original mold pattern is concave. However, the profile of the mold pattern shown by "x" marks in the figures was reversed in the vertical and horizontal orientations, and plotted as a convex shape. Moreover, the dotted lines in the figure are the profiles of the designed microlenses. In the case of a curvature radius of 12 μm, softened Pyrex glass was not completely filled into the mold pattern. This fact can be confirmed even at the low filling rate in Fig. 19. It is believed that the mold pattern was destroyed by the softened Pyrex glass before filling, and the entrance of the mold pattern was forcefully expanded outward. On the other hand, when the curvature radius was 20 μm, the depth of the mold pattern was shallow. In the case of either a 12 or a 20 μm curvature radius, the profile of the mold patterns and the microlenses correspond well. Therefore, it can be said that the transcript was well executed. The effect of the content ratio of P on the transcript was not particularly noticeable.

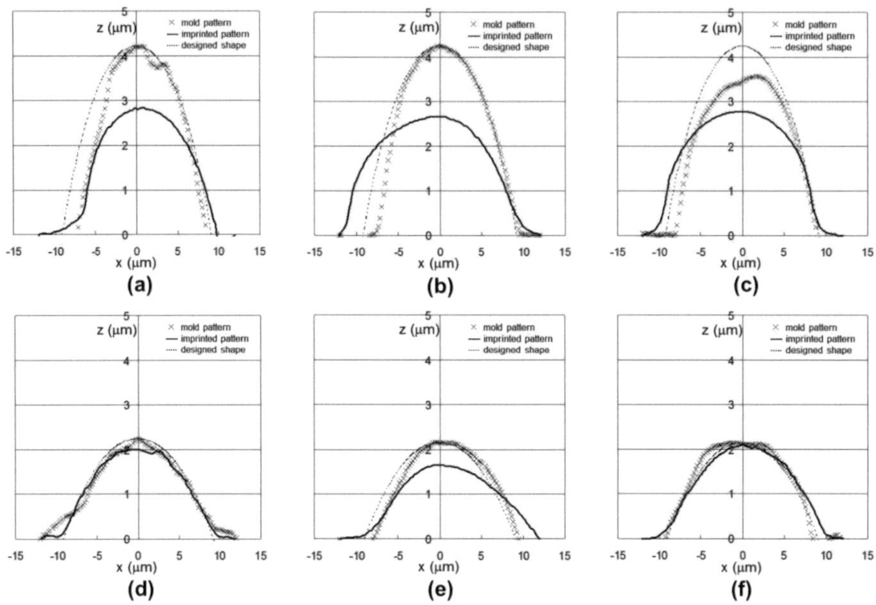

Figure 21. Measured cross-sectional profile of: (a) Ni–4P/polished-Inconel-600 mold and Pyrex glass microlens with a curvature radius of 12 µm, (b) Ni–8P/polished-Inconel-600 mold and Pyrex glass microlens with a curvature radius of 12 µm, (c) Ni–16P/polished-Inconel-600 mold and Pyrex glass microlens with a curvature radius of 12 µm, (d) Ni–4P/polished-Inconel-600 mold and Pyrex glass microlens with a curvature radius of 20 µm, (e) Ni–8P/polished-Inconel-600 mold and Pyrex glass microlens with a curvature radius of 20 µm, and (f) Ni–16P/polished-Inconel-600 mold and Pyrex glass microlens with a curvature radius 20 µm.

4.4. Application to large-area imprinting

Lastly, we report the results of imprinting a pattern that is larger than the microlenses on Pyrex glass. The pattern shape was an AIST logo of two different sizes. The size of the large logo pattern was about 250 x 50 mm^2, and for the small logo, the pattern was about 125 x 25 mm^2; a 50% reduction was also applied to other features in the vicinity. Figure 22(a) is an SEM image of a Ni-16P/unpolished-Inconel-600 mold, and Fig. 22(b) is that of a Ni–16P/polished-Inconel-600 mold. Figure 22(c) is the result of imprinting on Pyrex glass using the Ni-16P/unpolished-Inconel-600 mold. Regarding the molding conditions, the heating temperature, the contact time, and the loading force were kept at 640 °C, 20 min, and 700 N, respectively. In regard to the width, the character "I" in the large AIST logo pattern was measured. The width of the mold pattern was measured to be 8.3 µm, and the width of the imprinted pattern was also measured to be 8.3 µm. In the case of the small logo pattern, the width of the mold pattern was 4.5 µm, and the width of the imprinted pattern was 4.4 µm, which was almost same, considering the margin of error. On the other hand, Fig. 22(d) is an SEM image of the logo pattern when Pyrex glass was imprinted using the Ni-16P/polished-Inconel-600 mold. The surface of

http://dx.doi.org/10.5772/60408

the pattern appeared very smooth. When the width of the logo was measured, there was a difference between Figs. 22(b) and 22(d). The width of the imprinted pattern was 9.5 μm, while the width of the mold pattern was 8.7 μm in the case of the large logo. Moreover, with respect to the small logo pattern, the 4.9-μm-wide mold pattern was transcribed to the imprinted pattern at a width of 6.3 μm. In terms of a more comprehensive observation, the Ni-16P/unpolished-Inconel-600 mold seemed to be transcribed faithfully according to the pattern size. However, the surface of the imprinted patterns is quite different as is the microlens formation. Figure 23 shows expanded SEM images of a small AIST logo when observed at an inclined angle of 60^0. It is clear that the surface roughness of the molds influenced the imprinting accuracy. It seems to be important in precise Pyrex glass imprinting to add a polishing process to the Inconel-600 substrate by CMP technologies. When modifying one's views, for large irregularities on the surface of the Ni-16P/unpolished-Inconel-600 mold, it is assumed that the mold and the Pyrex glass are firmly fixed by an anchoring force, and that the extra flow of the softened Pyrex glass is suppressed. The depth of the mold pattern on the Ni-16P/polished-Inconel-600 mold measured with the NewView 5000 was 1.1 μm, and the height of the AIST logo imprinted on the Pyrex glass was 0.94 μm.

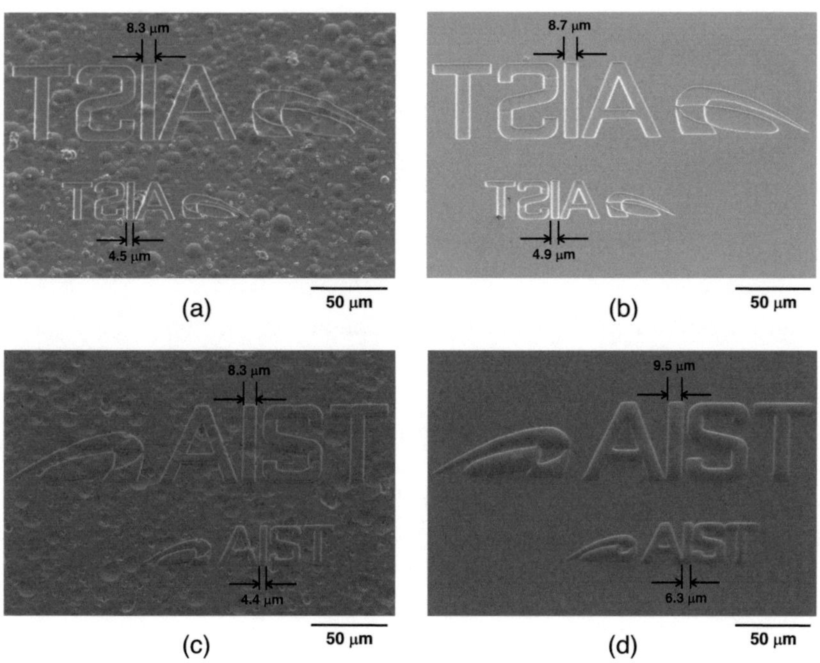

Figure 22. SEM images of AIST logo patterns on: (a) Ni-16P/unpolished-Inconel-600 mold, (b) Ni-16P/polished-Inconel-600 mold, (c) Pyrex glass imprinted using Ni-16P/unpolished-Inconel-600 mold, and (d) Pyrex glass imprinted using Ni-16P/polished-Inconel-600 mold.

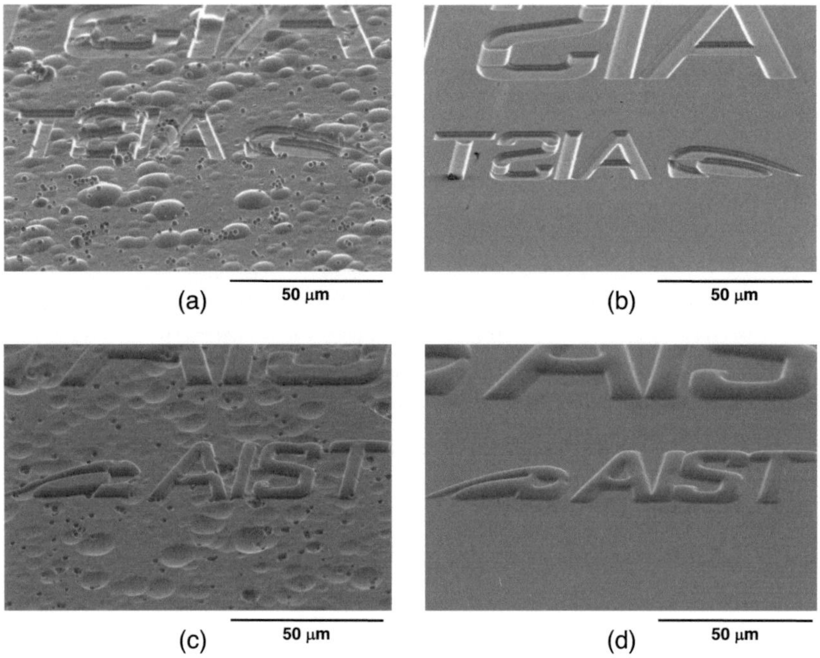

Figure 23. Expanded SEM images of the small AIST logo pattern from Fig. 22 when observed at an inclined angle of 60°.

5. Conclusions

At first, the author experimentally confirmed that an amorphous Ni–P alloy can be used as a mold material for nanoimprint lithography of Pyrex glass. Using X-ray diffraction, it was established that a P content of 8 wt.% was the right amount to make Ni an amorphous material. Moreover, it was shown that in fabricating the nanopatterns using FIB, and by making the Ni-P alloy amorphous it was possible to prevent the deterioration of process accuracy caused by the existence of crystal granularity. In addition, the hardness of the amorphous Ni-P alloy was rapidly improved by thermal treatment at 400 °C, which succeeded in creating characteristics more suitable for a mold material. And moreover, the author experimented with nanoimprint lithography using Pyrex glass and succeeded in molding patterns with linewidths of as little as 500 nm.

Next, the author improved and perfected a low-cost high-accuracy mold for Pyrex glass imprinting. In this process, an amorphous Ni-P plating layer was deposited on a Ni/Ti/Si substrate by electroless-plating, and then a microlens pattern was created on the surface by using an FIB system equipped with a 3D-CAD software. Compared to pure Ni mold, the amorphous Ni-P mold showed a superior processing accuracy by FIB. Especially, in the case

http://dx.doi.org/10.5772/60408

of large microlens with a curvature radius of 20 μm where amorphous Ni-P alloy mold was not used, an imprinted pattern could not be observed on a Pyrex glass. The Ni-P/Ni/Ti/Si mold could be heated up to 620 ºC, whereas the Ni-P/Cu/Cr/Si mold could not be used at this high temperature.

Finally, the author successfuly completed a low-cost high-accuracy mold for imprinting Pyrex glass. A plating layer of amorphous Ni-P was formed on an Inconel-600 substrate by an electroless-plating technique, and on which a microlens patterns were formed with a FIB system equipped with a 3D-CAD software. The author learned that the hardness of a Ni-P layer can be increased by subjecting it to a thermal treatment at 400 ºC for 3 h. Experiments were carried out to compare the impacts of polished and unpolished surfaces of Inconel-600 substrates prior to electroplating Ni-P layers. It was found to be very important to have a smooth surface on the Pyrex glass and to add this requirement to the process whereby the surface of the substrate was polished before Ni-P electroless-plating. In addition, the content ratio of P = 16 wt.% seemed to be more suitable than of P = 4 wt.% or P = 8 wt.% for achieving high hardness after the thermal treatment, and for obtaining smoothness over the rough surface of the mold. Moreover, raising the heating temperature improved the molding accuracy in the imprint process on Pyrex glass more than could be achieved by simply extending the contact time. Finally, microlenses with a 20 μm radius of curvature; and an AIST logo with a character width of 10 μm (or less) were successfully imprinted on Pyrex glass. We also proved experimentally that a Ni-P alloy can be used as a mold material to imprint on Pyrex glass at high temperature around 640 ºC.

Acknowledgements

The electroplated pure Ni and Ni-P substrates were supplied by Dr. Manabu Yasui of Kanagawa Industrial Technology Center and Takeshi Kitadani of Sawa Plating Co., Ltd, respectively. The author received valuable technical support from Chieko Okuyama of National Institute of Advanced Industrial Science and Technology and Michiru Yamashita of Hyogo Prefectural Institute of Technology. For FIB etching and e-beam 3D surface roughness analysis the author received technical supports from Tomoyuki Tsuchida and Dr. Jun-ichi Uegaki of Elionix Inc. The author thanks them for their advice and cooperation.

Author details

Harutaka Mekaru*

Address all correspondence to: h-mekaru@aist.go.jp

National Institute of Advanced Industrial Science and Technology (AIST), Tokyo, Japan

References

[1] Chou SY, Krauss PR, Renstrom PJ. Imprint of sub-25 nm vias and trenches in polymers. Applied. Physics Letter. 1995; 67(21) 3114-3116.

[2] Haisma J, Verheijen M, van den Heuvel K, van den Berg J. Mold-assisted nanolithography: A process for reliable pattern replication. Journal of Vacuum Science and Technology. B 1996; 14(6) 4124-4128.

[3] Colburn M, Johnson SC, Stewart MD, Damle S, Bailey TC, Choi B, Wedlake M, Michaelson T, Sreenivasan SV, Ekerdt J, Wilson CG. Step and flash imprint lithography: a new approach to high-resolution patterning: proceedings of the International Society for Optics and Photonics: Vladimirsky Y. (ed.) Emerging Lithographic Technologies III, 15-17 March 1999, Santa Clara, California: the International Society for Optics and Photonics; 1999; 3676 379-390.

[4] Sotomayor Torres CM. Alternative Lithography: Unleashing the Potentials of Nanotechnology. New York: Kluwere Academic/Plenum Pubkishers; 2003.

[5] Takahashi M, Sugimoto K, Maeda R. Nanoimprint of Glass Materials with Glassy Carbon Molds Fabricated by Focused-Ion-Beam Etching. Japanese Journal of Applied Physics Part 1. 2005; 44(7B) 5600-5605.

[6] Mekaru H, Okuyama C, Ueno A, Takahashi M. Thermal imprinting on quartz fiber using glasslike carbon mold. Journal of Vacuum Science and Technology B. 2009; 27(6) 2820-2825.

[7] Takahashi M, Maeda R. Large-Area Micro-Hot.Embossing of Glass Materials with Glassy Carbon Mold Machined by Dicing. Journal of the Japan Society for Technology of Plasticity. 2006; 47(549) 963-967 [in Japanese].

[8] Takahashi M, Goto H, Maeda R, Maruyama O. Desktop Nanoimprint System –Prototype and Performance: proceedings of the Japan Society for Precision Engineering Conference. 15-17 March 2006, Noda, Chiba: the Japan Society for Precision Engineering ;2006; 737-738 [in Japanese].

[9] Ikeda K. Bisai Tensha: Kako Gijutsu Zenshu (The Complete Works of Precise Transfer and Processing Technology). Tokyo: Technical Information Institute; 2008 [in Japanese].

[10] Takahashi M, Maeda R. Sugimoto K. Micro/nano-Hot Embossing of Quartz Glass Materials with Glassy Carbon Mold Prepared by Focused Ion Beam. Journal of the Japan Society for Technology of Plasticity. 2006; 47(549) 958-962 [in Japanese].

[11] Masui K, Maruno S, Yamada T. Heat-Induced Structural Changes in Electrodeposited Ni-P Alloys. Journal of the Japan Institute of Metals and Materials. 1977; 41 1130-1136 [in Japanese].

[12] Hou K, Jeng M, Ger M. A study on the wear resistance characteristics of pulse electroforming Ni-P alloy coatings as plated. Were. 2007; 262(7-8) 833-844.

[13] Yasui M, Takahashi M, Kaneko S, Tsuchida T, Hirabayashi Y, Sugimoto K, Uegaki J, Maeda R. Micro Press Molding of Borosllicate Glass Using Plated Ni-W Molds. Japanese Journal of Applied Physics Part 1. 2007; 46(9B) 6378-6381.

[14] Inoue A, Takeuchi A. Recent Progress in Bulk Glassy Alloys. Materials Transactions. 2002; 43(8) 1892-1906.

[15] Takenaka K, Togashi N, Nishiyama N, Inoue A. Structure, mechanical properties and imprint-ability of Pd-Cu-Ni-P glassy alloy thin film prepared by a pulsed-laser deposition method. Journal of Non-Crystalline Solids. 2010; 356(31-32) 1542-1545.

[16] Wheeler JM, Raghavan R, Michler J. *In situ* SEM indentation of a Zr-based bulk metallic glass at elevated temperatures. Materials Science and Engineering A. 2011; 528(29-30) 8750-8756.

[17] Tan H, Gildertson A, Chou SY. Roller nanoimprint lithography. Journal of Vacuum Science and Technology B. 1998; 16(6) 3926-3928.

[18] Morikawa T, Yokoi M, Eguchi S. Amorphous plating and the applications of this. Science and Industry. 1991; 41 213-221 [in Japanese].

[19] Mekaru H, Kitadani T, Yamashita M, Takahashi M. Glass nanoimprint using amorphous Ni-P mold etched by focused-ion beam. Journal of Vacuum Science and Technology A. 2007; 25(4) 1025-1028.

[20] Mekaru H, Tsuchida T, Uegaki J, M, Yasui M, Yamashita M, Takahashi M. Micro lens imprinted on Pyrex glass by using amorphous Ni-P alloy mold. Microelectronic Engineering. 2008; 85 873-876.

[21] Katayama K. The Japan Society of Mechanical Engineering Data Book Heat Transfer 4th Ed. Tokyo: the Japan Society of Mechanical Engineers; 1986 [in Japanese].

[22] INCONEL alloy 600. Huntington: Special Metals Corporation. SMC-027; 2008.

[23] Taguchi Y, Omata Y. Electrodeposition of the Composite Film of Nickel and Hydrophobic Graphite Particles. Journal of the Surface Finishing Society of Japan. 2006; 57(8) 564-583 [in Japanese].

[24] Mekaru H, Okuyama C, Tsuchida T, Yasui M, Kitadani T, Yamashita M, Uegaki J, Takahashi M. Development of Ni-P-Plated Inconel Alloy Mold for Imprinting on Pyrex Glass. Japanese Journal of Applied Physics. 2009; 48 06FH06-1-06FH06-10.

Section 5

Application of MOSSBAUER Spectroscopy in Alloy Development

Chapter 6

Mössbauer Spectroscopy of Radiation — Induced Processes in Metallic Systems

K.M. Mukashev, A.K. Shokanov and F.F. Umarov

Additional information is available at the end of the chapter

http://dx.doi.org/10.5772/60748

Abstract

Nuclear power engineering development is held back by both the nuclear reactor safety reasons and the problems related to creating materials suitable for using in the reactors. These materials must be resistant to radiation, able to stand high temperatures, and stable to the corrosive environments. In this work, the general regularity of the interstitial atoms and vacancy interaction with impurity substitutional atoms of $^{57}Co(^{57}Fe)$ in bcc lattice metals has been systematically investigated for the first time. The electron states and structure of "impurity-interstitial" atom, "impurity-vacancy" systems, and their mössbauer parameters are defined. For the first time, by the Mössbauer effect study, the complexes annealing stages from isochronal annealing temperature have been defined. The mössbauer imputity atoms vibrations rms amplitude values and their binding energy are determined. It has been experimentally established that atom mobility considerably increases in radiation-damaged zones created by high-velocity charged particles, fission fragments, or ionized displaced atoms. The compound dumbbell state in bcc metals was investigated, and it was shown that unlike fcc metals, in the bcc metals the considerable quadrupole splitting was revealed, which enables us to separate them on different interstitial configurations around ^{57}Co impurity. It was also established that non-cubic charge distribution around a mössbauer atom leads to the electric field gradient that causes the nuclear levels hyperfine splitting owing to quadrupole interaction.

Keywords: Mössbauer effect, positron spectroscopy, point defects, irradiation by nuclear particles, isochronal annealing, quadrupole splitting, "impurity-interstitial" complexes, "impurity-vacancy" complexes

1. Introduction

Most of the materials that are being used in innovative technologies have complex composition, structure, and phase. They can be seriously affected by the irradiation of nuclear particles with high energy. Irradiation causes an excess formation of defects at an atomic level. The defects of structure can accelerate diffusion processes in materials and create centres of allocation of new phases in alloys. Moreover, if nuclear reactions appear by irradiation processes, there are more changes in the internal structure of materials. As a result, the new structural units appear, which are absent in the original matrix. The most important processes that occur at the interaction of high-energy radiation with a solid are ionization, the displacement of lattice atoms from their sites, and the occurrence of fission products from nuclear reactions. Ionized atoms are quickly neutralized in metals with high electroconductivity. As a result of the elastic collision with bombarding particles, the lattice atom can obtain energy E_p that may exceed the threshold of displacement energy E_d. Then, the atom leaves its initial place in the lattice. If the knocked-out atom has enough energy, it goes through several interatomic distances to a complete stop, which shows the formation of a pair-displaced atom and a vacancy. For most solids, E_d has a value of about 20-30 eV. If E_p is not much more than E_d, then during each collision, only one pair of such defects is formed. For large E_p, two or more "vacancy–interstitial atom" pairs are formed.

If irradiation is created by heavy particles, the energy Ep obtained by the lattice atom can be very large. Then, the average distance between collisions can be equal to the interatomic distance. In these conditions, every atom on the way of the primary knocked-on atom moves from its place. As a result, an area of strong lattice distortion appears. This area is called displacement spike. Primarily, a knocked-on atom can appear at very high temperatures. This is called thermal spike area. However, fission fragments can cause stronger local damages. The thermal and displacement spikes can occur in materials that were affected by irradation in the reactor or the irradation by heavy particles, except in large cascade displacements.

The most important characteristic of the collision is the energy, which is transmitted to the impinged atom. This energy that can change from zero in collisions at small angles to a maximum E_{pmax} at elastic head-on collision is given by

$$E_{p\max} = 4EMm / (M + m)^2 ,$$

where E and m – energy and mass of incident particle, M – mass of the target atom.

This ratio is used for nonrelativistic particles such as neutrons, protons, etc. For electrons with energy of the order of MeV, relativistic effects need to be considered. Then, this formula is modified:

$$E_{p\max} = 2E\left(E + 2m_e c^2\right) / Mc^2 ,$$

where, m_e – rest mass of electron, c – speed of light.

The minimum energy E_{min} needed for E_d energy transmission is equal to

$$E_{min} = (M+m)^2 E_d / 4Mm.$$

This reasoning is not applicable for electrons and gamma-rays. In the case of electrons, the E_{min} value is large, and there is a need to consider relativistic mechanics:

$$E_d = 2E_{min}\left[E_{min} + m_e c^2\right] / 4Mc^2$$

Gamma-rays do not directly displace atoms. They transmit their energy to the electrons of the atom. During collision with the lattice atoms, they lead them to displacement. Minimal energy of gamma rays, which is required for the displacement of atoms, is determined from the relation:

$$E_d = 2E_{e\,min}\left[E_{e\,min} + m_e c^2\right] / Mc^2$$

When materials are irradiated by reactor neutrons or heavy particles with high energy, the crystal lattice undergoes a huge number of elementary damages. With this, free energy of the system increases, and activation energy of the system decreases. These processes are related to the movement of atoms and defects. As a result, a mobility of atoms and defects increase. Depending on the physical parameters of the material, nuclear, and other factors, their complexes of vacancies and interstitials may be formed. These formations under certain conditions can lead to the formation of dislocation loops.

At a sufficiently high temperature, defects undergo a series of transformations. Some of them cancel each other out. Another part of the defects can reach the surface of the metal or grain boundaries. If defects are absorbed by dislocation, they cause the pinning of the latter. If absorbed defects are too many, they move along the dislocation line. When they come together, they form the teeth, inhibiting the movement of dislocations. As a result of the absorption of defects, dislocation is the pinning and hardening of the material.

Experimental methods for direct observations of structural defects on the atomic level have undergone significant developments. They include the direct resolution of individual atoms on a field ion microscope, determination of the atomic configurations of point defects by x-ray diffusion scattering, Mössbauer and positron spectroscopy, and others. Field-ion microscope is used to study the phenomena on the crystal's surface. The X-ray diffusion scattering is very useful to study clusters, which are formed from point defects and vacancies. Positron annihilation is sensitive to single vacancies and their clusters. The Mössbauer spectroscopy is sensitive to changes in the energy shift of the nuclear levels. These changes lead to the hyperfine interactions. In the Mössbauer spectrum, it is detected by several parameters such as the isomer shift, quadrupole, and magnetic hyperfine splitting, as well as the value of the Mossbauer effect probability.

This work presented the results of the studies of metals and their alloys by nuclear gamma-resonance spectroscopy and x-ray structural analysis. In order to carry out these works we have designed and developed a set of tools. The Mössbauer effect study requires spectrometer, electric vacuum samples, annealing furnaces, helium and nitrogen cryostates, installation for samples' hardening, and devices for low-temperature irradiation and measurement.

2. The Mössbauer effect essence

The Mössbauer effect is a physical phenomenon observed in solids. It consists of the emission of a gamma-ray by a radioactive atom, without any loss of energy, for this atom must be in a bound state. Then, the gamma ray emitted at the expense of the recoil effect of the atom nucleus is also in the bound state. This discovery was made in 1958 by the German physicist Rudolf Mössbauer and won the Nobel Prize in Physics [1]. In his work, he used the radioactive nuclide osmium-191 which undergoes β-decay with the radioactive iridium-191 formation.

Nuclear-gamma resonance spectroscopy, which is based on the use of the Mössbauer effect, is one of the unique methods in nuclear-physical experiments. The scope of the application of the method extends from about nuclear and atomic phenomena and structures to a wide range of analytical problems, including the study of the short- and long-range order in solids. High sensitivity to the bare shift of the energy level of the nucleus allows it to apply to the study of the defect structure of metals. In this case, the Mössbauer atom, emitting and absorbing resonance gamma rays, serves as a probe. With the help of specially designed experiments, sensitivity to iron-containing materials can be reduced to ~ 10^{-4} at.%. Information about the object under study is contained in the Mössbauer spectra. Among these, we must be able to extract the parameters characterizing the state of the environment and the resonance atoms in the crystal lattice. The most important of these are the probability (value) of the Mössbauer effect and the width of the spectral line splitting. In practice, the magnitude of the effect is evaluated by the relation:

$$\varepsilon = \frac{I_\infty - I_0}{I_\infty} = \frac{N_\infty - N_0}{N_\infty},$$

where, I_∞,, N_∞ – the intensity of the radiation passing through the absorber at the source momentum, when the resonance absorption is virtually absent; I_o, N_0 the same with full overlapping emission and absorption.

In connection with the registration of gamma-rays in an environment where the solid angle of detection **Ω << 4π,** instead of parameters $\boldsymbol{I_\infty}$ and I_o, are limited to the counters $\boldsymbol{N_\infty}$ and N_o, respectively. Typically, the center of gravity in the abscissa of the spectrum does not coincide

http://dx.doi.org/10.5772/60748

with the zero point value of the velocity. This shift can be dismissed as a positive and a negative side, which is called the isomer shift. The reason for this shift is the Coulomb interaction of the nucleus with radius R and its electron shell. The extent of this interaction is estimated by the change in the potential energy of the nucleus and is determined by the density **|ψ(0)|** electrons of s-level:

$$\Delta E = -\frac{2}{3}\pi Z e^2 \left|\psi(0)\right|^2 \cdot < R^2 >,$$

where Z – the atomic number of the nucleus, e – electron charge, and <R> – the mean square radius of the nucleus.

However, the core radius in the ground state is different from the radius of the excited state. In addition, the density of s-electrons at the nucleus of the source and absorber can be different. In view of these factors for the isomer shift, it is possible to write a more acceptable formula in terms of velocity:

$$\delta = K \cdot \frac{\Delta R}{R} \cdot \Delta\left|\psi(0)\right|^2$$

where K – a constant determined by isotope used, $\Delta \mid \psi(0) \mid ^2$ – the difference between the electron densities of s – levels of the source and absorber, and ΔR/R – radius of the nucleus relative change at the transition from the excited state to the ground state.

The difference in density $\Delta \mid \psi(0) \mid ^2$ occurs due to differences in valence of s-level. Consequently, the isomer shift contains information about the chemical bond. Since this value is relative, you must always specify the origin (usually relative to a standard absorber – the potassium ferrocyanide).

The width of the nuclear transition is determined by the Heisenberg uncertainty relation. Its value is sufficiently small. Therefore, there is a spectral width at half height used. It is the sum of the corresponding parameters of the line source r_S and absorber r_a. The experimental spectrum is always wider than that amount. The line broadening is stipulated by several causes, such as the nuclei environment heterogeneity, presence of inclusions, final thickness of the source, and absorber, lines hyperfine splitting, etc. On the Mössbauer, spectra magnetic and quadrupole splitting can be observed. The first splitting occurs when the resonant core is in ground or the excited state has a magnetic moment, while the core has an effective magnetic field. Quadrupole splitting arises from the interaction of the electric quadrupole moment of the nucleus with the gradient of the inhomogeneous electric moment. As a result of this interaction (e.g., in an excited state), the nuclei ^{57}Fe level, with the spin Ia = 3/2, splits into two sublayers. In the ground state, quadrupole moment is absent, and there is no splitting. Under these conditions, instead of a single line in the spectrum, there are two adjacent lines. They are called doublets. The distance between them is proportional to the electric field gradient (EFG) and is recorded as the measure of this index. Thus, in all changes in Mössbauer's parameters, there is always a relationship between the characteristics of the nucleus and the properties of the object studied.

In Mössbauer experiments, the principle of the Doppler shift of gamma-ray energy by periodic motion of the source or absorber relative to the other is used to measure the absorption spectrum. If the experiments on the transmission of gamma-ray, then the absorber is investigated and the Doppler shift is provided by the source movement. In this case, the geometry of the experiment is called absorptive. When an object under study is the source itself, the movement is subjected to the absorber and the source remains stationary. The geometry of the experiment is then called emissive. The sensitivity of this method for mössbauer atoms is four-five orders higher in magnitude than in absorption. The choice of mode is determined by experimental conditions. The mode with constant acceleration is basic in the majority of Mössbauer measurements. In this case, the relative velocity of the source and the absorber changes linearly at a predetermined interval ($-V \div +V$).

Since some of the investigated materials (e.g., Mo) did not have their Mössbauer isotope, the process of creating a source for emission investigations was realized in the following way. First, radioactive cobalt chloride solution $^{57}CoCl$ is transferred onto the surface of the pre-annealed material and evaporated. Upon reaching the required activity, samples were thermodiffusion annealed in a hydrogen atmosphere at a high temperature. As a result, chloride solution was completely reduced to metallic cobalt. The ^{57}Co atoms embedded in a matrix occupy regular lattice sites, which is evidenced by obtaining a symmetric single resonance line for such materials and the absence of the electric field gradient at the nuclei. Mathematical processing of the experiment of the Mössbauer effect study spectra aims to obtain physical parameters that determine the shape of the resonance line. For sufficiently thin samples with a high range precision, the Lorentzian type can be described:

$$L(X) = a / \left[1 + 4 / b^2 (x - c)^2 \right]$$

where b – the half-width, c – the center of gravity position, and a – the amplitude of the resonance spectrum.

When processing, each experimental spectrum is represented as a superposition of several such lorentzians. The primary means of adjustment has been selected with the least squares method. This method describes precisely enough the resonance line shape with fitting quality assessment by χ-square criterion [2]. Between the parameters of the resonance when removing the degeneracy in the magnetic, electric fields, there are certain connections. Hence, components of the experimental spectrum can be divided into a number of characteristics. The isotope ^{57}Fe can introduce three types of components: single peak (degenerate line), quadrupole doublet (removal of degeneracy on the interaction of the ^{57}Fenuclear quadrupole moment with electric-field gradient), and sextet that arises in the splitting of the resonance line in a magnetic field.

http://dx.doi.org/10.5772/60748

To the fitting operation, performing a special program ALFA has been developed. It is intended to use the average power of the personal computer. Besides the mandatory background value N_∞, the varying parameters are as follows: for single peak-center position – the line width and amplitude; for doublet-center position – amplitude and width of the components in pairs 1 and 6, 2 and 5, and 3 and 4; and for the position of center of gravity – quadrupole and magnetic splittings. As a result, the spectrum processing are extracting parameters of lines and components, χ-square, the sum of the deviations of the theoretical values $N_T(v)$ from the experimental $N_i(v)$, a confidence interval for obtained fitting, and some others as appropriate.

The Mössbauer effect study spectroscopy can help us to determine the hyperfine structure and energy shifts of γ-line on ^{57}Fe nuclei in solids. Shifts are caused by the interaction of nucleus with electric and magnetic fields. They form externally to the nucleus charges [2, 3]. In Mössbauer spectroscopy, energy of hyperfine interaction does not depend on methods of measurement. It is determined by the product of two values. One of this corresponds to the nucleus or external field characteristic, while another one determines the parameters of nucleus. If it is needed to determine nucleus characteristics, the values of field that affect the nucleus need to be known. If the moment of nucleus is known, then we can determine the values of internal fields in solids.

In practice, energy shifts of γ-line is determined by an experiment in the position of the resonance Mössbauer spectrum with respect to any of the samples. This is expressed in mm/s. The calibration spectra is performed to change into eV. For this purpose, the metallic iron Fe or the potassium ferrocyanide – ($Na_2[Fe(CN)_5No]\cdot 2H_2O$) with standard ^{57}Co source in Cr were used. Quadrupole and magnetic moments of excited states of Mössbauer's nuclei and changes of charge's distribution radius in some nuclei were determined with the help of the Mössbauer effect study method. These characteristics of nuclei cannot be obtained with the help of other methods that is why they represent a scientific interest in nucleus physics. The method of the Mössbauer effect study is successfully applied to measure internal fields and electric density on the nuclei. This direction in the Mössbauer effect study spectroscopy is very important. Mössbauer nuclei can be introduced into a very large number of different compounds and groups of metals. In cases when the compounds are studied for absorption, along with the absorption method, one can also use the emission measurement technique on the basis of the Mössbauer effect. The Mössbauer effect study spectroscopy allows to study hyperfine interactions at extremely low concentrations of impurities. Thus, the Mössbauer effect study spectroscopy is a powerful tool for the study of hyperfine interactions in both single crystal and polycrystalline, and in the compounds.

3. The Mössbauer effect probability

The probability of the Mossbauer effect is determined by the dynamic properties of crystals. In crystals, atoms stay in oscillatory movement with respect to the equilibrium position. These fluctuations depend on the masses of the atoms and force constants. The latter are an important characteristic of the interaction between them. They determine the macroscopic properties of the crystal, as heat capacity and elastic characteristics. Probability of effect **f** is defined for a monatomic cubic crystal by the formula:

$$f=\exp\left(4\pi^2<x^2>/\lambda^2\right),$$

where, $<x^2>$ – mean square thermal displacements of atoms in the direction of the γ-ray radiation.

The probability of recoilless process will be greater if the wave length λ of the radiation is larger compared with the square root of the mean square displacement. The stronger the atom is bound in the crystal, the greater the probability of the Mössbauer effect. For determination, $<x^2>$ is needed to set up the absolute value of **f**. Thus, it is necessary to take into account the various factors that influence the measurement results. A simpler method of determining **f** is the measurement of the relative values of **f** in studies of the temperature dependence of this quantity.

4. The isomer shifts

The expression of the isomer shift can be obtained for a system consisting of a nucleus with charge distribution ϱ (r), which is in the electric field produced by the electron cloud e/ψ $(\vec{r})$ ^{2}, where, ψ (r) – the wave function of the electrons. Electrostatic interactions are determined through

$$v=\int\rho(r)\varphi\left[(r)dr\right]$$

where φ (r) – electrostatic potential of the electron shell. It should satisfy the Poisson equation

$$\frac{\partial\varphi}{\partial x}+\frac{\partial\varphi}{\partial y}+\frac{\partial\varphi}{\partial z}=4\pi edr$$

In practice, the shift in the emission line is determined relatively to the absorption line. At this time, the source and absorber are in different chemical states.

This value is the isomer shift. It is determined by two factors: the change in the density φ^2 (o) which is created by electrons in resonant nuclei in different compounds, and in the charge distribution, the change at the nucleus transition from the ground to the excited state. If the source and absorber are composed of different chemical compounds, the maximum difference of the absorption line is not at zero source velocity relatively to the absorber, and at some velocity is determined from $\delta E=v\frac{E}{c}$,

http://dx.doi.org/10.5772/60748

or

$$v = \frac{\delta E \cdot c}{E} = \frac{4\pi f e^2 r^2 c}{5E}\left(\Psi^{2}(o) - \Psi_{11}^{2}(o)\right)\frac{\delta r}{r},$$

where, $\delta r = r_B - r_o$ – the nucleus excited and ground stated radiuses difference.

Positive velocity corresponds to the motion of the source to the absorber $\frac{\delta r}{r}$ and expressed through

$$\frac{\delta r^2}{r^2} = (r_B^2 - r_o^2) / r^2$$

The electron charge density at the nuclei of φ^2 (o) is determined by s-electrons of various shells. Small contributions can also give $p_{1/2}$ electrons.

5. Quadrupole interaction

As a result of the quadrupole interaction in Mössbauer spectrum, it shows a hyperfine structure. Its individual components correspond to γ-transitions between split levels. For ^{57}Fe spin of the excited state 3/2 and the ground state 1/2. These hyperfine structures have the form of the doublet. Doublet is due to the splitting of the upper level into two states with 3/2 and 1/2. The ground state with spin ½ is not split (Fig. 1). Each level is shifted by the amount

$$W_{3/2} = +e^2qQ/n \text{ and } W_{1/2} = -e^2qQ/n,$$

where, Q – quadrupole moment of the nucleus.

There is a degree of deviation from the distribution of the averaged charge density of the nucleus in the m = 1 state from the spherical symmetry where, q – quadrupole interaction constant or the electric field gradient. The value of the quadrupole interaction depends on the product of the quadrupole moment and the electric field gradient.

The distance between the doublet components determines the quadrupole interaction constant. If the quadrupole interaction energy does not exceed the width of the nuclear level G, then instead of the individual components we can observe a broadened line:

$$\Delta E = W_{3/2} - W_{1/2} = \frac{e^2 qQ}{2}$$

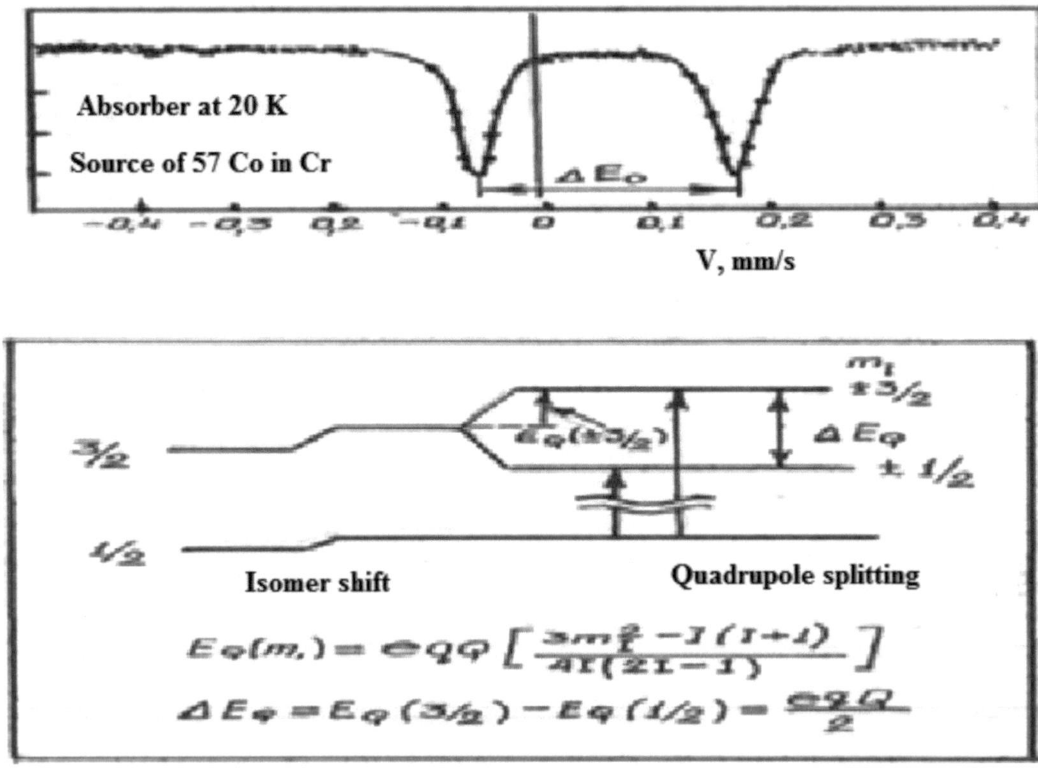

Figure 1. Quadrupole splitting of ^{57}Fe

The asymmetry of the hyperfine structure transition intensity occurs at the presence of a preferred direction. It is given by a magnetic field or electric field gradient.

6. Magnetic hyperfine structure

The magnetic hyperfine structure appears in the spectrum as a result of the interaction of the nucleus dipole moment **μ** with the its magnetic field **H**. This field is generated by its own atom electrons (the Zeeman effect) or by external sources. The magnetic structure is always absent at the nuclear levels whose spins are zero. In the Mössbauer effect, the γ-transitions between the two nuclear levels are observed. Any of these levels can have a magnetic hyperfine structure. Each γ-quantum corresponds to a transition from a magnetic level of the excited state of the nucleus to the ground state sublevels. The selection rules are defined by radiation multipolarity. The respective energy levels are described in terms of:

$$E_M = \mu_n H m_1 / I = g\mu_n H m_e,$$

where, μ_n– nuclear magneton; g-gyromagnetic ratio; m_1= I, I-1,....-I.

http://dx.doi.org/10.5772/60748

Picture of the splitting of the excited levels I = 3/2 and the ground state I = 1/2 is widely used in the Mössbauer effect study spectroscopy. The upper level is split into four sub-layers, and the bottom is split into two sub-layers. Between the upper and lower levels, there are six transitions. They are allowed by the selection rules. The lines of the magnetic structures are arranged relative to the center of the spectrum. For these lines to be well resolved, the magnetic interaction value must be greater than the natural line width (Fig. 2).

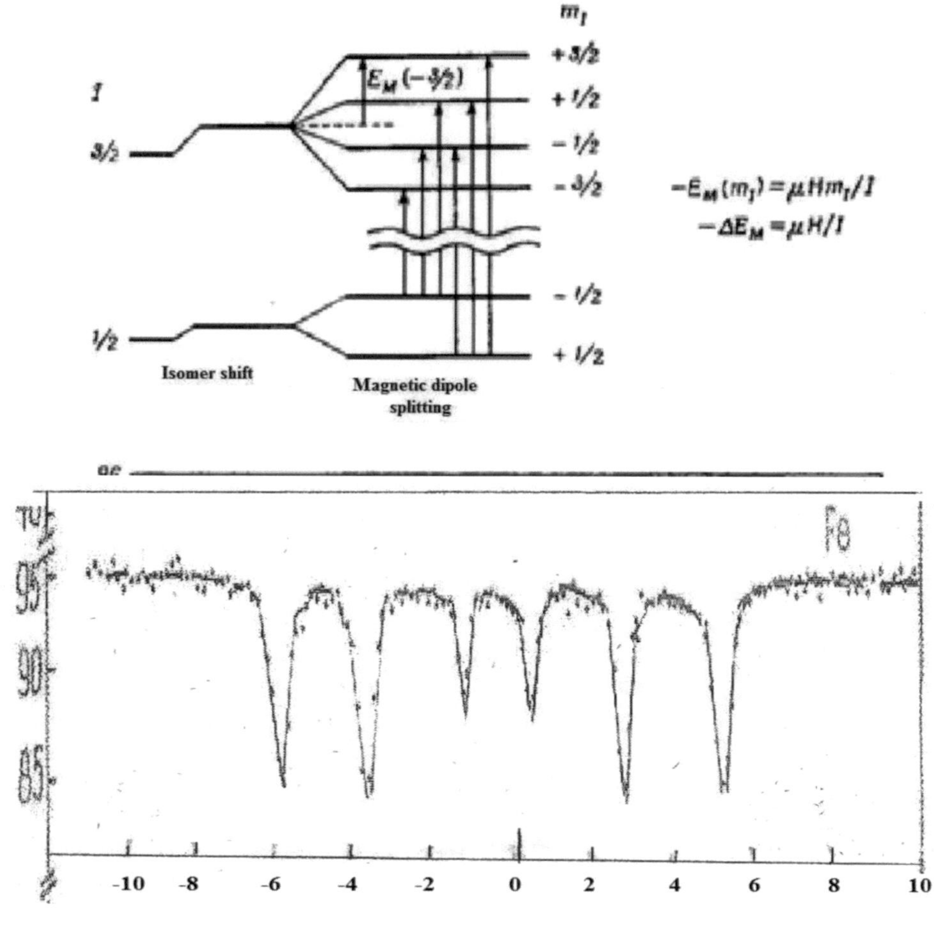

Figure 2. Magnetic hyperfine splitting of ^{57}Fe spectrum

The method of nuclear gamma-resonance spectroscopy based on the Mössbauer effect, due to the high sensitivity and the small shifts of the energy level of the nucleus, has been used successfully for the study of the defect structure of metals. In this case, all the information about the state of the defect region is extracted using the Mössbauer atoms emitting or absorbing resonance gamma-rays. In recent papers on the use of the Mössbauer effect to study of point radiation defects interaction with substitutional impurity atoms on the basis of the Mössbauer isotope, ^{57}Fe has appeared [4].

7. The Mössbauer effect study spectrometers

Experiments to investigate the state of the impurity atoms in metals before and after irradiation by different particles were carried out on the Mössbauer effect study spectrometers. Schematic diagram of the spectrometer is shown in Figure 3.

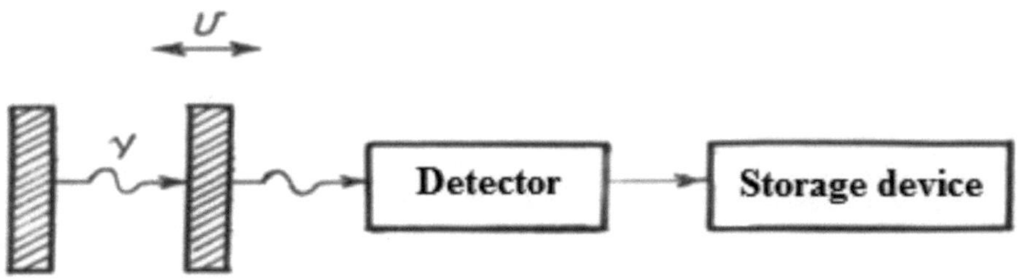

Figure 3. Schematic diagram of the Mössbauer effect study spectrometer

It consists of a γ-radiation source, an absorber as a sample under study, a detector, and a storage device. The absorber imposed some movement velocity **v** relative to the source of γ-rays. Consequently, there is a Doppler shift in the absorption line relative to the emission line. By varying the rate of absorption, one can obtain the intensity of radiation transmitted through the absorber from the absorber velocity. This relationship represents the absorption spectrum. The nuclei of the Mössbauer isotope in the source and in the absorber may be in the same state and chemical environment. Then, when the value of the absorber relative velocity v = 0, the γ-rays resonant absorption occurs. By increasing the source or absorber velocity, the emission and absorption lines overlapping area is gradually reduced. At a certain value of the velocity v, resonance condition is completely broken. In the process of the Mossbauer absorption the integration occurs over the area of absorption and emission lines. Then. the experimental line width determined as G'exp > G source + G absorption.

8. Irradiation by neutrons and charged particles

The VVR-K nuclear reactor of the Nuclear Physics Institute of the RK Nuclear Centre was used for irradiation of samples by neutrons. On the average, 2.5 neutrons were fitted in one fissionable nucleus in the nuclear reactor at ^{235}U fission. The neutrons' energy was distributed in the 0.1-15 MeV interval:

$$n(E)dE = \sqrt{\frac{Z}{\pi e}} sh\sqrt{Ee^{-E}dE}$$

http://dx.doi.org/10.5772/60748

The neutrons energy average value in the nuclear reactor channel was about 2 MeV. Consequently, collisions with surrounding atoms slow down the neutrons and decrease their energy. Therefore, they are much slower. The neutrons play a significant role in the creation of radiation defects in crystals. The samples' temperature are in the 80^0 C limits at the neutrons' irradiation.

The samples' irradiation by the charged particles was conducted on U-150 isochronous cyclotron of the Nuclear Physics Institute of the RK Nuclear Centre. At charged particles irradiation, the generation rate of radiation defects in metals is high by several orders than at the neutrons' irradiation. Thus, by the beam intensity regulating, it can receive, for a short time. such a damage rate which corresponds to a long-term neutron irradiation. The major portion of the bombarding particles energy is spent, meanwhile, on the sample radiation heating. Consequently, the target temperature increases to a few hundred centigrade degrees against the beam intensity. In order to prevent the annealing of radiation defects, we perform intensive cooling of the target during irradiation with the temperature of cooling that reaches that of liquid nitrogen.

9. The Mössbauer spectra processing and analysis

For the useful information extraction from complex Mössbauer spectra and its interpretation, it is necessary to apply computer processing methods. Generally, the resonance of the Mössbauer line by Lorenz curve description is presented at an acceptable accuracy:

$$L(x) = \frac{a}{1 + \frac{4}{\gamma^2}\left(x - c^2\right)}$$

where, a – amplitude; γ – half-width; and c – the resonance line center of gravity position. For the experimental spectra fitting by the Lorenzian set, the least squares method is used, that is the following functional is minimized:

$$F(A) = \sum_{i=1}^{n} \omega \left[N_{\text{ÿĕñ}} - \mathrm{N}\left(\mathrm{A}, \mathrm{x}_{\mathrm{i}}\right)^2 \right]$$

where, A is the varying parameter of adjustable function; Xi are the velocity values in the i_{th}-channel; $N_{i\text{экс}}$. is a count in i_{th}-channel of analyzer; ω *i* is the weighting factor (usually equal to 1); and N is a number of dots of the experimental spectrum. Taking into account the relations between the Mössbauer spectrum, individual parameters can reduce the number of variable parameters.

10. The Mössbauer effect at the point defects in the hardened molybdenum investigation

The sufficient vacancy concentration in metals can be obtained by the hardening method. The vacancy formation energy in metals is less than interstitial atoms formation energy [5]. At heating up to the melting temperature, a considerable quantity of vacancies appears in metals, and their concentration is determined by following expression:

$$n = \exp\left(-\frac{E}{kT}\right),$$

where, E – vacancy formation energy; T – heating temperature; k – Boltzmann constant.

At high-speed hardening, 10^4-10^6 degree/s is succeeded to preserve the considerable part of vacancies non-equilibrium concentration in metals. The Mössbauer effect study permits to observe the selectively of vacancies, divacancies, and small clusters with ^{57}Co Mössbauer impurity atoms interaction. To this effect, the virgin sample with ^{57}Co impurity heats up to 1200^0C temperature in 10^{-7} тorr vacuum. Subsequently, they rapidly cool with 10^4 degree/s velocity.

The observation of a single vacancy interaction with ^{57}Co impurity atom permits to determine the overpatching around impurity atoms and hyperfine interaction parameters. Simultaneously, the root-mean-square amplitude of Mössbauer impurity atoms vibrations in substitution lattice sites, before and after vacancies capture, has been determined. Thus, the experimental works on hardening are very important for the Mössbauer spectra definition and identification. The spectra measurements are carried out at a room temperature on the installation, with uniform acceleration, while using a resonance detector with the FeAl converter. As can be seen (fig.4) in the ^{57}Co in molybdenum spectra, the additional defect line appears after hardening. The spectra processing permits to determine the individual spectrum components. They consist of an undisplaced line corresponding to ^{57}Co states in substitution lattice sites and quadrupole doublet. The last of ^{57}Co state connected with divacancy is determined. The line half-width is Γ=0.34 mm/s, the single line isomer shift is δ=-0.24mm/s, the quadrupole doublet isomer shift is δ=0.13 mm/s relatively to the undisplaced line, and the quadrupole splitting value is ΔE= 0.90 mm/s.

The quadrupole doublet positive isomer shift relates to the electron density of 4s electrons on the nuclei decreasing by 20%. Meanwhile, the environment symmetry on ^{57}Co Mössbauer nuclei is disturbed and the electric-field gradient appears. Its interaction with the nucleus quadrupole moment is brought to hyperfine splitting on the emission spectra appearing. Consequently, the "defect" line which was described by the quadrupole doublet appears in the spectra. In order to determine activation energy of the vacancies related to ^{57}Co impurity atoms we have conducted isochronal annealing of the samples in the vacuum equal to 10^{-7} torr.

http://dx.doi.org/10.5772/60748

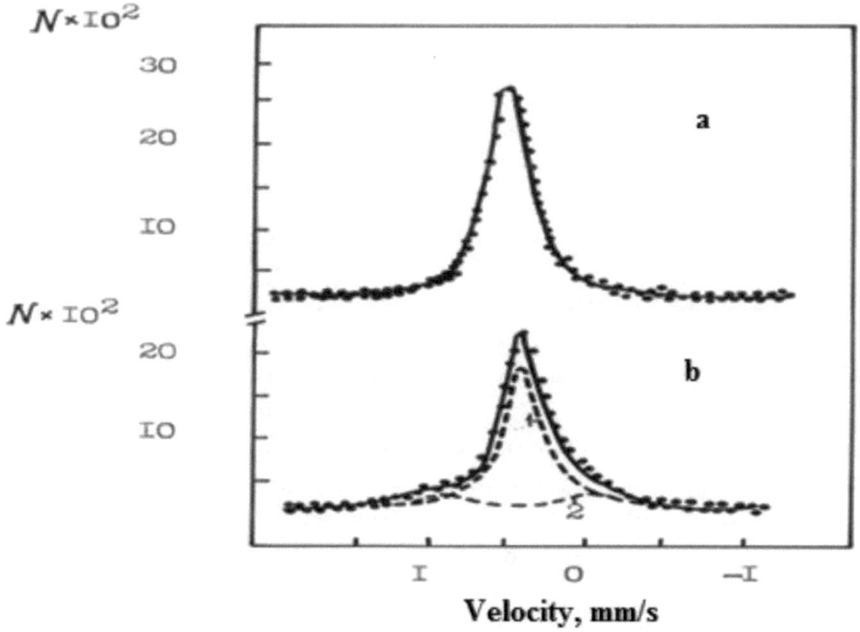

Figure 4. The Mössbauer emission spectra of ^{57}Co in molybdenum before (a) and after (b) hardening

The dependence of the "defect" line square relationship to the spectrum general square (Δs/s) from isochronal annealing temperature is determined (fig.5).

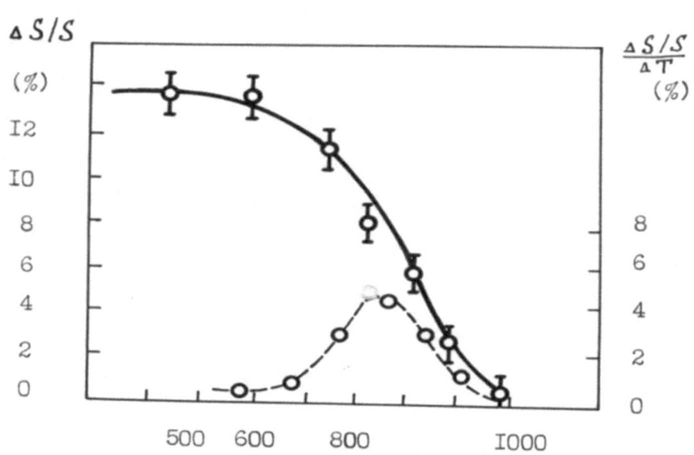

Temperature, °C

Figure 5. The molybdenum isochronal annealing after hardening

Δs/s parameter at a temperature interval from room temperature up to 300^0 C, shows a negligible change. Above of this temperature, a drastic decrease occurs. The "defect" line completely disappears at 980^0C temperature. The vacancies migration activation energy is determined by the following equation:

$$E_a = kT_0 \ln\left(\nu k/\alpha E_a\right),$$

where, $\nu=10^{13}\,s^{-1}$–Debye frequency; $\kappa=2.6\cdot10^{-5}$ eV/K-Boltzmann constant; T_0-mean temperature; and $\alpha=\Delta(T^{-1})/\Delta t$ –design coefficient.

The vacancy migration activation energy obtained value is 2.2eV in molybdenum. It is seen that at temperatures above 400^0 C, the Δs/s value is decreased. It is connected with the ^{57}Co impurity part associated with vacancies in the "impurity-vacancy" complexes decreasing. In temperatures above 400^0 C, these complexes begin to break down. The unallocated vacancies, migrate in the regular sinks. The free interstitial atoms return to the regular lattice sites.

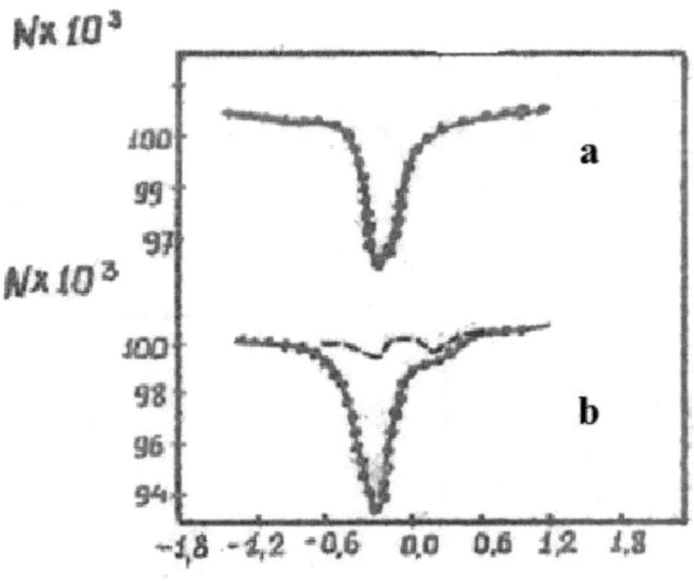

Velocity, mm/s.

Figure 6. The Mössbauer spectrum of ^{57}Co in niobium: before (a) and after (b) hardening

11. ^{57}Co impurity atoms interaction with vacancy defect in niobium

The Mössbauer spectrum of impurity ^{57}Co atoms state in niobium after hardening was a single symmetrical circuit with δ=-0.15mm/s. The standard sample of potassium ferrocyanide served as an absorber. The line width of Mössbauer spectrum was made up Γ=0.30 mm/s. The samples were annealed in 10^{-6} mm. of mercury column vacuum at 1200^0 C temperature. The hardening was conducted in vacuum oil. After the hardening, an additional "defect" line appeared in the right of the spectrum (fig. 6). The parameters of this line are best described by the quadrupole doublet with the width equal to Γ=0.45 mm/s. The isomer shift is equal to δ=0.18 mm/s relatively to the undisplaced line. The quadrupole splitting is ΔE= 0.96 mm/s. The positive isomer shift of the quadrupole doublet in the emission spectrum testifies reduction of the electron density on the nuclei of Mössbauer impurity. The quadrupole splitting appearance indicates symmetry disturbance around the impurity atoms' environment. Consequently, on the nuclei of the Mössbauer atoms, the electric-field gradient appeared. Such changes connected with ^{57}Co impurity are caused by vacancies. The impurity content in niobium did not exceed 10^{-6} in %.

The dynamic behavior of the ^{57}Co impurity by the isochronal annealing method in temperature interval has been investigated up to 1000^0 C (fig.7). The Δs/s parameter does not change up to 300^0 C temperature. Therefore, in this temperature interval, the "impurity-vacancy" complexes are the stable. At 750^0 C temperature, the Mössbauer emission spectrum is recovered thoroughly up to its initial state. With the annealing temperature growth, the single vacancies are left step-by-step as the ^{57}Co impurity atoms and crystal lattice is recovered up to its initial state. The vacancy migration energy in niobium made up in accordance with literary data [6], 1.8 (4) eV.

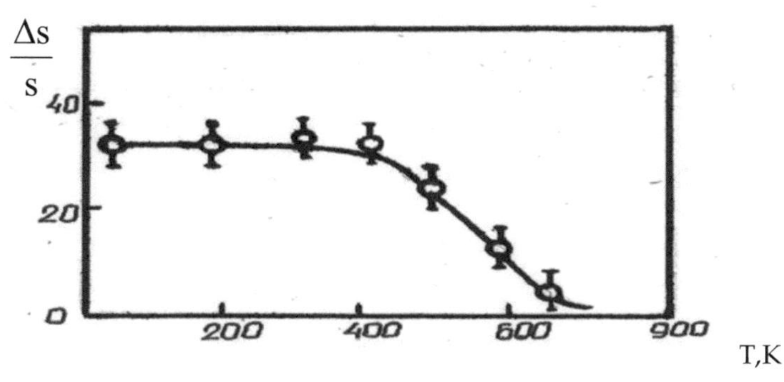

Figure 7. Isochronal annealing of hardened niobium ($\frac{\Delta s}{s}$)

12. Defects interaction with impurity atoms in molybdenum irradiated by electrons and neutrons

The interaction of defects with impurity atoms is one of the fundamental mechanisms in the field of metal radiation damage. The particular investigations by the instrumentality of electrical resistance measure showed that impurity atoms can efficiently capture the migrate interstitial and bring the annealing stage to decay or absolute suppression [7, 8]. This problem finds its confirmed experiments with molybdenum irradiated by electrons. The required concentration of vacancy and interstitial atoms in the sample of $Mo^{57}Co$ at the temperature of 100K, was created by electron irradiation (E=3 MeV) to up to $1.07 \cdot 10^{19} cm^{-2}$ fluence. At this temperature, interstitial in molybdenum are mobile but vacancy remains stationary. Therefore, interstitial can proceed recombination with vacancies if they are captured by ^{57}Co impurity atoms. After irradiation at spectrum positive Doppler velocity, a sharp "defect" line and additional convexity from the left side of the substitution line appears (fig.8b). This indicates their accessory to the quadrupole doublet with large splitting (fig.8c).

Defect line	1	2	3	4	5
σ, mm/s	0, 48	0, 45	0, 45	0, 47	0, 47
E, mm/s	1, 82	1, 14	1, 48	0, 81	0, 39

Table 1. Values of the "defect" lines centers shifts

The Mössbauer spectrum can decompose on 4 quadrupole doublets with about + 0.46 mm/s central shifts relative to the substitution line (1-4 lines). The evidence of these lines was obtained from annealing experiments after the fast neutrons' irradiation at 4.6K (fig.9). The configuration 1 most clearly appeared after annealing at 124K. After annealing at 214K, it disappeared and line 3 are dominated. This process is connected with the reorientation of the defect in configuration 1 to configuration 3 [9]. The successive annealing at higher temperatures is accompanied by the captured interstitial release. The defect line 3 becomes unstable. On III (470 K) stage, the defect line 4 appeared clearly on the spectrum (fig.9e). At the same time, the fifth component with small quadrupole splitting appear (line 5, table 1). The annealing above stage 3 brings new defect lines appearing and disappearing in 1-5 configurations. The new lines are characterized by half of centre shift, as compared with 1-5 defect lines. The full annealing of the irradiated samples is completed in stage 5 at about 1300K temperature.

Thus, the direct evidences of interstitial atoms capture in $Mo^{57}Co$ system around ^{57}Co impurity atoms have been found. Five variants of defect lines have been found with well-defined shear centre. They have different electric-field gradients. All 5 combinations were at the expense of the captured single, double or triple interstitials [10]. The observed stage 3 of the configuration existed at the expense of large cluster defects generated on ^{57}Co impurity atoms.

The substitutional impurities are always present in metals in small quantities. They constitute a significant part at the crystals' defect structure formation. The radiation defects' physical-mechanical properties are dependent on their configuration and electron structure.

The molybdenum polycrystal, with 99.98% purity, irradiated by reactor neutrons investigations, has been carried out by emission and absorption method. In the first event, the radioactive^{57}Co was used as an impurity atom, and in the second event, stable isotope of ^{57}Fe was used for absorption experiments, where the samples enriched by ^{57}Fe isotope up to 60% were used. The ^{57}Co samples in the Mössbauer spectra were at a symmetrical circuit, with

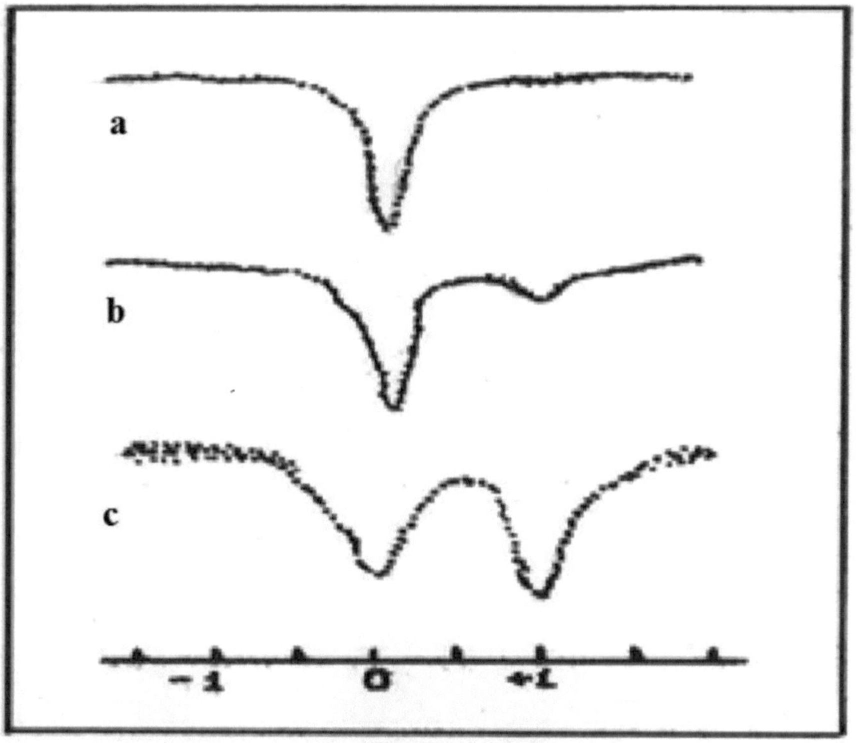

Velocity, mm/s.

Figure 8. Mössbauer spectra of ^{57}Co in Mo, absorber ^{57}Fe in Pd: a) before irradiation; b) after electron irradiation, E= 3 MeV at 100K temperature up to $1.078 \cdot 10^{19} cm^{-2}$ doses; c) after substitution line deduction.

$\delta = 0.30(3) mm/s$ isomer shift relatively to a standard absorber (fig.10a). The absorption spectra measurement was conducted employing a ^{57}Co cosource in copper.

The conversion electrons resonance counter on the basis of FeAl was used as a detector. The irradiations of samples were conducted in the 10^{22}-10^{23}neutron/m^2 fluence interval. The emission spectra of irradiated samples change in accordance with the Mössbauer nuclei state in substitution lattice points. In the positive part of the spectrum, the additional "defect" line appears (fig.10 b, c, d) and the principal line is widened. In the sample spectra on absorption, if it is irradiated up to $5 \cdot 10^{22}$ neutron/m^2 doses along with undisplaced line, it may be separated in addition of at least two quadrupole doublets (Fig.11). These quadrupole doublets have different isomer shift relatively to the undisplaced line: δ_1 =0.13(3) mm/s and δ_2 =0.46(3) mm/s. The quadrupole splitting values are ΔE_1=0.75(3) mm/s and ΔE_1=0.82 (3) mm/s respectively. At fluences above $5 \cdot 10^{22}$ neutron/m^2, the third quadrupole doublet with δ_3 =0.46 (3) mm/s and ΔE_3=0.82 (3) mm/s parameters appeared in the spectra. With the irradiation doses increasing, the total intensity of the additional line increased and principal line position did not change. The sample spectra on ^{57}Fe absorption in molybdenum after irradiation also changed. Along with the principal lines, the "defect" line appeared in the negative part relative

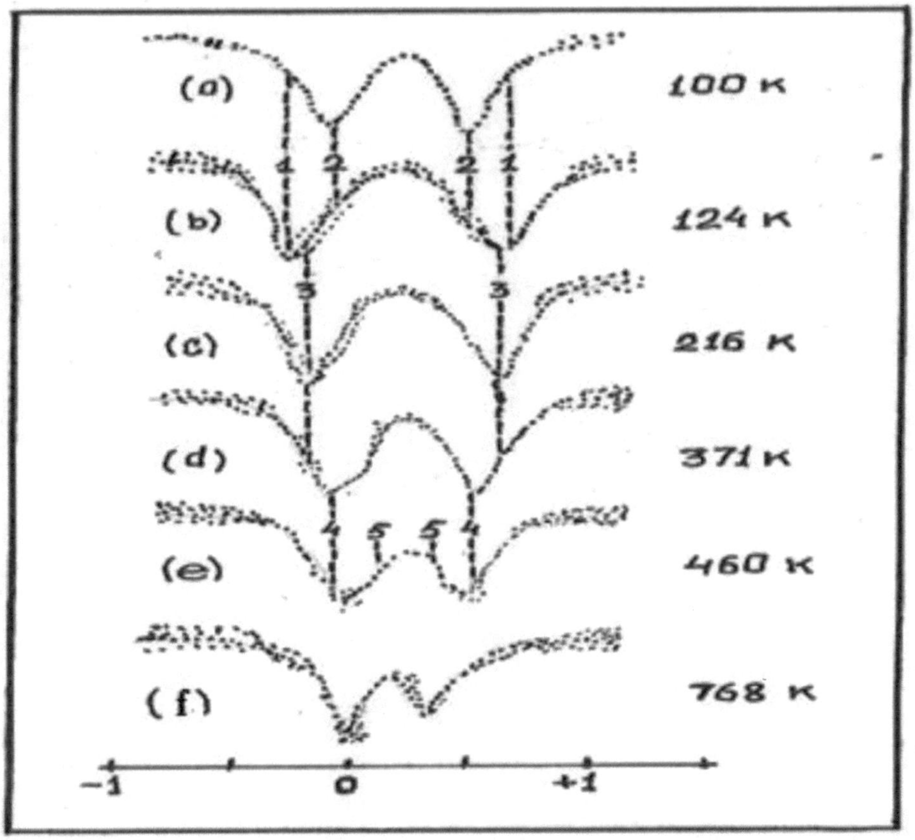

Figure 9. The Mössbauer spectra of ^{57}Co in Mo, irradiated by electrons with E=3 MeV (T= 4.2 K): a) after electron irradiation up to $1.07 \cdot 10^{19} cm^{-2}$ doses at 100K temperature, b) after fast neutrons irradiation up to $6.9 \cdot 10^{18} cm^{-2}$ doses at 4.6 K temperature, c) after isochronal annealing at 214 K, d) 371 K, e) 460 K and f) 768 K

to the undisplaced line. It has been described by two quadrupole doublets: δ_1 = -0.11(3) mm/s and δ_2=0, 76(3) mm/s. The quadrupole splitting values of these doublets were ΔE_1=0.95(3) mm/s and ΔE_1= 0.72(3) mm/s, respectively.

The "defect" line appearance in the sample spectra after irradiation signifies the radiative single defects captured by the Mössbauer impurity atoms. The present stage of the residual electrical resistance measurement after irradiation corresponds to the migration of the interstitial type radiation defects in molybdenum [11]. The interstitial atoms connected with ^{57}Co impurity atoms in molybdenum formed the stable dumbbell shaped configuration.

These complexes have been annealing at higher temperature as compared with the free interstitial atoms. In emission spectra, such complexes appear in two configurations and in only one configuration – in absorption spectra. In addition, the neutron irradiation leads to the appearance in the absorption spectra of yet another impurity state of cobalt atom. It is characterized by the quadrupole doublet with δ = 0.13(3) mm/s isomer shift. A doublet with

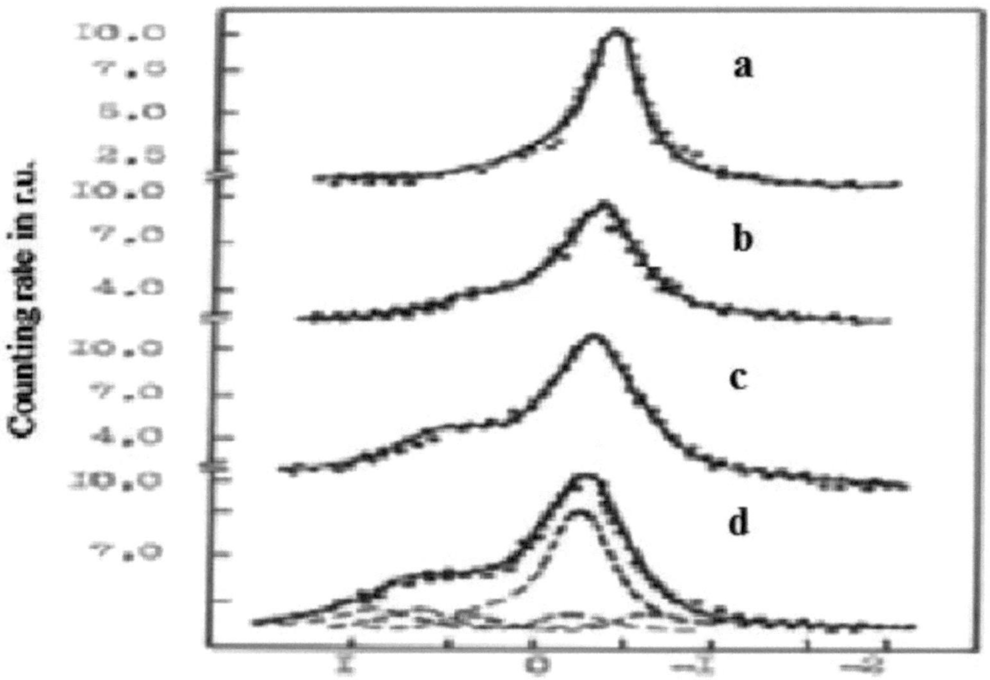

Velocity, mm/s

Figure 10. The emission Mössbauer spectra of ^{57}Co in molybdenum before (a) and after irradiation at doses: 10^{22} (b), $5 \cdot 10^{22}$ (c) and 10^{23} (d) neutron/m^2. The calculated spectrum decomposition are indicated by the dotted line.

such parameter has been detected in the samples subjected to hardening. Thus, these complexes can relate to the connection with vacancies cobalt impurity atoms. Therefore, the simultaneous formation of "impurity-interstitial" and "impurity-vacancy" complexes are possible in molybdenum samples after irradiation. The most likely orientations of ^{57}Co impurity atoms and interstitials mixed dumbbells are directed along <110> and <111> axis (fig. 12).

The isochronal annealing has allowed to determine the annealing stages of "impurity-interstitial" and "impurity-vacancy" complexes. In fig.13, the isochronal annealing results of the samples irradiated by neutrons of up to 10^{22}(a), $5 \cdot 10^{22}$(b) and 10^{23}(c) м$^{-2}$ fluences are presented. In the temperature interval of up to 600K, the ΔS/S relationship remains constant. At an annealing temperature of above 600K, the vacancies become free from fixed sinks and begin to migrate along the crystal. In the case of "impurity-interstitial" complexes encountered, the vacancies are annihilated with these complexes. The Mössbauer impurity atoms become free and recover in the lattice points. The Mössbauer impurity atoms' environment symmetry are recovering. At 950K temperature, the irradiated sample spectrum is completely recovered to the initial state.

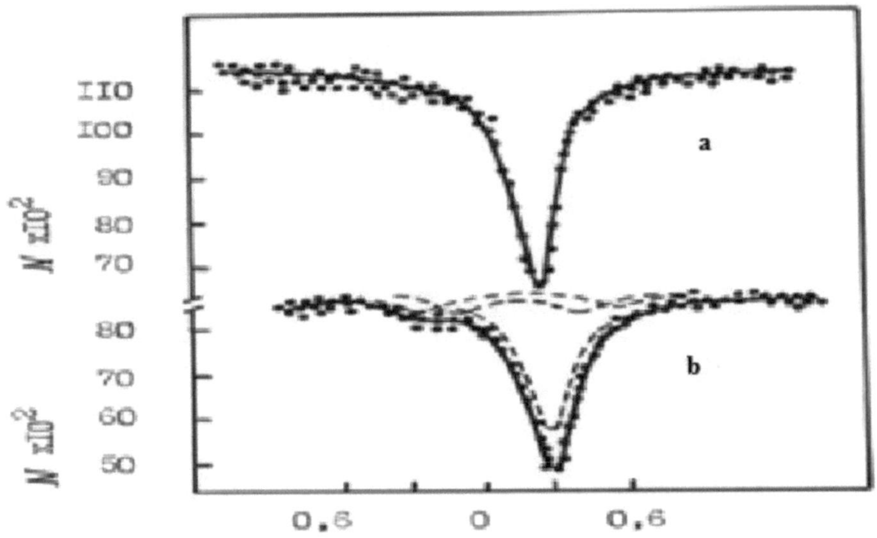

Velocity, mm/s.

Figure 11. Spectra of ^{57}Fe absorption in molybdenum before (a) and after (b) neutrons irradiation up to 10^{21} neutrons/m^2 doses

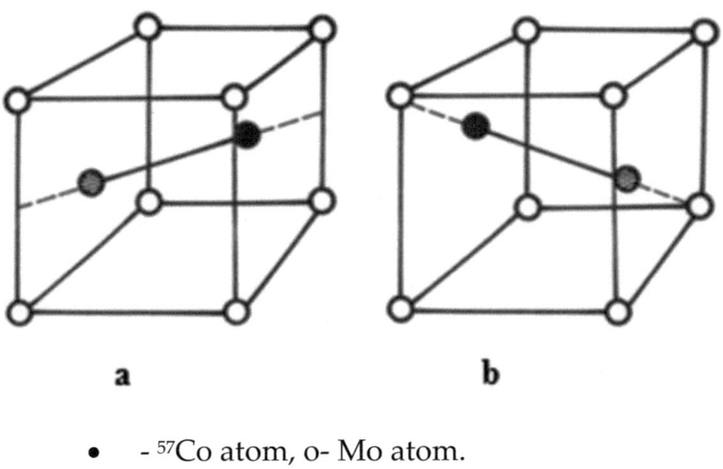

Figure 12. The orientation direction of mixed dumbbell in molybdenum lattice: a- <110> and b - <111>

13. The point defects atomic configuration in the chromium irradiated by high energy protons

The atomic configurations of irradiated metals with bcc lattice have been investigated commonly by numerical methods with molecular dynamics methods [12]. The current metal electron theory is qualitative. It rightly describes the peculiarities of interatomic interactions.

http://dx.doi.org/10.5772/60748

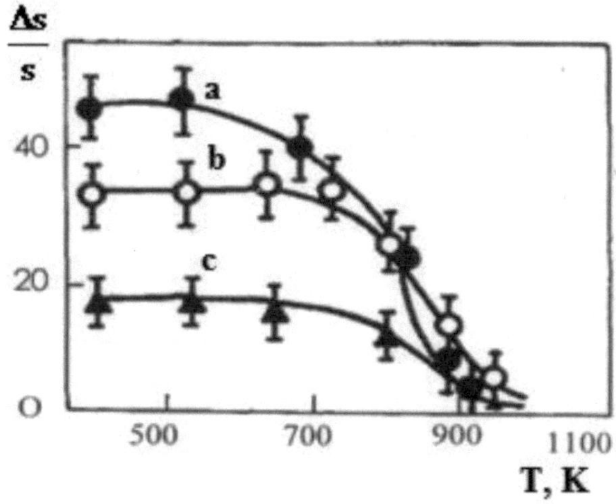

Figure 13. The molybdenum samples isochronal annealing curves, irradiated by different neutrons doses

But quantitative adjectives of these interaction receptions present a considerable difficulty. Therefore, the ^{57}Co impurity atoms' interaction with interstitial point defects in chromium was investigated here by the emission method. The initial Mössbauer spectra of samples consist of symmetrical single lines. The line width is $\Gamma = 0.28$ mm/s, the Mössbauer effect value is $\varepsilon = 23\%$, and isomer shift $\delta = 0.07$ mm/s relatively to the standard absorber.

The sample irradiation has been realized by 10MeV protons at $10^{21} - 3 \cdot 10^{22}$ m^{-2} fluences. The sample Mössbauer spectra have been measured at liquid nitrogen temperature. After irradiation, the absorption line becomes asymmetrical. The spatters on the line wings are observed. The spatters' appearance is testified that the "defect" line spectrum consists of more than one quadrupole doublet (fig.14). Along with a single substitution line, the additional "defect" line described by three quadrupole doublets has been observed (Tabl.2).

Parameters	1	2	3	4
δ, mm/s	0, 07	0, 45	0, 43	0, 38
Δ E, mm/s	0	1, 14	1, 48	0, 86
ε, %	2000%	82	34	56

Table 2. Defect" line with quadrupole doublets

The first line corresponds to the substitution line of ^{57}Co nuclei in chromium regular lattice site. The presence of a single line and absence of quadrupole doublets indicate ^{57}Co environment symmetry by matrix atoms. The large values of the observed shows that the Mössbauer effect probability in chromium is very high.

The other spectra components correspond to quadrupole doublets. They characterize the different stages of cobalt in irradiated chromium defect structure. The new three quadrupole doublets appearance in the spectrum is probably stipulated by the three different complexes formation. They correspond to the Mössbauer "impurity-interstitial" atom mixed dumbbells structure. These dumbbells are formed as a result of interstitial capture by Mössbauer atoms.

There is a high probability of the formation of the above configurations. However, the dumbbells along [110, 111] and [100] directions are energy-optimal. The three quadrupole decomposed lines appearance in the Mössbauer spectra can attribute to these mixed dumbbell three configurations. The equal shift (δ=0.45 mm/s) of doublet lines is indicated on their centres' location in the regular lattice sites. The quadrupole splitting large value is probably stipulated by the electric-field gradient at the expense of the considerable displacement of impurity and interstitial atom from the equilibrium position.

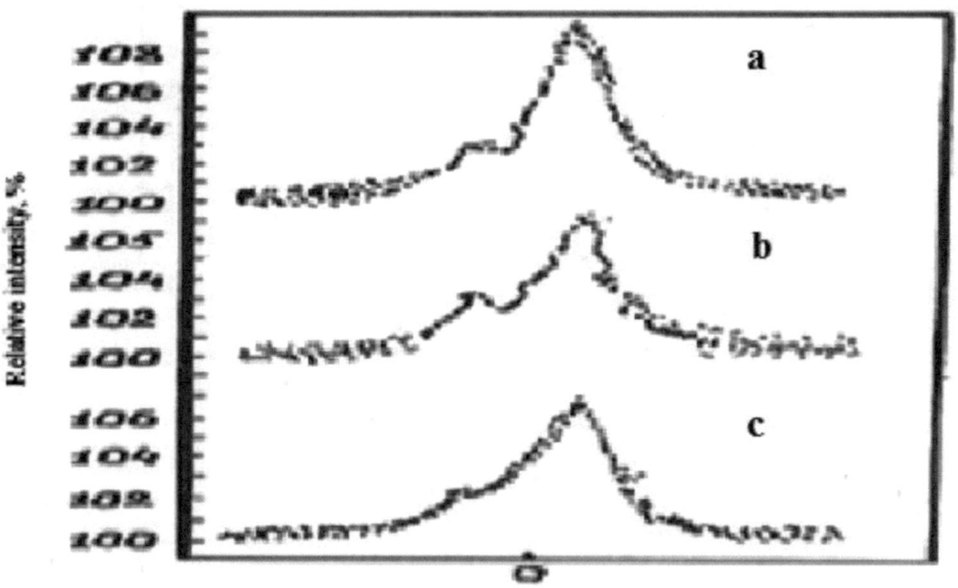

Velocity, mm/s

Figure 14. Emission Mössbauer spectra of ^{57}Co in chromium, irradiated by protons at fluences: a) $5\cdot10^{21}$, b) $7\cdot10^{21}$, c) $3\cdot10^{22}$ m^{-2}

The mixed dumbbell configurations splitting value in chromium is six times higher than in aluminium. The Mössbauer impurity atoms in aluminium represent an interstitial dumbbells structure. The ^{57}Co impurity atoms can pass from one position to another. We thereby can witness a generation of the equivalent configurations in most likely directions in the fcc structure.

Such behaviour of ^{57}Co impurity atom in aluminium is confirmed by the Mössbauer effect for the "defect" line decreasing from 0.9 up to 0.2, which is described by one quadrupole doublet.

http://dx.doi.org/10.5772/60748

14. The point defects interactions in neutrons irradiated tungsten, vanadium, and chromium

The majority of bcc structure metals are the perspective materials in the nuclear technology and space engineering. Among them, the molybdenum and niobium are the most investigated in details [13-18]. It has been established that the point defects generated in crystal structure, after irradiation, interacted between themselves like impurity atoms. The Mössbauer nuclei application as impurity atoms allow to consider the electron state and structure of the "impurity-interstitial" and "impurity-vacancy" complexes. For the molybdenum and niobium cases, the "impurity-interstitial" complexes have a dumbbell configuration. Their Mössbauer parameters depend on the bombarding particles type, energy, and irradiation temperature [12, 19, 20]. For annealing temperature interval of 300-900K in molybdenum, three different configurations of "impurity-interstitial" complexes have been determined. They are moving from one state to another against irradiation conditions and bombarding particles type. From the spectra analysis we can distinguish several annealing stages of the "impurity-interstitial" systems. In the temperature range of 300-550 K, the state, which corresponds to the quadrupole doublet, with the following parameters δ=0.13mm/s and ΔE=0.75mm/s, is stable. At the annealing temperature of about 450K, a new state occurs along with the aforementioned. It is characterized by the quadrupole doublet with δ = 0.46 mm/s and ΔE = 0.82 mm/s. At a higher temperature of 800K, another state with δ= 0.46 mm/s and ΔE= 1.01 mm/s appears on the Mössbauer spectra. All these complex states are completely annealed at temperatures higher than 1150 K, and the sample Mössbauer spectra recover up to the initial state [21, 22]. In niobium, the bound states of impurity atoms with interstitials and vacancies formation occur by an analogous scheme, but the Mössbauer spectra parameters have their peculiarity [23, 24]. However, some questions on this problem, connected with radiation defects interaction with impurity centers are still not answered. To address this problem, initial samples of the polycrystalline foils by ~99.99% cleanness have been used. Preliminarily, the samples undergo a high-temperature annealing at 1500 K and the vacuum of 10^{-8} torr in the course of 20 hours. This is followed by the ^{57}Co radioactive isotope spread on the sample surface by electrolysis in about 3-5 hours. As an isotope sourc, e the $^{57}CoCl_2$ radioactive solution has been used. The emission spectra is measured by a resonance detector on the basis of the $^{57}FeAl$ alloy. Potassium ferrocyanide was used as a standard absorber.

In the initial state, the radioactive^{57}Co impurity atoms are located in tungsten substitution sites. Their concentration does not exceed 10^{-6} %. At such concentrations, the impurity atoms do not interact among themselves. They also do not influence the crystal structure. Therefore, all derived information reflect the processes taking place in crystal as a consequence of radiation point defects interaction with ^{57}Co impurity atoms. Such method application allows to identify the different stages which are formed by the Mössbauer atoms capture on interstitial atoms and vacancies.

The initial spectra of all investigated samples correspond to the single Lorentz symmetrical circuit (fig.15). This indicates the absence of electric-field gradients on the Mössbauer nuclei. The Co^{57} isomer shifts in W, V, and Cr are observed respectively with 0.41 mm/s, 0.08 mm/s, and 0.07 mm/s relatively to the potassium ferrocyanide. The sample irradiation was conducted by fast neutrons at a temperature lower than 80⁰C. The neutrons fluence reached 10^{23}neutron/ m^2. All spectra were measured at room temperature on the Mössbauer installation with uniform acceleration. The sample isochronal annealing was conducted in 10^{-4} Torr vacuum. In the irradiated sample spectra along with the substitution line, we can observe the appearance of the additional "defect" line (Fig.15). Processing of the spectra enabled us to distinguish two different quadrupole doublet lines in W, besides the substitution line. These doublets can be provided by the defect structure formation nearby of the Mössbauer impurity atoms. The first quadrupole doublet with δ= -0.15 mm/s isomer shift and ΔE=0.80 mm/s quadrupole splitting, probably, characterize the "impurity-vacancy" complexes state. The negative isomer shift indicates the decrease of electron density of complexes. On Co^{57}nuclei we can observe an appearance of the electric-field gradient.

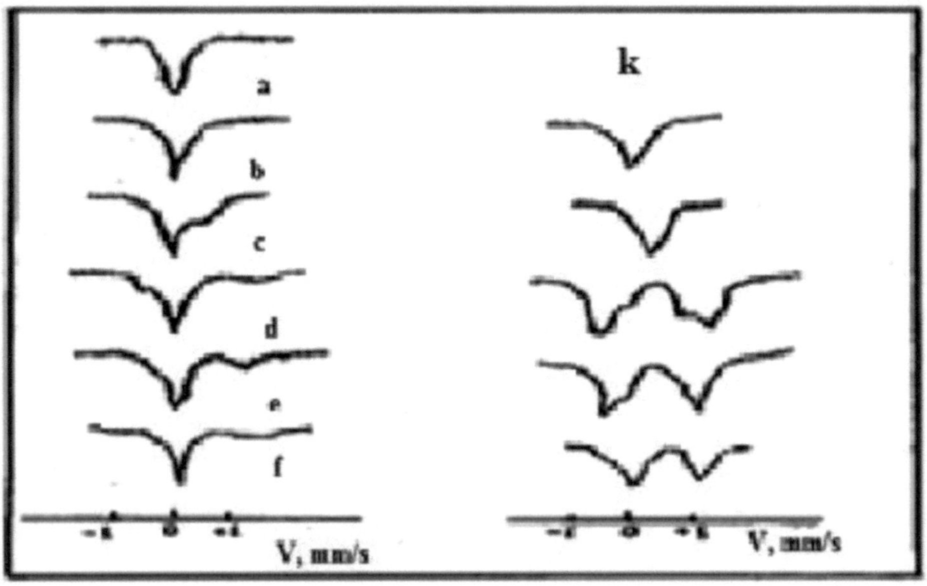

Figure 15. The ^{57}Co inW Mössbauer spectra before (a) and after irradiation (b) by neutrons up to fluences: 10^{22} (b), $2\cdot10^{22}$ (c), $5\cdot10^{22}$ (d), $7\cdot10^{22}$ (e), 10^{23} (f). The spectra after undisplaced line deduction are shown in (k)

The second quadrupole doublet shows a positive isomer shift. It shifted on δ = 0.45 mm/s relatively to the substitution line. The quadrupole splitting of the present doublet is equal to ΔE=1.14 mm/s. This doublet appearance is probably stipulated by the Mössbauer impurity atoms interstitial capture. The latest have become more active readily at irradiation temperature. As a result of the interaction, the ^{57}Co impurity atoms are displaced from crystal lattice sites and generated "impurity-interstitial" bound states of the dumbbell configuration. By computer simulation data in desired conditions, the most likely direction of the mixed dumbbell is along the <110> axis. Such assumption is confirmed by the electron density increasing on the Mössbauer nuclei. The electric-field gradient appeared on^{57}Co nuclei. On the right of Fig.15, the spectra after substituion line deduction are shown (k). At 800 K isochronal annealing, the "impurity-vacancy" complexes are annealed completely and in their place, the other "defect" line appears. It is described by the third quadrupole doublet with δ= 0.43 mm/s and ΔE= 1.14 mm/s parameters.

At the mentioned temperature, the interstitial atoms become free from the different traps and crystal structure irregularities. The activation energy of interstitial atoms migration in tungsten makes up 0.19 eV. The migrating interstitial atoms are recombined with vacancies in meeting with the "impurity atom-vacancy" complexes and the impurity atoms released. The latter further return to the regular crystal lattice. In addition, a part of the interstitial atoms are, probably, interacted with impurity atoms. Consequently, the new formation which presents also the "impurity atoms-interstitial" bound states are formed.

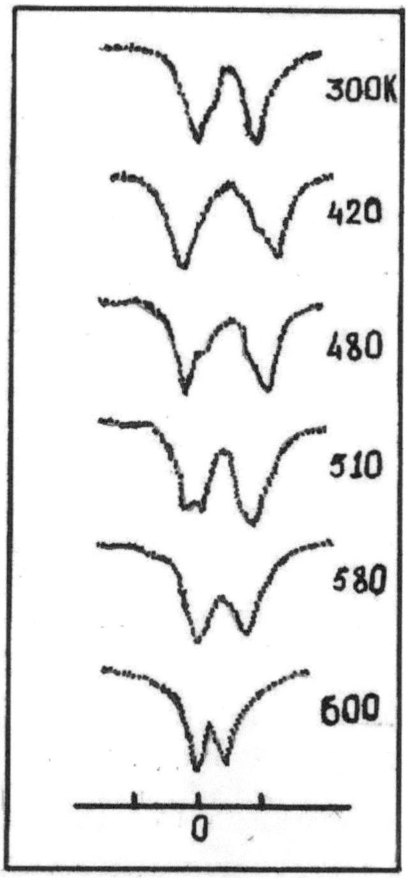

Velocity, mm/s.

Figure 16. The ^{57}Co Mössbauer spectra in V, irradiated by neutrons up to 10^{23}neutron/m^2 fluence after the substitution line deduction at different isochronal annealing temperatures

The "defect" line with positive isomer shift appearance in the Mössbauer spectra of vanadium samples after isochronal annealing at 800K temperature is confirmed of this assumption (fig. 16). Such "defect" line is well described by the quadrupole doublet with δ=0.41 mm/s relatively to the substitution line and $\Delta E = 1.48$ mm/s. The calculation analysis has shown that such states corresponds to the dumbbell configuration along <111> axes. At a further increase on isochronal annealing temperature, the relative area under the spectrum decrease, but δ and ΔE values do not change. The "defect" lines in the spectra are completely annealed at 1200K temperature. The Mössbauer "impurity-interstitial" atom complexes annealing at temperatures above 1200K is probably realized by the single vacancies releasing, migration and their recombination with interstitial atoms. The vacancies migration activation energy is 1.3 eV in vanadium.

The chromium irradiation is at the same conditions. In the Mössbauer spectra, along with undisplaced line, the one 'defect" line also appears. It is described by one quadrupole doublet with $\delta = 0.38$ mm/s and $\Delta E = 0.86$ mm/s. Such quadrupole doublets are relevant for the tungsten samples also. Thus, one can assume that in metals with bcc structure, after neutrons irradiation, both "impurity-vacancy" and "impurity-interstitial" complexes are formed. In the Mössbauer spectra, the quadrupole doublets with electric-field gradients large values on ^{57}Co nuclei are appearing.

15. Investigation of recoil nuclei states in molybdenum, are generated by protons and α-particles irradiation

The recoil nuclei are generated as a result of the elastic and inelastic interactions of incident particles with target atoms, and the investigation of their states is a quite relevant problem of the radiative study of materials. The Mössbauer effect application allows to investigate the recoil nuclei states in crystals after irradiation. A new formed nuclei as a result of nuclear reactions can correspond to the Mössbauer radionuclide. Then, it can register selectively with such recoil nuclei by emission spectroscopy [25, 26]. By absorption and emission spectra comparison for these nuclei, it is possible to determine the state of the primary knocked-on atom (PKA). It is known that the atom displacement from the lattice minimum energy is equal about 25 eV. The recoil nuclei energy at (n, γ)-reactions may reach to a value of some 100 eV. At (d, p)-reactions, the recoil nuclei energy may reach no more than 10 MeV. Even so, the atom leaves its initial position in the crystalline lattice site. The stored energy is used for elastic collisions with other atoms. Over 10^{-12} c time, the PKA energy decreases below the displacement threshold energy. The life time of the Mössbauer level is considerably greater(10^{-7}c for ^{57} Fe) than the PKA stopping time. Therefore, the Mössbauer effect offers the unique opportunity for the PKA new location detection. At the same time, its immediate environment and electron state can determined.

To study this effect, the experiments are conducted on polycrystalline molybdenum samples with 99.98% purity. The two sets of samples, enriched up to 80% by ^{57}Fe isotope, were prepared. The isotopes are deposited on the molybdenum samples surface by $2 \cdot 10^{-3}$ Pa vacuum coating. The enriched ^{57}Fe evaporated layer thickness constitutes 0.2 mg/cm^2 and native iron -20 mg/

http://dx.doi.org/10.5772/60748

cm^2. The evaporated layer enriched by ^{57}Fe samples was irradiated by protons with the energy of 50 MeV to up to $4.2 \cdot 10^{21} m^{-2}$ fluence. The sample with native iron were irradiated by α-particles with 50 MeV energy, up to $2.7 \cdot 10^{21}$ m^{-2}. Liquid nitrogen was used for cooling, so that the samples temperature at irradiation do not exceed 100K. The beam current density was $8 \cdot 10^{-3} A/m^2$. The samples' spectra before irradiation corresponded to sextet. The spectrum parameters corresponded to the native iron spectra parameters (fig. 17, 18). The absorption spectra did not change after irradiation. In the emission spectra along with the sextet, a single symmetrical line appears (fig.17a, 18a). The ^{57}Co Mössbauer nuclei in irradiated samples are generated by $^{57}Fe(p, n)^{57}Co$ and $Fe(\alpha, p2n)^{57}Co$ nuclear reactions. The recoil nuclei mean energy at elastic interaction is defined by following equation:

$$<E> = M_1 m E_0 / \left(M + m\right)^2,$$

where, m – bombarding particle mass; M – target nucleus mass; M_1 – the nucleus mass obtained after nuclear reaction; E_0 – the bombarding particle energy.

The recoil mean energy <E> at elastic collision of impinging particles with target atoms is shown in Table 3. Below, the corresponding path lengths in iron and molybdenum are presented.

Particles	Energy, MeV	Recoil nuclei mean energy, eV	Path length, μm	
			Fe	Mo
Protons	10	389	$5 \cdot 10^{-4}$	$5 \cdot 10^{-3}$
α-particles	50	506	$7 \cdot 10^{-4}$	$7 \cdot 10^{-4}$

Table 3. Recoil energy and ranges in Fe and Mo at elastic interactions of impinging particles with target atoms

The ^{57}Fe enriched iron layer thickness is 0.15 μm and native iron's is 0.71 μm. The recoil nuclei with 5-7Å free path length incipient at elastic collisions remains on iron matrix. The ^{57}Co Mössbauer radionuclides are generated consequently by $^{57}Fe(p, n)^{57}Co$ and $^{56}Fe(\alpha, p2n)^{57}Co$ nuclear reactions. Therefore, the sextet appears in the absorption spectrumand corresponds to the ^{57}Fe spectrum. The model which is described [27] has been used for recoil nuclei mean energy calculation at inelastic interaction. The mean energy values of the particles inelastic interactions with target atoms are given in Table 4.

Particles	Energy, MeV	Mean recoil energy, эВ	Path length, μm	
Fe	Mo			
Protons	10	0, 16	0, 43	0, 04
α-particles	50	3, 16	0, 93	1, 88

Table 4. Recoil energy and ranges in Fe and Mo at inelastic interactions of impinging particles with target atoms

It is seen that the recoil nuclei path length in the targets are comparable with the iron evaporated layer thickness. Therefore, the part of ^{57}Co nuclei penetrates into the target substrate. In emission spectra, a symmetrical line appears along with the sextet (fig.17, 18). It is characterized by the ^{57}Co impurity atom state in molybdenum crystalline lattice. It is stipulated, thereby that at inelastic interactions, the Mössbauer nuclei obtain the energy sufficient for the matrix lattice penetration.

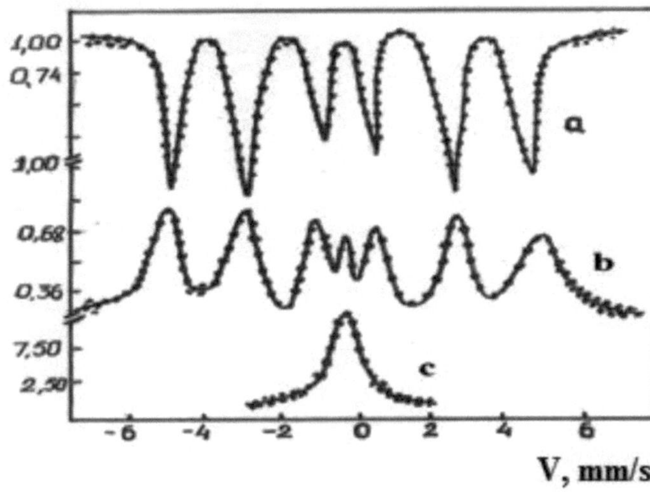

Figure 17. Absorption (a) and emission (b) spectra of ^{57}Fe, is evaporated on molybdenum surface after protons irradiation. Emission standard spectrum of ^{57}Co in Mo (c) taken from [28].

By Mössbauer spectra character, one may assume that the ^{57}Co isotope atoms predominately occupy the substitution sites in the molybdenum crystalline lattice lines, broadening what is probably stipulated by the fast hardening at ^{57}Co recoil nuclei slowing down and with the thermal spikes formation in the substitution region. The sextet appearance in the emission spectra after irradiation is characterized by the ^{57}Fe atoms state in matrix of metallic iron. The calculations have shown that the recoil mean energy for ^{57}Co atoms is 016 MeV at 10MeV protons irradiation and 3.16 MeV at α-particles irradiation. The recoil nuclei path length in target constitutes 0.04 and 1.80 μm, respectively [28].

The nuclei recoil means that energy is in the 400-500eV range. The path length of the nuclei recoil in molybdenum is about $(5\text{-}7)\cdot 10^{-4}$μm. Such nuclei recoils must remain practically and completely in the range of the evaporated iron layer on the molybdenum samples surface.

16. The Mössbauer spectroscopy of magnetic nanoparticles

Currently, the research activities and development of new diagnostic and medical technology in oncology appears as one of the important areas in medicine [29]. In spite of the sweeping

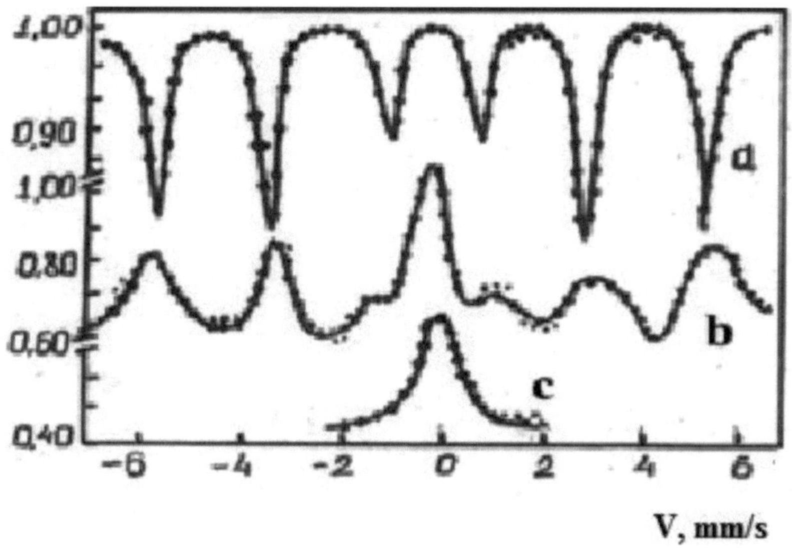

Figure 18. Absorption (a) and emission (b) spectra of the native iron, evaporated on molybdenum surface after α-particles irradiation. Standard emission spectrum of ^{57}Co in Mo (c) taken from [26, 28]

progress in modern medicine and biology development, cancer remains as an intractable disease which has taken million of humans life yearly. Each 30 seconds, one oncological patient in the world dies. The traditional treatment mode application such as the radiotherapy, surgical, chemotherapeutical, and other methods are limited by the difficulty of accessing the tumor, propagation of the cancerous cells to the whole of the body, and lack of selectivity of these methods against cancerous cells. Although radiotherapy is the unique means in fight against cancer, it very often affects the sound tissue.

In this section, the brand-new innovation technique on the basis of marked magnetic nanoparticles application has been proposed for use. This method of cancer therapy differs from standard therapy methods, whereby metal nanoparticles can be delivered directly to the cancerous cells. Hereupon, it is possible to make a direct impact on the malignant tumor without damage to the neighbouring healthy cells. The different magnetic nanomaterials production is present as a practical interest. It is stipulated, thereby that during nanodimensional state transition, a number of fundamental properties are changed. The nanoparticles are possessed by a highly-developed active surface and by a high absorptive capacity respectively. Because of their small dimensions they are comparable with cells, viruses, proteins and DNA dimensions. The nanoparticles can approach, interact and connect with the biological object. Nano-dimensional materials on the base of ferrites are very important and in disperse aqueous and non-aqueous solvents. Currently, investigation and engineering applications of the broad-spectrum of magnetic nanoparticles on the base of Co, Fe, and Ni metals, ferric oxides, $MgFe_2O_4$, $CoFe_2O_4$, $MnFe_2O_4$, $LiFe_5O_8$, and CoPt, FePt, MnAl, $SmCo_5$ etc., have been investigated [30].

In Fig.19, the scheme of the magnetic nanoparticles target delivery, diagnostics, and controlled hyperthermia are shown. The main point of the proposed method is in the following. Due to the fact that the nanoparticles can function both on tissues and cells levels, they can undergo endocytosis or phagocytosis. Consequently, the nanoparticles are absorbed by cells. In the course of this process, the magnetic nanoparticles can penetrate over the cytoplasmic membrane and in some cases in the nuclear membrane also.

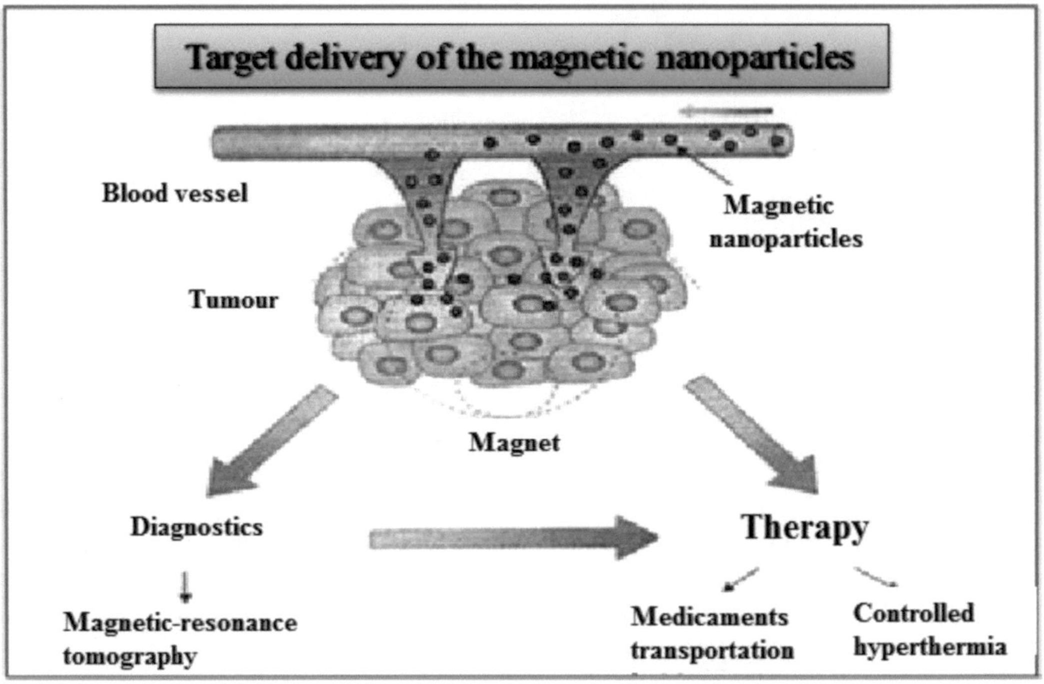

Figure 19. The magnetic nanoparticles transportation scheme for the diagnostics and therapy

As summarized above, the magnetic nanoparticles are finding increased applications in oncology. Owing to newly developed drugs, free transportation by the intended users and the malignant tumor destruction with minimal side effect and minimal damage to surrounding tissues are also finding increasing applications. To achieve this goal nanoparticles obtaining the special biological markers are necessary to be designed, in possession of specific properties characterized by the different tumors. They correspond to the magnetic nanoparticles and antibodies special mixtures which identify the tumor location and penetrate in it. Since the nanoparticles are penetrated in the tumor structure, it is necessary to excite them by high-frequency magnetic fields.

Under this magnetic field, the whirling currents appear in the nanoparticles, which causes local heating. By the magnetic field intensity change, it is possible to enhance the nanoparticles temperature and consequently change the pathology region temperature. As the malignant tumor temperature increases to up to 43°C and higher these malignant tumors are disintegrated and later on disappear completely [31]. Currently, the perspective in oncology magnetic nanoparticles on the ferric oxide base (hematite, magnetite, and maghemite) were approved by the Food and Drug Administration (FDA) in the many countries of the world.

The Mössbauer spectroscopy can give valuable information on ferric oxide nanoparticles super paramagnetic behaviour. The Mössbauer effect study enables us to control the hyperfine structure and the electron state of a magnetic nanoparticles on the base of ^{57}Fe nuclei ferric oxides. By researching results, the surface layer thickness can be calculated also. In addition, the initial samples, which are used at magnetic nanoparticles synthesis, are sufficiently and accurately identified. The Mössbauer spectroscopy can give selective information about the inside and surface properties of the nanoparticles. As an example, the hematite, magnetite, and their aggregation with FeO and natural iron Mössbauer spectra are given below (fig.20). One of the important problem at the ferric oxides particles synthesis process is the FeO, hematite – Fe_2O_3, magnetite – Fe_3O_4 and maghemite – γ-Fe_2O_3 phases identification and separation. If the FeO has a cubic crystal structure and the hematite has a corundum-type structure, that both magnetite and maghemite have the spinel type structure, thus, is not distinguishable by the x-ray method. As shown by Fig.20, the hematite spectrum consists of one sextet. Also, the line is widened.

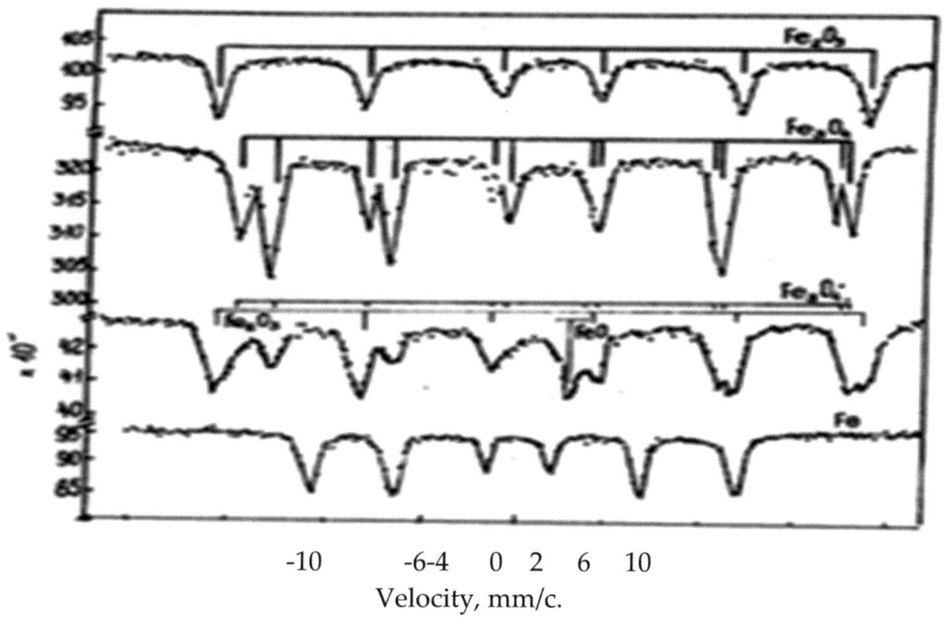

Figure 20. The Mössbauer spectra of the ferric oxides– hematite Fe_2O_3, magnetite Fe_3O_4, their aggregation (Fe_2O_3, Fe3O4) with FeO and natural iron

The line broadening denotes that the spectral line represents a superposition of two unresolved lines, which cause the respective states of the ^{57}Fe atoms isotope. The magnetite spectrum corresponds to two clearly defined sextets for the ^{57}Fe atom states located in the octahedron and tetrahedron sublattices with the spinel structure. On the sample spectra, which possessed all the ferric oxides components, the separate components are good and visible. The last spectrum corresponds to the pure metallic iron.

17. Conclusion

- In this work, the general regularity of the interstitial atoms and vacancy interaction with impurity substitutional atoms of ^{57}Co(^{57}Fe) in bcc lattice metals have been systematically investigated for the first time. The electron states and structure of "impurity-interstitial" atom, "impurity-vacancy" systems, and their Mössbauer parameters were defined.
- The irradiation influence on complexes on the base of ^{57}Co impurity atoms and interstitials formation stages, the dependence of electron state, internal crystalline electric fields from dose and irradiation energy of neutrons, fission fragments and accelerated charged particles were investigated. In the investigation, the Mössbauer impurity atoms vibrations rms, amplitude values, and their binding energy were defined.
- For the first time, through the Mössbauer effect study, the complexes annealing stages from isochronal annealing temperature have been defined. The processes of Mössbauer ^{57}Co impurity atoms thermal diffusion in bcc metals has been investigated. In iron-beryllium and iron-titanium laminated systems, the phase formation, which depends on the beryllium concentration in titanium and also on the temperature of isochronal annealing, was investigated. Using different experimental methods, the peculiarities of the impurity atoms accumulation process at high dose ion implantation in molybdenum have been shown.
- It has been experimentally established that the atoms mobility considerably increases in radiation damaged zones created by high-velocity charged particles, fission fragments, or ionized displaced atoms. As a consequence of the interaction of radiation defects with impurity atoms, their electronic states, nearby environment, and hyperfine structure of nuclear levels undergo changes. These were selectively registered by the Mössbauer spectroscopy method.
- The systematical investigation of point radiation defects with substitutional impurity interaction were performed. The "impurity-interstitial" complexes' stability temperature interval is indicated and their annealing stages have been determined. The Mössbauer parameters of the compound dumbbells composed of impurity and interstitial atoms have been determined. The technology of gamma radiation Mössbauer source production for emissive investigation is reinvestigated.
- In the fcc metals, the efficiency of the method was demonstrated on Al on impurity ^{57}Co atoms. Unlike the fcc metals, the bcc metals' considerable quadrupole splitting has revealed the ability to separate them on different interstitial configurations around ^{57}Co impurity. By following the process of formation and decay of these configurations after irradiation by different particles, as well as after isochronal annealing, it becomes possible to obtain important information about the defects' mobility and "impurity-defect" complexes state.

http://dx.doi.org/10.5772/60748

- - It has been established that non-cubic charge distribution around a Mössbauer atom leads to the electric field gradient. This causes the nuclear levels' hyperfine splitting, at the expense of quadrupole interaction, which emerges as two adsorption lines in case of ^{57}Fe. Quadrupole splitting is proportional to the electric field gradient on Mössbauer nuclei. As a result of systematic investigations on $Al^{57}Co$ diluted alloys, it has been shown that one can observe the interstitial atom captured by ^{57}Co impurities directly by the appearance of an additional "defect" line in the Mössbauer emission spectrum.
- In the course of investigating radiation processes in metallic systems, the basic idea about atoms displacement mechanisms and defects stability was developed. The diffusion mechanism of the interstitials movement models was realized. Also, the impurity atoms role in the process of formation of "impurity-interstitial" complexes was studied.
- As a result of systematic investigations on $Al^{57}Co$ diluted alloys, we have shown the interstitial atom captured by ^{57}Co impurities directly by the appearance of an additional "defect" line in Mössbauer emission spectrum.
- At the same time, for the purposes of its application to transportation of medicinal products to the fields of cancerous cells focused in human, the Mössbauer spectroscopy capability for nanoparticles properties was thoroughly investigated.

Acknowledgements

The authors would like to thank Dr. Umarov T.F. and Ospanov Y.O. for their useful help.

Author details

K.M. Mukashev[1], A.K. Shokanov[1] and F.F. Umarov[2*]

*Address all correspondence to: farid1945@yahoo.com

1 Department of the Theoretical and Experimental Physics of the Kazakh National PedagogicalUniversity named after Abai, Almaty, Kazakhstan

2 Department of Geology and Earth Physics of the Kazakh-British Technical University Almaty, Kazakhstan

References

[1] Mössbauer R.L.//J. Phys. 1958. V. 151. P. 124.

[2] Spinel V.S. γ -rays resonance in Crystals.-M.: Nauka, 1969.

[3] Vertheim G. Mössbauer Effect. Foundations and Application. M.: Mir, 1966.

[4] Erich U.Z. Phys. 1969, 227, №1 p. 25-44.

[5] Herries D.R. et.al. Te Third Intern. Conf. on Peaceful user of Atomic Energy. Geneva, 1964, № 162.

[6] Orlov A.N., Trushin Yu.V. The Point Defects Energy in Metals. M.: Energoatomizdat, 1983.

[7] Mansel W., Vogl G., Fast neutron radiation damage in aluminum studied by interstitial trapping at^{57}Co Mossbauer atoms.- J. Phys. F.:Metal Phys., 1977, V7.N.2 P. 253-271.

[8] Vogl G., Mansel W., Petra W., Groger R. A comparison of bindtails in Aluminium and Silver. – Hyperfine Interact., 1978. V.4. P. 681-683

[9] Vogl G. and Mansel W., Hroc. Conf. on the Fundamental Aspects of Radiation Damage in Metals, ed. by M.T.Robinson and F.W. Young, Jr., ERDA Report, conf 1976, 751006, P. 349.

[10] Schilling W. Self-interstitial atoms in metals.- J.nucl.Mater., 1978, v. 69, №. 1-2, P. 465-489.

[11] Vogl G. and Mansel W., Prog. Int. Conf. On the Fundamental Aspects of Radiation Damage in Metals, ed. by M.T. Robinson and F. W. Young, Jr., ERDA Report, CONF1975 1006, p. 349.

[12] Hyperfin Interact.1978, V.4, P. 681.

[13] Zhetbaev A.K., Ibragimov Sh.Sh., Shokanov A.K., Ozernoi A.N. Mossbauer effect study of radiation defects interaction with impurity iron atoms in neutron irradiated niobium. Proc. Indian Nat. Sci Aced. New Dehli 1982.

[14] Zhetbaev A.K., Kupchishion A.I., Shokanov A.K., Mukashev K.M. Radiation effect in neutron irradiated studiet by gamma - resonanse and positron methods. Radiation effect 1989, v.106.

[15] Zhetbaev A.K., Ibragimov Sh.Sh., Shokanov A.K. A Mossbauer study of molybdenum oxidation kinetics. J. dePhys. 1980.v.41 p.

[16] Zhetbaev A.K., Ibragimov Sh.Sh., Shokanov A.K. A Mossbauer study of neutron irradialed Fe-57 in molybdenum. Int. Conf. Mossbauer spectroscopy. Portorog, Abstracts, 1979.

[17] Zhetbaev A.K., Ibragimov Sh.Sh., Shokanov A.K., Ozernoi A.N. Mossbauer effect study of radiation defects interaction with impurity iron atoms in neutron irradiated niobium. Int. Conf. Applie. Mossbauer effect Abstracts, India, 1981.

[18] Shokanov A.K., Salikhodjaev D.M. The investigation of the ^{57}Fe impurity states in molybdenum. In book: Transport processes in solid and gaseous mediums. Subject collection of the Kazakh Pedagogical Institute after Abai scientific studies.1989, P. 13-18.

[19] J. Phys.F.: Metall Phys. V7, № 2, 1977.

[20] Murray C.B., Norris D.J., Bawendi M.G., "Synthesis and characterization of nearly monodisperse CdE (E = sulfur, selenium, tellurium) semiconductor nanocrystallites.", J. Am. Chem. Soc., 1993, 115, 8706.

[21] Zhetbaev A.K., Mukashev K.M., Kupchishion A.I., Shokanov A.K. Gamma-resonance and positron diagnostics of radiation defects in irradiated by neutrons molybdenum. Izv. ANKazSSR, ser. phys. math., N4, 1989.

[22] Tompson M. Defects and radiation damages in Metals. M.: 1971.

[23] Zhetbaev A.K., Ibragimov Sh.Sh., Shokanov A.K. The "impurity-interstitial" interaction observation in the irradiated niobium. DAN SSSR, V.265, N2, 1982.

[24] Jonhson R. A., Lam N.Q.-Phys.Rev. Ser. B. 176, v. 13 h. 4364.88. Erginsoy C., Vineyard G.H., Englert A.- Pys. Rev. Ser. A, 1964, v.133, p.595.

[25] Garber R.I., Zavalishin A.A., Lukyanova A.N., Tkach V.I., Fedorenko V.I. Investigation of the structure and properties of tungsten and molibdenum irradiated in reactor at 100-1800° C. In book: VANT, ser. FRP and RM, 1976, V.1 (3), P.46-55.

[26] Shokanov A.K. The Mossbauer investigations of the radiation point defects interaction. Materials of the International theoretical and practical Conference initiated to Buketov E.A. Eighty years anniversary. Karaganda, 2009, V.3, P.703-707.

[27] Baldin A.M., Goldansky O.N., Rosental I.N. The nuclear reactions kinematics. M.: FM, 1959.

[28] Shokanov A.K. The Investigation by the instrumentality of Mossbauer effect of the recoil nuclei states in molibdenum generated after protons and α-particles irradiation. Materials of the International scientific symposium, devoted tocentenary of K.I. Satpaev "K.I. Satpaev academician and his role in Kazakhstan Science, Education, and Industry development", Almaty, 1999, P.68-69.

[29] Gubin S.P., Koksharov Yu.A., Xomutov G.B., Yurkov G. Yu. "Magnetic nanoparticles: production methods, structure, and properties", 2005, Chemistry progress, 74, 6, 539-574.

[30] Kechrakos D., Trohidou K.N., "Magnetic properties of dipolar interacting single-domain particles", Phys. Rev. B, 1998, 58, 12169–12177.

[31] Khedr M.H., Omar A.A., Abdel-Moarty S.A., "Magnetic nanocomposites: preparation and characterization of Co-ferrite nanoparticles", Colloids and Surfaces A: Physicochem. Eng. Aspects, 2006, 281, 8–14.

Section 6

Generation of Hydrophobicity in Alloys

Chapter 7

Hydrophobicity — A Green Technique for Enhancing Corrosion Resistance of Alloys

Zaki Ahmad, Asad U. Khan, Robina Farooq, Naila Riaz Mastoi and Tahir Saif

Additional information is available at the end of the chapter

http://dx.doi.org/10.5772/60815

Abstract

The corrosion phenomenon is as old as the age of the planet. The cost of corrosion has risen alarmingly with industrial progress and it is estimated to be around 300 billion dollars or 3 to 4.5% of the GNP of developed nations. Thousands of alloys have been developed to control corrosion, which is a major consideration in the development of new ferrous and non-ferrous alloys. Several corrosion control techniques such as inhibitor treatment, coatings, cathodic protection, alloying additions, and designing for corrosion protection have been developed to combat corrosion. Despite their merits, techniques such as inhibition treatment and coatings are limited by their adverse effect on the environment because of their volatile organic components. Due to an increasingly alarming carbon footprint, there is a growing global concern to keep the environment clean. Hence, a great need exists to replace the current control methods by eco-friendly methods. The potential of the green technology of hydrophobicity has therefore been exploited to control corrosion by fabricating hydrophobic surfaces on alloys and these surfaces have shown highly promising results. This technology offers a novel method to control corrosion of metals, alloys, polymers and composites.

Keywords: Hydrophobicity, Wet contact angle, Sol-gel Technique, Potentiodynamic Polarization, Surface Roughness

1. Introduction

With the dramatic growth in technology witnessed during the last twenty years, new alloys have been developed to fulfill specific requirements for components in the aerospace, automotive, shipping, transportation and metal industries. Researchers have focused their attention mostly on aluminum, copper, nickel, magnesium and ferrous alloys to enhance their mechanical strengths [1-16]. In recent years dramatic improvements have been made in enhancing mechanical, chemical and physical characteristics of ferrous and non-ferrous alloys by introducing nano-particles into the alloys. This is best exemplified by the nano-structured Fe. 20Cr alloy [17, 18] and the nano-crystalline surface of 304 stainless steel [19, 20]. The corrosion resistance depends on the formation of a passive and homogeneous film. Work on 304 stainless steel showed that the capability of passivation is enhanced, accompanied by enhanced corrosion resistance on a nano-structured surface [20].

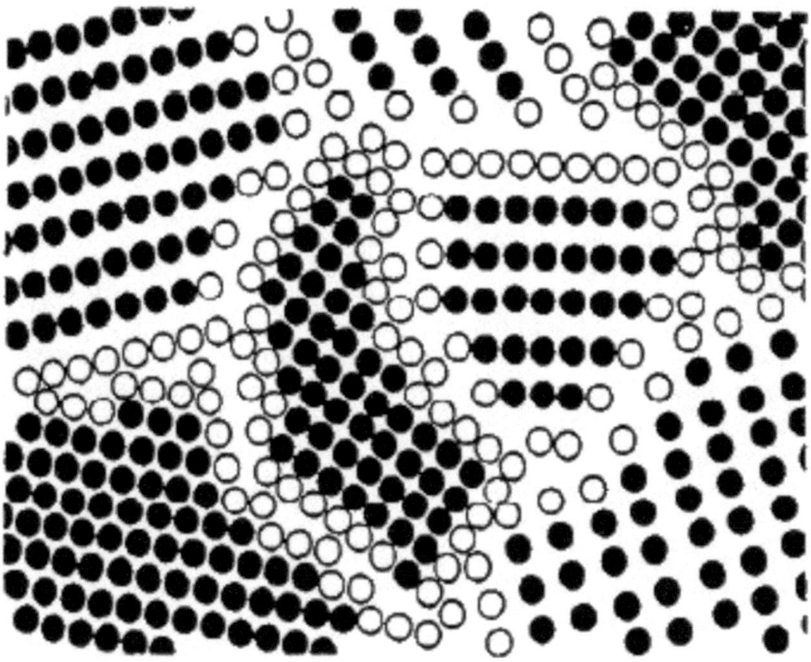

Figure 1. Schematic representation of the atomic structure of a two-dimensional nano-crystalline material distinguishing between the atoms associated with the individual grains (solid circles) and those constituting the grain boundary network (open circles) [24].

In both examples, the grain size cited ranged in the nano-scale regime. A study of 304 stainless steel prepared by equal channel angular processing with a grain size of 80-120 nm showed an improved tendency for passivation; studies on light alloys such as magnesium alloys showed the beneficial effect of nano-structured cerium oxide film. Potentiodynamic and electron impedance spectroscopy studies showed improved corrosion resistance of Mg alloy AZ31 [21]. The corrosion resistance of copper based alloys is attributed to the formation of cuprous oxide (CuO) and cupric oxide (Cu_2O) protective layers. They are excellent condenser tube materials for application in desalination plants [22]. The nickel development institute has done pioneering work on nickel and its alloys. Investigations have been conducted on nano-crystalline nickel alloy [23]. Enhanced overall current density is shown by nano-crystalline nickel and its

alloys because of high defects arising from the high surface fraction of grain boundaries and triple junctions. The high defect concentration provides increased number of potential sites, increased overall current density but reduced current density on individual sites which inhibits localized corrosion in nano-crystalline materials as shown in Figs. 1 and 2 [24, 25].
The grain growth is limited by triple junction (Fig 4) [25].

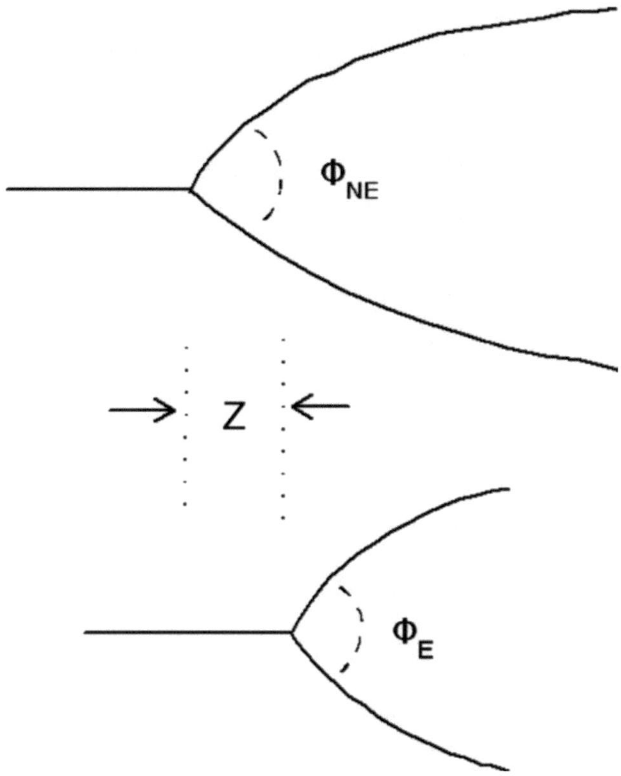

Figure 2. Schematic illustration of triple junction with (upper) non- equilibrium dihedral angle (Φ_{NE}) and (lower) equilibrium dihedral angle (Φ_E); migration of triple junction by distance z enables the retention of equilibrium dihedral angle [25].

Another factor is even distribution defect sites as compared to poly-crystalline materials. Nano-crystalline nickel shows uniform corrosion compared to severe localized corrosion in poly-crystalline nickel. The variation of anode/cathode current density in nano-crystalline materials depends upon the surface morphology and the surface modification as shown in Figs. 3 and 4 [26].

The decrease in corrosion resistance is attributed to the matrix/inter-crystalline (m/L) ratio and anode-cathode area ratios, which may decrease to 20 for grain size smaller than 100 nm. The uniformity of grain size may reduce penetration current density $i_a = i_c$. Improved corrosion resistance has been reported for nano-crystalline $Al_{90}Fe_5Gd_5$ and $Al_{87}Ni_7Y_{43}$ and $Cu_{90}Ni_{10}$ alloys [27, 28]. A nano-structured copper alloy showed less pitting corrosion attack than poly-

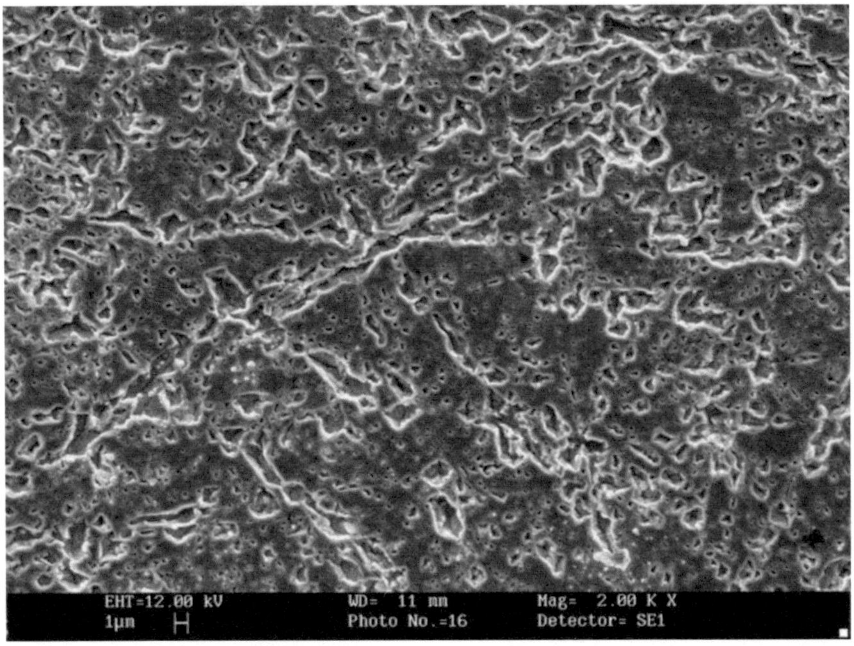

Figure 3. Surface morphology of nanocrystalline nickel [26].

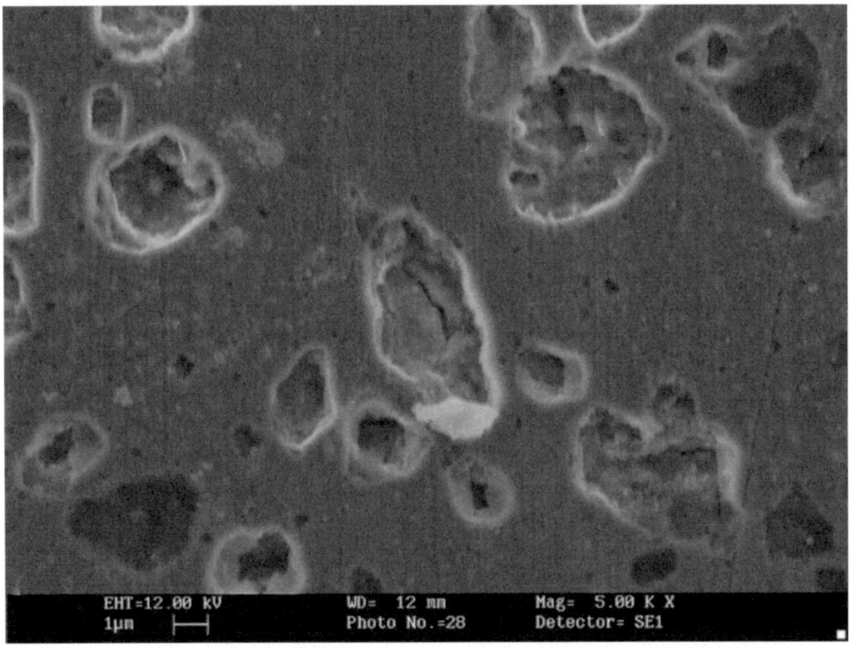

Figure 4. Surface modification to minimize corrosion [26].

crystalline copper electrode after polarization from 0.8 to 0.1V at 0.2 mv/S in a 3.5 wt% NaCl solution because of the formation of passive films of CuCl and Cu_2O [29, 30]. Studies on electrochemical behavior of nano-crystalline aluminum fabricated by plastic deformation showed enhanced passivation ability of nano-crystalline aluminum [31].

In a work, the addition of 0.3-0.4% scandium with 0.14% zirconium signally suppressed corrosion due to a passive layer created by the formation of nano-precipitates of Al_3Sc. A strong tendency of pinning of grain boundaries was observed. This shows how the grain size and nano-precipitates affect the passivation behavior of Al-Sc-Mg alloys [32, 33].

2. Cost of corrosion and bio-fouling

Although reliable figures and the most recent breakdown of corrosion losses have not been reported yet, from the recent past estimates the cost of corrosion is estimated to be around 85 billion dollars. In U.K, the cost is estimated to be 4-5% of GNP [34]. The corrosion free life of automobiles in the estimated cost of Saudi Arabia is only 6 months [35]. Corrosion costs are huge in land, sea and air transportation, oil and gas transmission pipelines and nano and micro-devices. In marine corrosion bio-fouling is encountered which affects the speed of ships and causes excessive consumption of fuels. Marine bio-fouling has manifold effects on the corrosion and life cycle of marine vehicles and plants using sea-water such as dual system desalination plants. The condenser tubes become the target of corrosion bio-fouling. Bio-fouling affects marine vehicles such as ships and boats by the adhesion of layers of micro-organisms and deposition of bio-films. Their attachment to the substrate also leads to the formation of differential aeration cells and onset of localized corrosion. The active organisms which attack are generally bacteria, diatoms, algae and invertebrates causing increased hydrodynamic drag resulting in increased fuel consumption [36].

Both chemical and physical methods have been used to control bio-fouling. The circulation of sponge rubber ball with cooling water has been employed conventionally. The use of a suitable polymers has been quite successful. Biocides have been used to kill micro-organisms. Glute-aldehyde as well as proprietary mixtures of complex organic compounds has been extensively used. Despite all research, the development of a bio-film repellent surface appears to be a new remedy which is being resolved by super-hydrophobic surface. The damaging effects are not only limited to corrosion and bio-fouling, but they also affect the environment. The international maritime organization estimates that fuel air emission could increase between 38 and 72% by the year 2020. Anti-fouling coatings have been developed from natural resources; however their behavior is not conclusively known. Low surface energy nano-SiO_2 coatings have the capability to reduce the adhesion strength of micro-organisms. The micro, nano-layer structures on the coating films could be obtained when an appropriate mass ratio of resin, nano-SiO_2 and other pigments are approached. Sea-water tests have shown that lower surface energy and the elastic modulus of coatings are less in adherence of the bio-films [37].

The technology of super-hydrophobic surfaces has come to the rescue of bio-fouling. Super-hydrophobic surfaces are defined as surfaces with wet contact angle greater than 150°. These surfaces are water repellent and due to their hierarchical nature of surface and deposition of low energy compounds, they repel water. This phenomenon will be fully described under "Super-Hydrophobicity theory" to understand super-hydrophobic coatings which lower the surface energy of a substrate. In a work reported, a super-hydrophobic surface attached significantly less micro-organisms after nine days on silicon coated enamel, fluorinated ethylene propylene (FEP) abutments and modified silicon rubber voice prosthesis with fluoroalkyl-silanes showing a significant reduction in bio-film formation [38-40]. A few strategies have been adopted to mitigate bio-fouling. The hyper-branched fluoropolymers (HBFP) and polyethylene glycol (PEG) film showed a strong resistance to proteins, serum, albumen, lectin, and lipopoly-saccharides [41].

In another study amphiphilic block polymers, co-polymers based on polystyrene (PS) and polyacrylate (PA) were designed. The PS block acted as a binder and PA block exposed the amphiphilic side chain to the exterior. These materials showed a good resistance against ulva and cells of diatoms [42]. The application of hydrophobic surface exposed to marine environment has shown a dramatic change since 1926, when Navy used plastic paints to prevent bio-fouling. The use of anti-adhesive water repellent surfaces is a departure from polymer technology practiced in the past.

3. Corrosion protection methods

Conventional corrosion protection methods include alloying, coating, designing for corrosion protection, cathodic protection and surface modification, inhibition and selection of materials for a particular environment. There is a vast amount of literature available on these conventional methods in books and journals on corrosion such as "NACE Corrosion and Materials Protection" published by NACE, Houston Texas, U.S.A. and other corrosion journals [43-48]. All the conventional methods have their merits and they are highly productive. It is very well known that water, moisture, humidity, acid and basic electrolytes and marine environment induce corrosion in different forms. It was not until the lotus effect was discovered by Barthlott and Neinhuis [49] on lotus leaves that research was focused on hydrophobicity and hydrophilicity as a tool to mitigate corrosion.

Super-hydrophobicity is now the focus of attention because of its promising application potentials on increasing life cycle of metal and alloys, cleaning of environment, reduction of drag in ocean going vessels, anti-icing of aircrafts, properties minimizing reduction of drag speed of ships, corrosion of materials, control of bio-fouling, enhancement of heat transfer on condenser surface and conservation and conversion of energy [50-52].

4. Hydrophobicity

Hydrophobicity or water repellency is an inherent property bestowed by nature on some plants, the most significant being the lotus. While the lotus plant stays in dirt, it remains clean

http://dx.doi.org/10.5772/60815

as its leaves repel water which is deposited in the form of droplets and flows away. A similar phenomenon is attached to several other plants and insects such as the desert beetle (steno cava). Thus a natural phenomenon has been mimicked and hydrophobic engineered surfaces have been fabricated on metallic and nonmetallic substrates. Hydrophobicity is defined by wet contact angle formed between the droplet and the substrate. If this angle is greater than 90°, the surface is called hydrophobic and if greater or equal to 150°, the surface is called super-hydrophobic. Super-hydrophobic surfaces also show low contact angle hysteresis. As demonstrated by lotus leaves, there are basically two requirements for a surface to be hydrophobic;

a. The surface must have a roughness in the range of 4-6 μm.

b. The surface energy must be very low or made lower by deposition of low energy compounds like polydimethyl siloxanes (PDMS).

A hierarchical surface is the key to hydrophobicity. A hierarchical surface is characterized by a structure comprising of microhills and valleys with epicuticular crystals of wax as found in the lotus flower and nano-hairs found in the gecko. The grain size is also an important feature as the grains need to be in the nano-regime, less than 100 nm. The hills are also called asperities. The surface area of asperities being very small, water drops cannot be attached and consequently they roll down. Increasing the roughness of a surface increases its wettability. When the wet contact angle between substrate and liquid becomes greater than 90°, the surface becomes hydrophobic. When a liquid drop is deposited on a flat surface as shown in Fig. 5 [53], the wettability is determined by the chemical composition of the surface; the intrinsic water contact angle $\varnothing_{flat}$, can be co-related to interfacial free energies, that is free energies at the solid air (γ_{SV}), solid liquid (γ_{SL}) and liquid air (γ_{LV}). By Young equation [54],

$$Cos\,\varnothing_{flat} = \frac{\gamma_{SV} - \gamma_{SL}}{\gamma_{LV}} \tag{1}$$

the expression γ_{SL} can b estimated by

$$\gamma_{SL} = \gamma_{SV} + \gamma_{LV} - \sqrt[2]{\gamma_{SV} \cdot \gamma_{LV}} \tag{2}$$

Equations 1 and 2 indicate that hydrophobicity increases with decreasing the surface energy of the solid-air (γ_{SV}) interface. The lowest surface energy is possessed by trifluoromethyl (CF_3) group terminated surface (~ 6 mN/m) with an $\varnothing_{flat\ of}$ 120° [55].

It may be noted that low surface energy as the major factor and roughness only enhances it. The contact angle on an ideally rough surface is shown in Fig. 6 [53]. Wetting on rough surface was discovered by Wenzel [56] and later followed by Cassie and Bexter [57]. Wenzel suggested that if the droplet filled the roughness counters, it should enhance hydrophobicity with a linear relationship between contact angle and roughness given by;

$$Cos\,\varnothing^{W}_{rough} = \gamma cos\varnothing_{flat} \tag{3}$$

where $\varnothing$ = static contact angle

and $\varnothing^{W}$ = apparent contact angle enhanced by roughness

$$\text{Roughness}\gamma = \frac{True\ surface\ area}{Horizontally\ projected\ surface\ area} \tag{4}$$

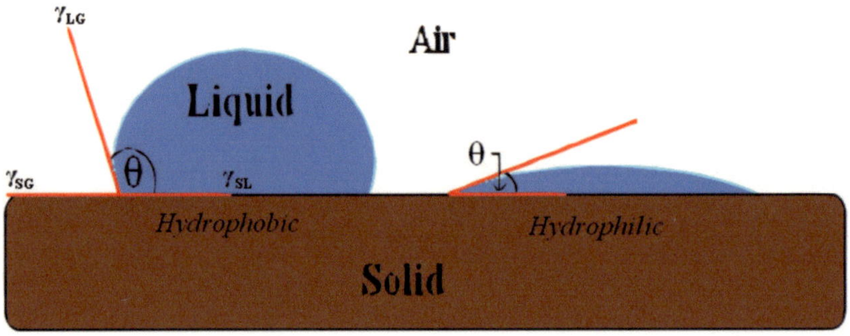

Figure 5. Contact angle made by droplet on hydrophobic (Left) and hydrophilic surface (Right) [53].

As surface roughness is greater than 1, the equation predicts that, $\varnothing_{\text{flat}} > 90^{\circ} >$, $\varnothing_{\text{rough}} > \varnothing_{\text{flat}}$ and if $\varnothing_{\text{flat}} < 90^{\circ}$, $\varnothing_{\text{rough}} < \varnothing_{\text{flat}}$.

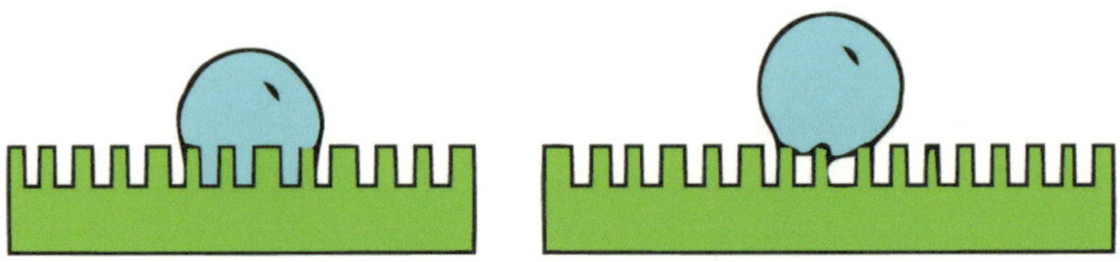

Figure 6. Theory of droplet formation on a rough surface. Fully wetting the surface contours (Left) and suspension on a solid/air composite (Right). These are the two basic models in SH theory [53].

Hence in the Wenzel state, increasing surface roughness will make surface more hydrophobic and intrinsically hydrophilic surface more hydrophilic. If, however, the surface roughness is increased, it becomes difficult for water to be in complete contact with the substrate. The air is likely to be trapped between water and the surface texture, which is called a composite surface. The water comes in contact with a composite surface of solid and air and forms droplets. In this case the contact angle is described by the Cassie Bexter equation [57]. The apparent contact angle described by Cassie Bexter is

http://dx.doi.org/10.5772/60815

$$\text{Cos } \varnothing_{\text{rough}} = \varnothing.\text{Cos } \varnothing_{\text{flat}} + \varnothing\text{V}.\text{Cos } \varnothing_{\text{LV}} = \varnothing_{\text{S}}.\text{Cos } \varnothing_{\text{flat}} - \left(1\text{-} \varnothing_{\text{S}}\right) \tag{5}$$

$\varnothing_S$ and $\varnothing_V$ are the fractions of solid and air contacting with water ($\varnothing_S + \varnothing_V = 1$). Since the contact angle of water on air ($\varnothing_{LV}$) is 180°, (Cos $\varnothing_{LV}$ = -1), that is, air entrapped would increase the surface hydrophobicity. As shown by Eq.(5), monotonic decrease of $\varnothing_S$ would increase $\varnothing_{rough}$ with an eventual increase in hydrophobicity. Figure 7 [58] shows the relationship between $\varnothing_{rough}$ and Cos $\varnothing_{flat}$. The two states are illustrated in Figs. 8 [58] and 9 [58].

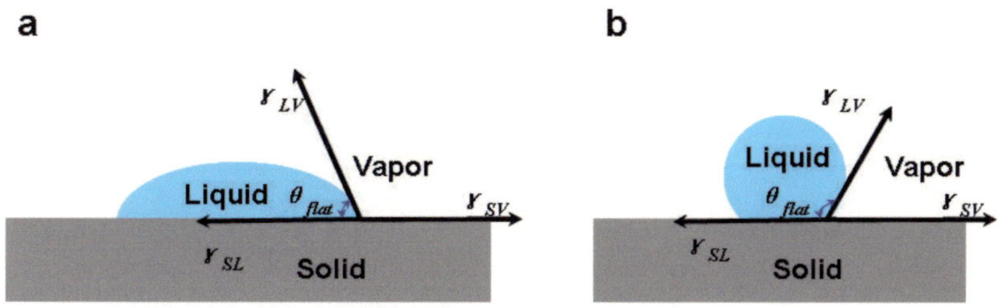

Figure 7. Contact angle of a liquid on an ideally flat surface (Young's model). For water, if θ_{flat} <90°, it is an intrinsically hydrophilic surface (a); if θ_{flat} >90°, it is an intrinsically hydrophobic surface (b) [58].

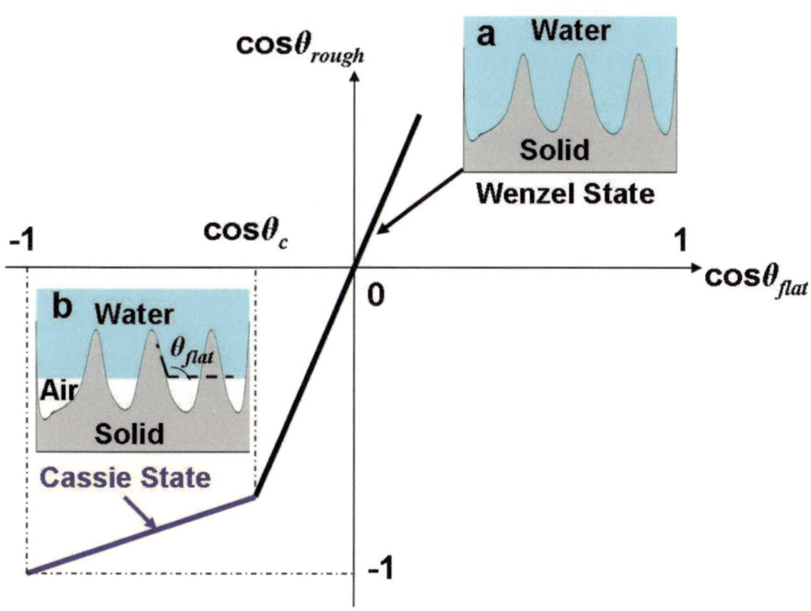

Figure 8. Relationship of cosθ_{rough} with cosθ_{flat}. The black solid and blue solid lines correspond to the Wenzel state and the Cassie state, respectively [58].

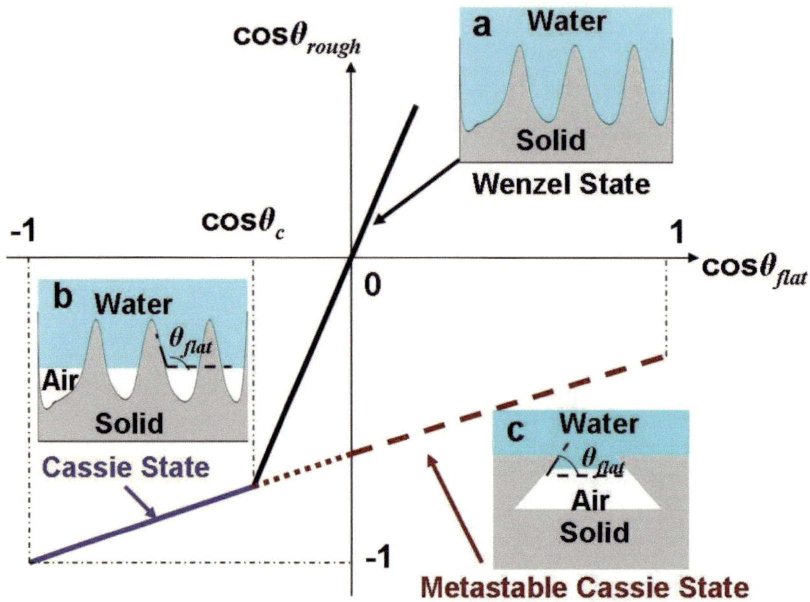

Figure 9. Relationship of *rough* cosθ with *flat* cosθ. The dotted and dashed lines correspond to the metastable Cassie state when cos> 90° *flat* θ and the metastable Cassie state when cos< 90° *flat* θ, respectively [58].

The research on super-hydrophobic surfaces faces several challenges such as the durability of a hydrophobic surface and stability. Another problem faced is the contamination of their surfaces by oil as such surfaces are not oil repellent. A few oil repellent surfaces have been reported; however, oil contact angles have been reported with oils of high free energy (γ_{LV}> 30 mN/m) such as polyols esters, arcnes and their mixture [55, 59, 60].

In recent findings on the role of super-hydrophobic surfaces in condenser tube, it has been shown that the heat transfer coefficient can be enhanced by water droplets falling on a condenser tube with a super-hydrophobic surface; however, the major problem is the withstanding of high temperature and stability of the surface [61]. Attempts have been made to produce high temperature surfaces using macro-porous silicon and to achieve a wet contact angle of 155° [62]. A lot of work remains to be done on corrosion wear, fatigue, and bio-fouling resistance of super-hydrophobic surfaces. Research in progress shows highly promising application potential of hydrophobic surfaces in the near future.

5. Methods of fabrication of hydrophobic and super-hydrophobic surfaces

Several techniques have been employed to generate hydrophobic surfaces. Modified paint technology has been used to produce super-hydrophobic surface. The paint (binder + pigment + particles + solvent) are the major constituents of paint. By keeping particles loading high and

http://dx.doi.org/10.5772/60815

binder content low, the binder evaporates and the particle adheres with the binder which roughens the surface. Major techniques employed to create hydrophobic surfaces are described below.

5.1. Layer by layer deposition

The fabrication of multi-composite films by layer by layer procedure means literally the nanoscopic assembly of hundreds of different materials in a single device. The film displays close to identical properties after deposition of first few layers. By utilizing the attraction between electrostatically utilized charged particles, one can form a multi-layer structure by dipping a substrate in a positive and negative charged polyelectrolyte. For instance, a micro-porous substrate can be generated by mixing polymer, polyalkylamine hydrochloride (PAH) and polyacrylic acid (PAP) and coating it with silica nanoparticles. This technique has several advantages, such as longer adsorption time which gives good adhesion, the desired thickness, and uniformity of surface. The cross-contaminate of deposition solution is also avoided as it may lead to depletion of the concentration of adsorbing molecules. It offers a very large processing window; hence, it is important to keep the processing conditions constant.

LBL technology allows us to design functional surfaces and surface based nano-devices in a build to order fashion. It exceeds simple self organization under equilibrium condition by its capability to arrange different materials at will in the presence of a nano-scale. In this procedure, self assembled technique enables deposition of multi-layer films on non-flat surfaces. By employing LBL technique, a super-hydrophobic surface was prepared by depositing silica nanoparticles on the surface of a 10 bilayer poly acrylic acid (PAP) coated ZrO nanoparticles polyallylamine hydrochloride (PAH) film followed by simple fluorination [63, 64]. LBL assemble allows deposition of multi-layers on non-flat surface. It has been shown that multi-layer films of poly diallyl dimethylammonium chloride (PDDA)/sodium silicate films can be deposited by LBL techniques. It is a simple and expensive way to create hydrophobic surface obtained by coating positive and negative polyanions in a sequence, the last layer being hydrophobic.

5.2. Electrochemical methods

They are widely used to form hydrophobic surfaces on rough structures. For example, Gold clusters have been electrochemically deposited on the matrix of polyelectrolyte multiple layers [65, 66]. Rough surfaces required for deposition of hydrophobic films have been fabricated by electro-deposition. Polymers can be deposited electrochemically with ease as they are inherently hydrophobic. Complete object can be coated by electrodeposition. Super-hydrophobic surfaces of silver, zinc oxides, and copper have been fabricated by electrochemical methods [67-69]. To make an aluminum surface hydrophobic, it is first anodized in an acid at specific voltage to create pores and the oxide layers, before applying a hydrophobic coating such as perfluoroallylsilane. A hydrophobic film of myristic acid can be deposited on copper by electrochemical technique.

5.3. Plasma and laser treatment

Plasma deposition can be used to produce hydrophobic surfaces directly as exemplified by deposition of polymers. Treatment by polymer leads to shrinkage and generates roughness on the surface. Some single step methods have been used to produce hydrodynamic surfaces as shown by treatment of silicone in CF_4. Laser treatment is more effective than plasma deposition because the vaporized material is re-deposited in cluster and the treated surface. The plasma technique provides a convenient tool to produce surface with special optical properties. The success of its method is shown by the preparation of fluorinated super-hydrophobic surface [70, 71].

5.4. Chemical and physical vapor deposition

Hydrophobic surfaces have been prepared by microwave plasma enhanced chemical vapor deposition (CVD) using organic silicon compounds and inert atmosphere using argon gas. By introducing CO_2 as an additive gas, ultra-pure water repellent films have been obtained. Super-hydrophobic surfaces using perfectly aligned carbon nano-tubes forest have been produced. Physical vapor deposition technique has also been used to produce super-hydrophobic surfaces of n-hexatriacontane [72-74].

5.5. Electro spinning

It is the use of electrostatic force to draw a thin fiber with a viscous polymer liquid on a substrate. For this purpose, polymer solutions, liquid crystals, suspensions of solid particles, and emulsions are electro-spun in an electric field at 1 KV. After the jet flows from the droplet in a straight line, it bends into a complex path which brings about changes in shape during which the electrical forces thin it. The solvent evaporates and bare fringent nanofibers are left.

Spinning can produce fibers down to micrometer and nanometer sizes from a wide range of materials including carbon fibers and carbon nano-tubes. Super-hydrophobicity is obtained without any modification. Cotton fibers have been coated with carbon nano-fibers and have shown good resistance against hot water [75, 76].

5.6. Sol-gel process

The sol-gel process involves the evolution of inorganic network through the formation of a colloidal suspension (sol) and gelation of the sol to form a continuous network. The pre-cursors of these colloids consist of a metal or metalloid element surrounded by reactive ligands. The most widely used aloxides and alloxysilanes, for example, tetramethoxysilane, metal oxo-polymer are obtained by polymerization reactions through hydrolysis and condensation of metal alkoxides $M(OR)_2$, M=Si, Ti, Zr, Al, An, Ce.

Super-hydrophobic surfaces have been created by combination of silicon particles, polymer binder, and hydrophobizing agents. Silica is used in the form of silica sols. The silica sol is used in sol-gel reactions where a combination of hydrolysis and condensation reactions of oxide structures takes place. The reactions are shown below:

http://dx.doi.org/10.5772/60815

$$\cdot\ (RO)_3Si - OR + H_2O \longrightarrow (RO)_3Si - OH + ROH \quad \text{(Hydrolysis)}$$

$$\cdot\ (RO)_3Si - OH + RO - Si(OR)_3 \longrightarrow (RO)_3Si - O - Si(OR)_3 + HOH$$

$$\cdot\ (RO)_3Si - OH + HO - Si(OR)_3 \longrightarrow (RO)_3Si - O - Si(OR)_3 + H_2O$$

Reaction 2 and 3 are condensation reactions. The silica properties can be made hydrophobic by treating with different silane. Surface modifications occur by a reaction between the silanol groups of the silica surface and the silanes, as shown in Fig. 10. Compounds like fluoroalkyl-silanes are excellent hydrophobic agents due to their low surface energy. Several different silanes made by Dow Corning have been used as hydrophobic agents [77-80]. The sol-gel technique provides a simple and cost effective method for the preparation of super-hydrophobic surfaces. Most of TiO_2 coatings with silanes have been prepared by sol-gel technique. The SiO_2- TiO_2 coatings in photocatalytic reactors have largely used this technique.

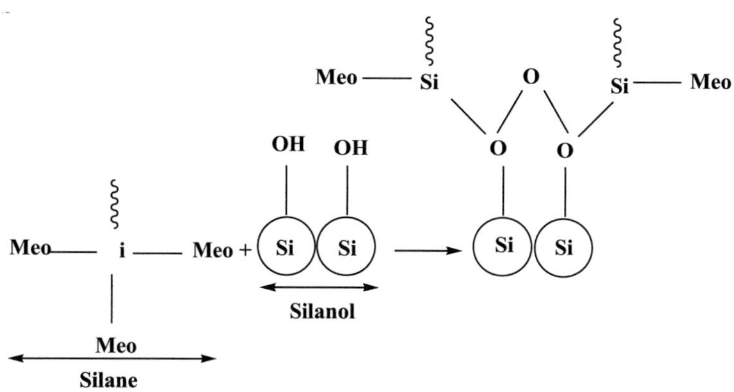

Figure 10. Surface modification of the substrate by silanol.

Water repellent, UV protective and anti-microbial coatings have been fabricated on textile fiber. Hydrophobic coatings on glass have been produced by sol-gel techniques using fluorinated compounds. Fluorinated silica nanoparticles have been produced by sol-gel process for production of super-hydrophobic films on glass, FAS 13 (fluoroalkyl-silane) was used to create hydrophobicity [81].

5.7. Self-Assembled Mono-layers (SAMS)

These are molecular assemblies formed by exposing the substrate in a solution of active surfactant in an organic solvent. It is best illustrated by alkanethiols which bears a hydrophobic alkyl chain and a thiol group which acts as an anchor. The SAMS of alkanethiol were applied on Golu surface [82]. Low surface energy coatings were developed on glass slides, silicon wafers, silicon nanoparticles, and zeolites by using organic silanes. A contact angle of 154° was obtained by modification of a SAM of octyldimethoxy silane. Fatty acids have also been employed for fabrication of super-hydrophobic surface such as copper [83-85].

5.8. Lithographic techniques

The basic idea is to expose a photo-resistant polymer to radiation through a photo-mask and remove the exposed or the unexposed paints depending on the nature of polymer. The patterned polymers can be directly used or can be used as a mask for deposition and etching. Nano-patterned lithography can also be used. One promising method is to directly use the dual scale surface roughness of surfaces found in nature such as lotus leaves or wings of cicada as stamps. The surface is pressed in PMMA at an elevated temperature. This is followed by transferring the pattern to a silica substrate.

Lithography is a well known technique for creating a rough surface. Patterns of mask were transferred onto silicon wafers followed by using hydrophobizing by silanes. The spin coating of mono-dispersed polystyrene beads for lithography is used to obtain large and periodic nano-sphere arrays. The nano-spheres are coated with gold film and modified with octadeca-nethiol. The hydrophobic surface can be tuned from 132° to 170° [86-88].

5.9. Phase separation

Due to coating or change of pressure, phase separating begins and a bi-continuous network is formed; if one of the phases becomes solid, the remaining liquid can be removed, forming a bi-continuous, three-dimensional network. In case the desired wet contact angle is not obtained, an additional super-hydrophobic coating can be applied.

5.10. Chemical etching

Besides lithography, chemical etching is another useful technique which has been successfully employed on steel and glasses. Rough surfaces on stainless steel were created by using hydrofluoric acid. After the deposition of fluorocarbon film, the etched samples of 304 and 316 stainless steel displayed a wet contact angle of 159° and 146.6° respectively [89]. Composition of etchant and time of exposure of substrate in the etchant have a significant bearing on the wet contact angle.

6. Super-hydrophobic surfaces and corrosion behavior

Corrosion is a serious problem in engineering systems such as automobiles, aircrafts, oil and gas transmission pipelines and ocean vehicles, reinforced concrete, and in micro- and nano-mechanical devices. Whereas traditional methods such as alloying, inhibition addition, cathodic protection, coatings, and designing for corrosion protection have been very beneficial,

http://dx.doi.org/10.5772/60815

the development of hydrophobic films have opened a novel route for corrosion mitigation. A brief review of the application of hydrophobicity to control corrosion is presented below. The application of hydrophobicity on steels, aluminum, magnesium, copper and their alloys is the focus of review.

6.1. Effect of hydrophobic surface on the corrosion resistance of aluminum alloys

Aluminum alloys 6061, 6013, 5052, 7075 and 2024 are well known for their excellent corrosion behavior in marine environments; however, the effect of hydrophobicity on their corrosion behavior is yet to be highlighted. Sulfuric acid, oxalic acid and phosphoric acid have been mainly used as electrolytes in anodizing to increase the wet contact angle.

Anodizing of aluminum alloys is also performed in chromic acid electrolyte followed by sealing in potassium dichromate to improve the corrosion resistance. Despite the use of several sealing products such as molybdate, cerium (IV), boric acid (V), Chromium (VI)N, and tartaric acid, these sealing agents need further improvement. A new procedure developed involves the use of carboxylic acids for post-treatment after anodizing. In an experiment conducted on Al 2024, anodizing experiments were carried out in 40 g/L in presence of 80 g/L of tartaric acid at a voltage of 14V for 20 min. Four different types of carboxylic acids, hexanoic acid, decanoic acid, myristic acid and stearic acid have been used. The contact angle increased from 10° to 110° by treatment with decanoic acid. This shows that carboxylic acid reacts with aluminum hydroxide to form an organic film which shows strong hydrophobic properties. Results of impedance spectra and salt spray tests showed decrease of corrosion due to the formation of hydrophobic films. Plots after and before and after post-treatment are shown in Fig. 11 [90]. Using myristic acid, a super-hydrophobic surface was created on aluminum alloy by anodizing and hydrophobization by myristic acid. The results showed that the super-hydrophobic coating with a contact angle greater than 160° decreased the corrosion rate tenfold compared to untreated aluminum. The contact angles measured on anode layer after treatment are shown in Fig. 12 [90]. The super-hydrophobic surface mainly affected the anodic reaction. The corrosion current (Icorr) was reduced by their orders of magnitude and the corrosion potential (Ecorr) shifted positively by 0.2V, indicating a significant reduction in corrosion rate [91, 92].

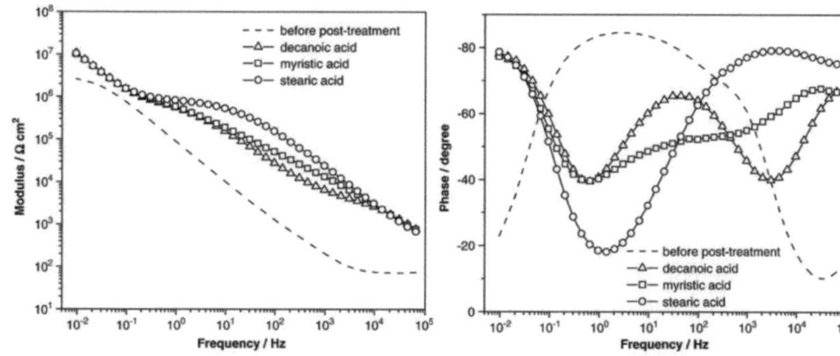

Figure 11. Bode plots of unsealed anodic layers obtained before or after post-treatment for 30 min 75 °C in pure molten acids (diagrams plotted after 2 h of immersion in 0.5 M Na_2SO_4) [90].

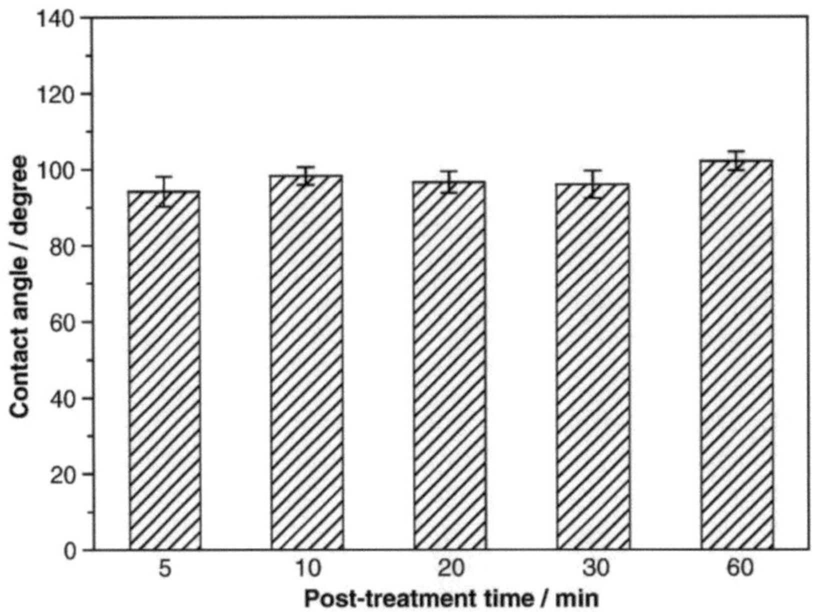

Figure 12. Contact angles measured on the surface of sealed anodic layers after post-treatment for 5-60 min at 75 °C in pure molten stearic acid [90].

It has been demonstrated that the super-hydrophobic surfaces prevent adherence of bio-fouling organisms. Work on scandium (Sc) doped Al-Mg alloys showed that addition of 0.3-0.4% Sc with 0.14% Zirconium (Zr) enhances their corrosion resistance due to the formation of Al_3Sc nanoparticles. These ultra-fine precipitates allow the formation of a highly uniform film of ScO_3 with boehmite. Different types of precipitates are shown in Fig. 13 [32].

Some advanced work showed that reaction of the substrate by myristic acid and perfluorosilane enhance the corrosion resistance three-fold as compared to untreated samples [32, 93].

In a work, water based silica films were applied on aluminum alloy 6082 and AZ31. The water based sol was prepared from 3-glycidoxypropyltrimethoxysilane (GPTM), triethylorthosilane (TEOS), water as a solvent and acid as a catalyst. The sol with TEOS showed that it acts as a barrier to water penetration. The hydrophobicity was enhanced by addition of GPTM. The cyclic voltammetry (CV) measurements showed pitting corrosion for uncoated aluminum and magnesium alloy in chloride solution, whereas no corrosion was observed in the substrate coated with silica film. The corrosion current density of coated sample was 3.96E-3mA/cm^2 for aluminum. Sample 6082-T6 was coated with silica film. The silica coated substrate enhanced the protection efficiency to 96.68%. This is attributed to the bonding of the silica film to substrate through covalent bonding thus ensuring its adhesion. The SiO_2 particles may react with OH^- ions which are generated from the cathode. The passive silica layer minimized the corrosion of aluminum alloy. This also applies to magnesium alloy AZ31. The corrosion current density of uncoated aluminum sample was 3.96 μA/cm^2 and 0.052 μA/cm^2 for sample coated with silica film. The protection efficiency is defined by:

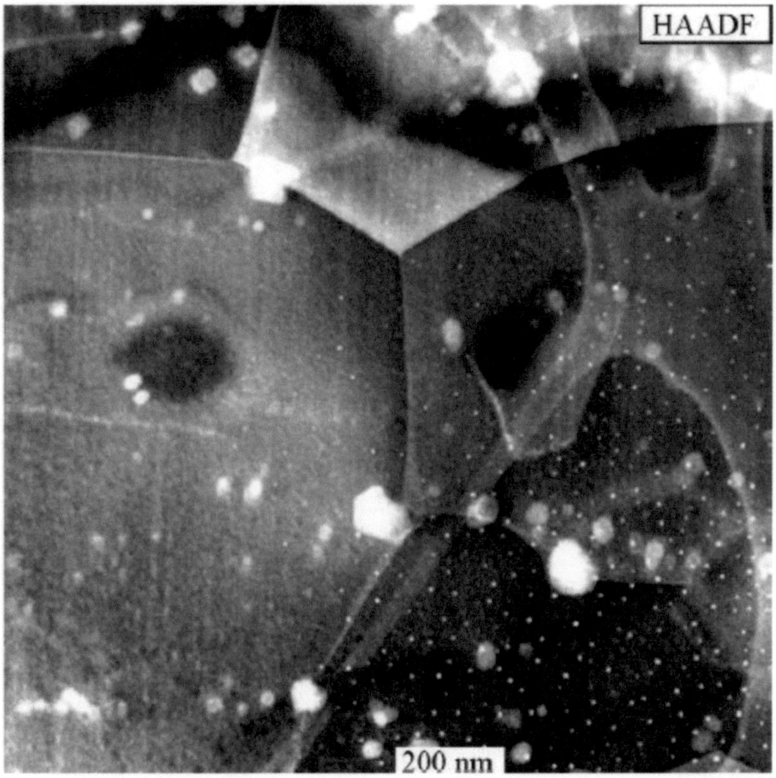

Figure 13. Different types of precipitates are clearly shown in HAADF STEM image. Al3Sc precipitates on the grain boundaries and homogeneous distributes of nano-precipitates are observed [32].

$$\text{P.E C\%} = \frac{\mathbf{I}_{\text{corr (uncoated)}} - \mathbf{I}_{\text{corr (coated)}}}{\mathbf{I}_{\text{corr (uncoated)}}} \times 100$$

The polarization curves are shown in Dig.14 [94].

Studies in super-hydrophobic treatment and corrosion protection for aluminum in sea-water were undertaken to evaluate its effectiveness [94]. Anodization was carried out at a current density of 0.32A/cm^2 in a 15 wt% of sulfuric acid at 27V, the time duration being 7 min. The sheets were modified by melting in myristic acid for 30 min at 70°C and washed in ethanol at 70°C and de-ionized water followed by drying in oven. Potentiodynamic electrochemical studies were conducted in the potential range of 600-60+ mV at a scan rate of 2 mV/s. Electrochemical impedance was performed by IMC electrochemical work station. Nano-spheres of diameter 50-60 nm was observed. The roughness was about 2 μm after hydrophobization with myristic acid, and a leaf like structure was obtained, as shown in Fig. 15 [94]. The potentiodynamic and bode plots of specimen under treated and untreated conditions are shown in Fig. 16 [94].

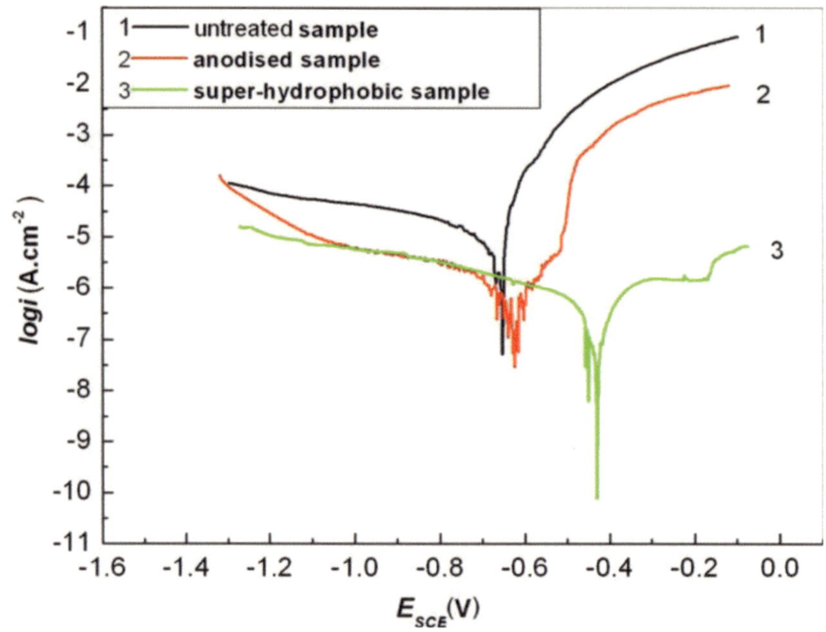

Figure 14. Potentiodynamic polarization curves of untreated, anodized, and super-hydrophobic samples for 24 h in sterile seawater at 2 mV s^{-1}[94].

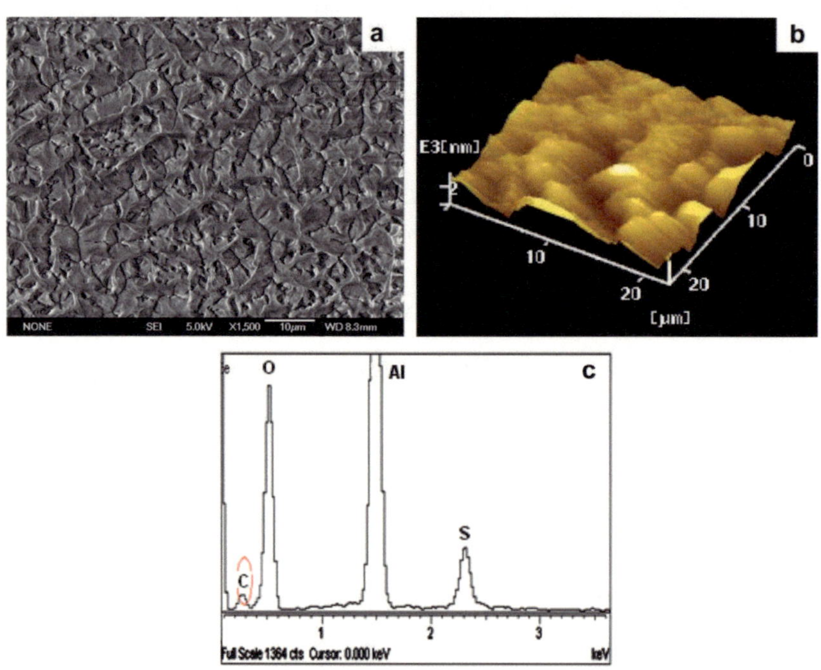

Figure 15. (a) SEM, (b) AFM, and (c) EDS of the super-hydrophobic aluminum surface [94].

http://dx.doi.org/10.5772/60815

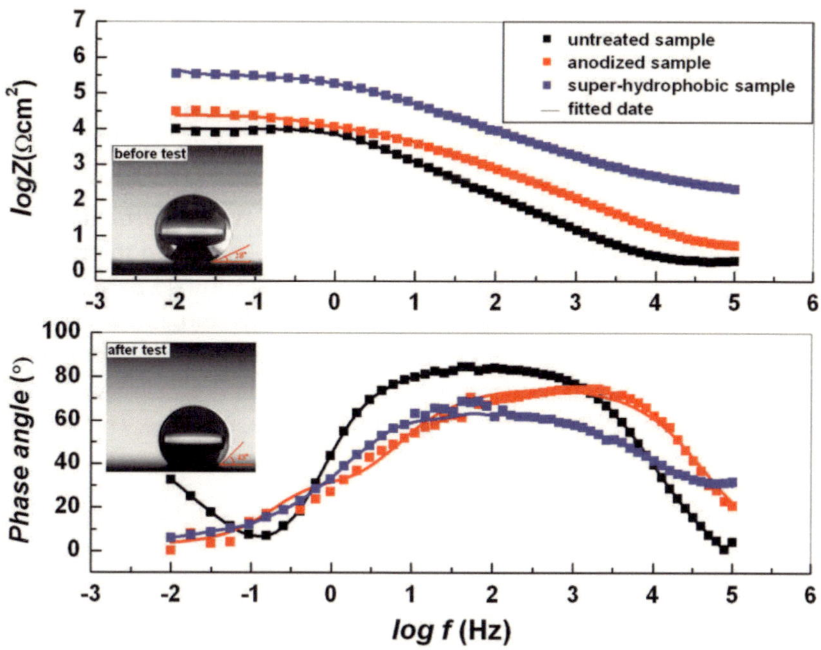

Figure 16. Bode plots of untreated, anodized, and super-hydrophobic samples after immersion in sterile seawater (32‰ NaCl, pH 8.02) for 24 h with the contact angle images of the super-hydrophobic surface before and after EIS test [94].

The porous anodic aluminum oxide was subjected to corrosion by penetration of chloride ions, where excellent corrosion resistance was shown by the surface modified with myristic acid. The basic requirement for corrosion prevention is to minimize the wetted area. It depends on whether the film is stabilized within the roughness grooves. It may be noted that air is trapped in the grooves which makes it a composite surface. The air entrapment would be metastable for the condition.

$$90 \leq \ \leq cos\frac{f_1 - 1}{\gamma - f_1}$$

where f_1= ratio of area of liquid beneath the drop in contact with solid

γ = roughness ratio of the surface

∅ = contact angle

If the ∅ is > 90°, the air would be entrapped. As a contact angle of 138° was observed, it showed that the film was stabilized within the grooves, which minimized the migration of chloride species to the surface and hence the substrate was protected. The corrosion behavior of magnesium alloys coated with fluoroalkylsilane molecule in corrosive NaCl has been investigated [21].

In the investigation, magnesium alloy AZ31 (29.8% Al, 0.88% Zn, 0.38% Mn, 0.0135% Cu, 0.002% Fe, and balance Mg) with a thickness of 1.5 mm was used. After cleaning, the alloy was immersed in 50 ml of distilled water containing 1.086 g of cerium nitrate hydrate ($Ce(No_3)_3.6H_2$). After final washing, the samples were immersed in 40 m toluene solution containing 400 μL of fluoroalkylsilane ($CF_3(CF_2)_7CH_3CH_2Si.(CH_3)_3$) and 40 μL of tetramethyl-siloxy titanium (TEST) for 30 min to deposit a hydrophobic film. No change in surface morphology revealed a porous structure after 1440 min; the nano-structured cerium oxide layer is comprised of a porous layer in contact with the substrate and intermediate compact layer and fibrous top layer. The averaged wet contact angles ranged between 136±2° to 152° after 360 min and indicated no significant change after immersion. However after 1440 min, significant change in wet contact angles occured which indicated enhancement of corrosion. The surface morphologies showed no significant change after immersion for 180-360 min. The formation of granular products increased with increased exposure time. The corrosion product was identified as $Mg(OH)_2$.

Potentiodynamic curves of untreated alloy in 5% of NaCl for 30 min and super-hydrophobic surface formed after 30, 60, 18 and 144 min was analyzed. The untreated alloy showed higher current densities compared to the treated alloy with hydrophobic surface. The EIS measurement also indicated as improvement in corrosion resistance. The studies present conflicting results for exposure periods exceeding 360 min. The substitution of self-assembled monolayers as an alternate for coatings on aluminum has been investigated in recent years [96, 97]. It is to be re-called that the molecules which constitute SAM have three major characteristics:

a. The functional group adheres to the substitute

b. The aliphatic chain acts in the self-organization process

c. A functional group is exposed on top of the layer

SAMS condensation over aluminum surface has shown that methoxy-silane monolayers, for example octadecylmethoxy silane (ODMS) and propyltrimethoxy silane (PTMS) protect aluminum base materials as these molecules form very compact coatings on oxalic surfaces. The SAM coated aluminum alloy 7075 surface showed that the SAM layers coated substrate showed a potential shift in the positive direction. The ODMS mono-layer showed superior properties over PTMS. The siloxane SAMs appear to hold a promising potential not only for Al 7075 but for most aluminum alloys used for marine applications. The major problem with Al 2024 is their enrichment in copper content which acts as a cathodic and initiates the possibility of a galvanic attack. SAMs were formed on 2024-T3, using decyltriethoxy silane (DTES) and n-octadecyltriethoxy silane (ODTES). The SAMs were created by converting ethoxysilanes ($R_3SiOCH_2CH_3$) to silanol (R_3SiOH). The ethoxy-silanes were hydrolyzed to silanols in solution of 0.02 silane, 0.006M HCl, and 0.28M H_2O in ethanol. The solution was hydrolyzed and diluted by a factor of 10. The alloy was immersed in the solution. This was followed by a condensation reaction resulting in the formation of M-O-Si bond. The curing temperature was 50°C. Water contact angles of 113° and 117° were obtained with decatriethoxy silane and n-octadecyltriethoxy silane, respectively [98]. As shown by electrochemical studies, the corrosion potential values (Ecorr) are shifted in the positive direction compared to the bare surface. The suppression of corrosion was clearly evident due to suppression of anodic dissolution. The results on anodic dissolution would depend on whether the SAM is attached to the oxide surface or directly to the substrate as alkanethiol SAMs on iron and copper showed no significant change in anodic currents [99, 100].

These methods yet remain to be applied on a large scale to fully explore their benefits. The effect of SAMs on the corrosion resistance of alloys like 6061, 6013 and 5052 needs more attention of researchers because of their increasing demand for industrial application in automotive, transport, desalination plants, and marine vessels. More work is required to understand the mechanism of corrosion of the hydrophobic surface for long term and short term duration.

6.1.1. Role of nano-pores on the hydrophobicity of aluminum alloys

Nano-pores play a crucial role in controlling hydrophobicity and corrosion resistance of aluminum alloys. Hence, the role of nano-pores on anodized aluminum needs to be better understood. More research needs to be focused on the role of nano-pores. The anodized aluminum is used as a template for the creation of ordered nano-structures. This requires ordering of the nano-pore as shown in Fig. 17.

The formation of nano-pores depends on the electrolyte used and the voltage applied. Anodizing potential has a significant effect on pore formation. An increase in temperature increased pore diameter. Pore diameter may be increased in an acid solution, such as phosphoric acid. A cross section of porous aluminum is shown in Fig. 18 [101]. The pores are hexagonal in shape. Important parameters of pores are pore diameter, inter-pore distance, wall thickness and barrier layer thickness. Pore diameter is directly related to anodizing potential [102].

D_P 1.20 $nmV^{-1}U$

D_P = pore diameter

U = anodizing potential

As shown by anodization in sulfuric acid, pore diameter depends on temperature for anodizing in sulfuric acid. An increased temperature causes increased pore diameter. The pore diameter can be widened in phosphoric acid. The pore diameter is also dependent on size, which is related to the steady state achieved during anodization when the current voltage curve becomes stationary. Another important factor is the inter-pore distance as it allows determining the number of pores per unit area. Figure 18 shows the inter-pore relationship to anodizing potential mechanism of oxide formation. The anions of the acid are transferred to the oxide/electrolyte interface. The acid anions can form OH^- or O_2^{-2}:

$$2Al + 6OH^- \rightarrow Al_2O_3 + 3H_2O + 6e^-$$

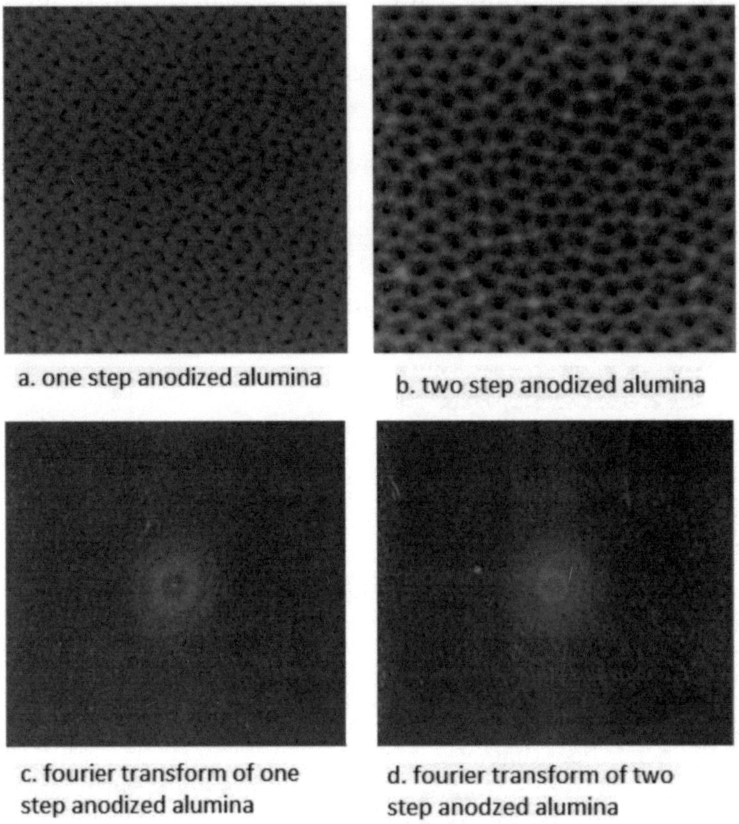

Figure 17. SEM images of one and two step anodization and their Fourier transforms.

$$2Al + 3O^{2-} \rightarrow Al_2O_3 + 6e^-$$

The oxides can also grow on oxide, electrolyte, and interface leading to the following reaction:

$$2Al + 6H_2O \rightarrow Al_2O_3 + 6H^+$$

The theory of formation of self-organized nano-pores is not well understood. It is popularly believed that there is a strong competition between the oxidation at the metal/oxide and dissolution of aluminum at the oxide/ electrolyte interface [102]. The process of anodization is followed by depositing a hydrophobic surface. The corrosion resistance of several aluminum alloys has been described in the earlier section; however, the effect of pore diameters, inter-pore spacing, shape of pores, hydrophobicity is not conclusively understood. The morphology of hydrophobic surfaces of aluminum alloys affects their corrosion resistance; however, the mechanism is not well understood.

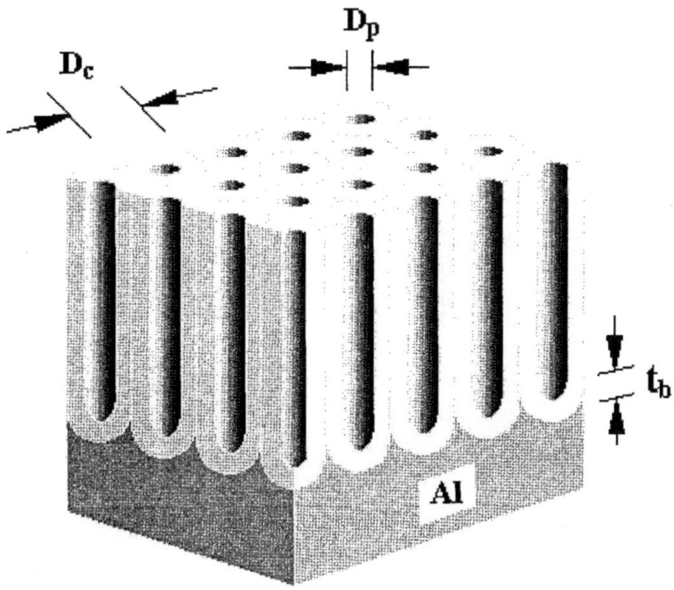

Figure 18. Ideal hexagonal "alumite" or PTF. Metallic Al, on the bottom, is covered by an impervious "barrier" oxide of thickness t_b and then by a hexagonal array of pores of diameter *Dp*, with a cell (repeat) distance *Dc* [101].

6.2. Effect of hydrophobic surface on corrosion resistance of copper alloys

It has been described earlier that the degree of corrosion of nano-structured materials is related to their micro-structure; in terms of crystallographic structure, porosity impurities, and triple junction [103], much depends on the homogeneity of the surface filmed formed. Whereas improved corrosion resistance has been reported for nano-crystalline $Al_{90}Fe_5Gd_5$ and $Al_{87}Ni_8Y_{43}$ and $Fe_{73}Si_{10}B_{25}C_{13}$ films [104], reduced corrosion resistance has been reported for $Cu_{90}Ni_{10}$ alloy [105]. In a study, nano-crystalline copper was deposited on titanium substrate by electro-deposition technique. The nano-crystalline copper electrode showed lesser localized corrosion than polycrystalline copper electrode in 3.5 wt% NaCl because of the formation of passive films consisting of CuCl and CuO_2 [106]. The uniform film formed on the nano-structured surface protects the nano-crystalline copper from corrosion.

Hydrophobicity has played a leading role in corrosion protection of copper and copper alloys in marine environment. The surface of copper was polished by SiC papers (400-1500 grade), degreased with acetone, washed with distilled water, dried, and etched in 7M HNO_3 for 30 s. To deposit a hydrophobic surface, the specimens were immersed in an ethanol solution of n-tetradecanoic acid 0.06 M at room temperature for 10 days. They were rinsed with de-ionized water and ethanol and dried in air. Samples of tough pitch copper were used (C1100). The surface of copper showed a flower like structure as shown in Fig. 19 [107].

Potentiodynamic polarization studies were conducted. Their distinct regions were observed with respect to the polarization curves, maximum corrosion current (I_{max}), minimum corrosion current (I_{min}), and limiting current (I_{limt}). The minimum corrosion current is attributed to the

Figure 19. SEM images of the anodic surface of the copper electrode after the application of 30 V DC voltage in an ethanolic stearic acid solution for (a) 0.5 h, (b) 1.5 h, (c) 3 h, (d) the morphology of a single particle as observed after 3 h [107].

formation of CuCl layer. Dissolution occurs by formation of cuprous ions leading to enhance current.

$$Cu \rightarrow Cu^{+} + e^{-}$$

This is followed by the formation of CuCl, leading to minimization of corrosion current. The corrosion current again increases upon further increase of potential due to the formation of $CuCl_2$. The formation of CuCl and $CuCl_2$ can be represented by:

$$Cu + Cl^{-} \rightarrow CuCl$$

$$CuCl + Cl^{-} \rightarrow CuCl_2^{-}$$

The values of polarization resistance for the treated specimens A, B, D, and E were 6607, 8000, 6105, 5000, respectively, compared to the values of 2857, 500, 218, 6200 and 1233, 3000 respectively for F, C, and H (bare copper). Specimens A, B, D were chemically modified with myristic acid where as C, F, and H were untreated.

http://dx.doi.org/10.5772/60815

A composite state (Cassie Bexter regime) was formed by the flower like structure where water droplets were entrapped within the grooves, thus enhancing hydrophobicity [108]. Studies on suppression of copper have been conducted by anodizing copper in triazinedithiol (TDT) solution. Pre-treatment with TDT showed a very high corrosion resistance. It was indicated that the corrosion reaction is suppressed by increased number of carbon atoms in the hydrophobic radicals [109].

Self-assembled mono-layer technique has been used to protect copper, like on aluminum. Studies were made to compare the effects of SAMs of hexanethiol and hexaneselenol on copper surface. Vibrational sum frequency spectroscopy showed that the surface of hexaneselenol was removed during corrosion due to galvanic effect; this was in contrast to hexanethiol because of its stronger bond with the copper substrate. The mechanism is not, however, clearly understood [110].

A novel and facile method was developed to successfully fabricate a regular hierarchical surface structure on copper sheet via combination of polydimethylsiloxane (PDMS) template and chemical etching. After cleaning a fresh leaf of lotus, a liquid mixture of PDMS was mixed and cast on the cleaned lotus leaf and cured for 24 h. The PDMS was peeled off the lotus to obtain a porous PDMS template as shown in Fig. 20 [111].

To mimic the hydrophobic surface of the lotus flower, a 10 g $FeCl_3$ solution was prepared in ethanol. Simultaneously, a copper sheet and a glass slide were ultrasonically cleaned by HCl. The glass slide was firstly immersed horizontally in $FeCl_3$ for 20 min. It was withdrawn vertically from the solution slowly. This was followed by wetting of PDMS template with $FeCl_3$ and pressing gently against the glass slide. The copper plate was corroded with PDMS template and a pressure of 2.5 KN/m^2 was exerted for 24 h. The copper sheet was finally immersed in stearic acid to obtain a hydrophobic surface. A contact angle of 153° was obtained. The surface mimicked the lotus leaf minimized corrosion. However, reliable data is not available and it offers a good scope for further work on the stability, adhesion, and corrosion resistance of a hydrophobic film [112].

Recently, attempts have been made to develop corrosion resistant hydrophobic grapheme oxide polymer composite coating on copper. The coating was deposited by electrophoretic technique. An aqueous dispersion of grapheme oxide was prepared (0.001-0.1g/L). A homogeneous solution of graphene and polymer was obtained by magnetic stirring and ultra-sonication for 20 min in an ultrasonic process. Electrophoretic deposition of polymer isocyanate associated with hydroxyl function acrylic adhesion was achieved as shown in Fig. 21 [113].

Two parallel plates of copper were used as electrodes, one acting as a cathode and the other acting as auxiliary electrode. Electrophoresis was performed at 10- 30 V. The deposition time ranged from 5 to 50 s. The deposit was treated with silicone to increase the hydrophobicity. Potentiodynamic polarization tests were conducted in 3.5 wt% NaCl to study the corrosion behavior of GOPC composite samples. The tafel plots of bare copper and GOPC coated copper are shown in Fig. 21. Images taken by FESEM shows that copper coated by composite and treated by silicone fluid shows no evidence of corrosion as shown by the uncoated copper

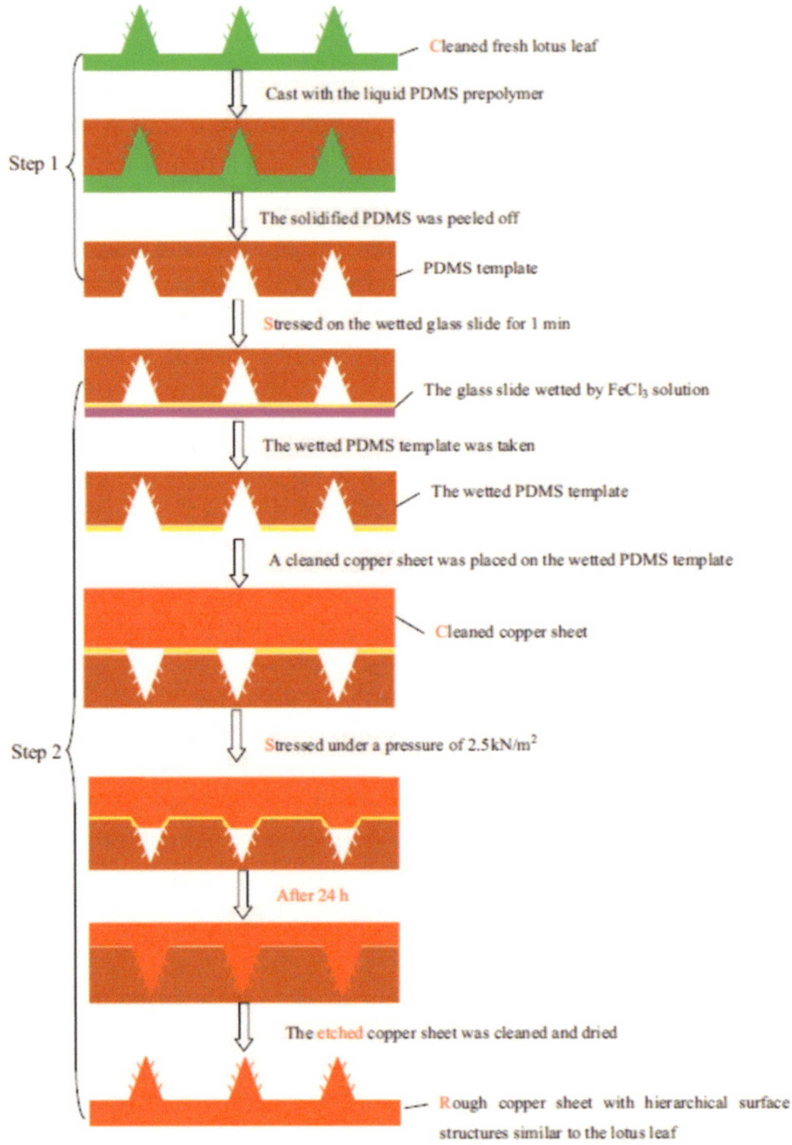

Figure 20. Procedure scheme for fabricating the lotus-leaf-like surface structures [111].

substrate. The corrosion resistance may be attributed to the formation of metal hydroxide at the electrode during electrophoresis deposition. The electrochemical impedance spectra showed the highest value for corrosion resistance for the sample coated with the GoPc [113] as shown in Fig. 22.

Enhancing the corrosion resistance of pure copper and copper alloys is extremely important because of a wide range of its application in marine, aggressive, and humid environments.

http://dx.doi.org/10.5772/60815

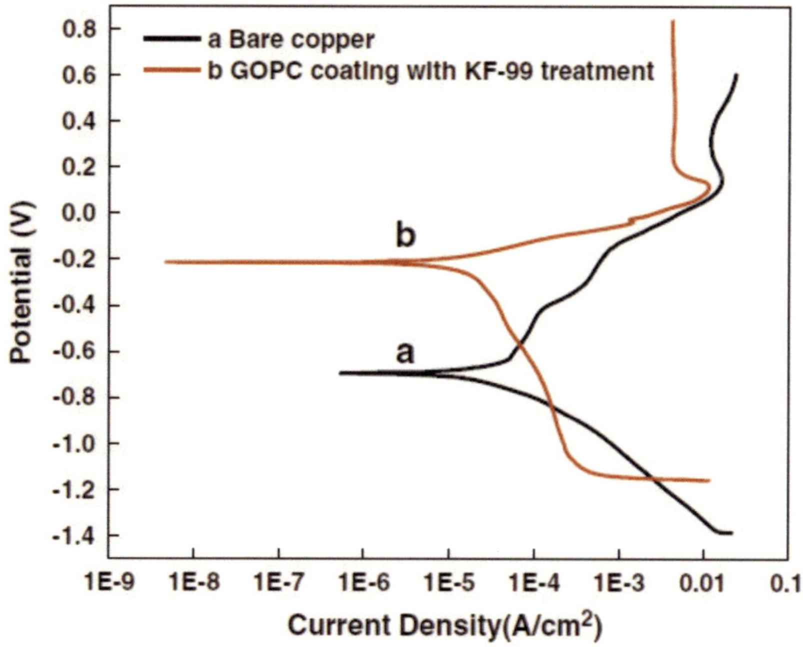

Figure 21. Tafel plots of (a) bare copper (b) GOPC coated copper after KF-99 treatment [113].

Cost-effective methods for industrial applications need to be developed to meet the increased industrial demand.

6.3. Effect of hydrophobic surface on the corrosion resistance of steels

Nano-structure plays a leading role not only in the creation of hydrophobic surfaces, but also on the corrosion resistance because of the grain size. TiO_2 thin films have been extensively used because of their applications in solar energy, photocatalytic reactors and anti-microbial properties [115-117]. Several methods have been used to fabricate thin TiO_2 films, e-beam evaporation, sputtering, chemical vapor deposition, physical vapor deposition, SAMs, and several other techniques [118-122]. The sol-gel technique has been the most popular because of its lesser equipment requirements and being a simple technique. In a work on preparation of hydrophobic surface on steel by sol-gel techniques, the optimum temperature for uniform distribution of nanoparticles was found to be 400°C. The average roughness was around 10.61 nm [123].

Chrome Moly steels are known to possess a useful combination of improved mechanical properties and high resistance to corrosion. These steels are widely employed in power generation plants. Samples of Cr-Mo steels were anodized to create roughness. The electrolyte was 1 M sulfuric acid and 0.16 M hydrofluoric acid and followed by pickling in hydrochloric acid. The samples were dipped for 10 days in myristic acid to obtain a hydrophobic surface. The AFM images showed a smooth and efficient coating. After exposing in myristic acid for

10 days, a wet contact angle of 148±4° was obtained. The contact angle depends on several factors such as surface roughness, size of crystallites (nano) and exposure period in solution used for making hydrophobic surfaces, such as fatty acids as in the above case. The electrochemical studies showed shifting of corrosion potential to more noble values, indicating the passivation of the surface. Electrochemical impedance studies showed an increase in charge transfer resistance (1.3×10^{11} Ω cm^2) and a lowering of capacitance values (2.0×10^4 F/cm^2s^{-1}), which indicated the enhanced corrosion resistance of 9Cr-1Mo steel due to its hydrophobic surface. Not only the corrosion resistance but the tendency to attack micro-organisms in sea-water is significantly decreased. Several studies on bio-fouling have suggested that the attachment of microbial organism is reduced whereas some studies claim complete detachment of bio-films and microbial attachment [124].

A uniform TiO_2 film was created on the surface of a 316 stainless steel by sol-gel technique. The specimens were ground with 1 and 0.3 μm Al_2O_3 powder and ultra-sonically cleaned in distilled water. This was followed by heating to 450°C for 30 min. In order to avoid cracking, the samples were subjected to heating in boiling water for 10- 50 min followed by treatment in an oven at 450°C. The specimens were treated with fluoroalkyl-silane (FAS-13) to obtain a hydrophobic surface. The specimens were put in a dark room to avoid any photochemical effects. The polarization curves for 316 stainless steel in oxygen- saturated Ringer solution are shown in Fig. 22 [125].

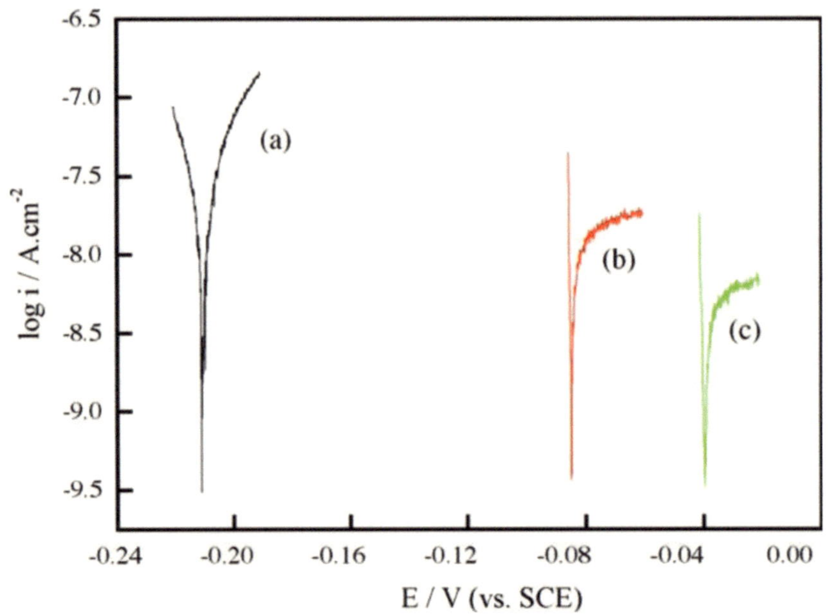

Figure 22. Polarization curves for bare 316L and films in oxygen-saturated Ringer solution. (a) Bare 316L; (b) TiO_2 / 316L coatings; (c) FSA/TiO_2 /316L coatings [125].

The nano-TiO_2 coated and FAS/nano- TiO_2 coated 316 stainless steel electrodes exhibited lower current densities. The corrosion potential (Ecorr) of FAS/nano- TiO_2 coated shifted to more noble values (-0.4V vs. SCE). This may be attributed to the formation of surface grooves, where gas is trapped and oxygen concentration is not disturbed and consequently the cathodic reaction is enhanced. The phenomenon is not clearly understood and more research is needed to find a conclusive mechanism. Work on impedance spectroscopy showed a low value of capacitance (10^{-8}) which reflects the insulating nature of the coating. The above work was pioneering for steels and it showed that the technique could be applied on a large scale for corrosion protection [114]. Super-hydrophobic surfaces on hastelloy with contact angle greater than 170° have been fabricated by sol-gel technique [126].

The hastelloy was pre-treated by ultra-sonic cleaning in acetone, alcohol and de-ionized water. The alloy was dip coated from TiO_2 pre-cursor containing TiO_2 nanoparticles (25 nm) followed by heat treatment. An aggregation of nanoparticles (150- 200 nm) was generated to create a hierarchical surface, a pre-requisite for the fabrication of a hydrophobic surface. The surface was treated with a 10 wt% ethanol solution of FAS-17, to generate hydrophobicity. The specimens were washed and dried at 140°C. The contact angle measured was 156.7°.

To evaluate the corrosion resistance, the alloy was exposed to hydrochloric acid, nitric acid, sulfuric acid, acetic acid, and sodium hydroxide. The bare hastelloys surface showed corrosion in the above media, as indicated by a color change of the substrate whereas no corrosion was observed on the hydrophobic surface. The wet contact angles and hysteresis showed that the magnitude of corrosion affects the degree of these angles. The higher contact angle before corrosion was 173°, which decreased with the degree of corrosivity. The high degree of wet contact angle and maintenance of hydrophobicity in highly corrosive environment suggests extended application of hastelloy in highly corrosive environment.

Hydrophobic surface on stainless steel may be obtained also simply by making the surface rough by etching, followed by electrolysis plating in a silane. A stainless steel plate was sandblasted with 200 μm silicon dioxide under a pressure of 0.4 mPa for 30 s and cleaned ultrasonically. This was followed by immersion in heptadecafluoro-1, 1, 2, 2, tetrahydrodecyl trimethoxy silane (HFTHTMS) to generate a super-hydrophobic surface prior to hydrophobic treatment, it was immersed in a plating bath of 1 mol/L $NiSO_4 + 6H_2O$ + 0.25mol/L N_2H_4 for 4 h at 80°C at a pH value in the range 8- 10. Corrosion studies on the hydrophobic surface were conducted by potentiodynamic polarization. The steel (nickel coated) with a hydrophobic surface showed a more noble potential. Potentiodynamic polarization curves for the stainless steel in 3.5 wt% NaCl solution are shown in Fig. 23.

The porosity value (1.29) was very small, which significantly showed the compactness of the hydrophobic film. The possibility of micro-galvanic corrosion cannot be ignored in view of small valueΔEcorr. The effect of direct immersion in HFTHTMS with corrosion behavior of stainless steel after surface roughening is not known. A contact angle of 150° was maintained even after three months of exposure [125]. The relation between corrosivity and contact angle is not well understood and it needs a better understanding.

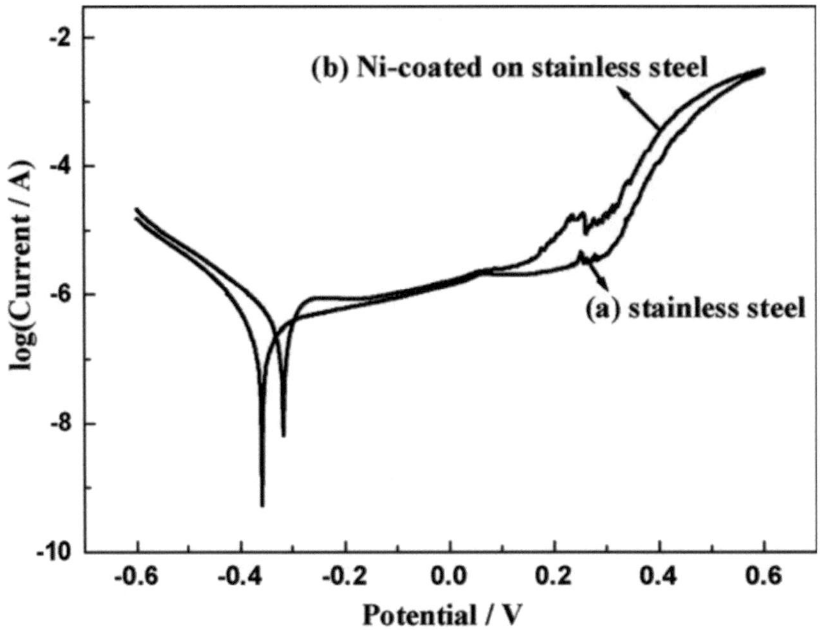

Figure 23. Potentiodynamic polarization curves for the stainless steel in 3.5 wt% NaCl solution (a) and the Ni-coated stainless steel (b) [125].

In a study, steel samples were coated with trimethoxypropyl silane (TMPS), polytetrafluor-oethlylene (PTFE), and silica nanoparticles (NPS) to determine the effect of hydrophobicity on the corrosion resistance of carbon steels in salt solution [127]. Samples of steel, steel/silane, steel/silane/PTFE, steel/silane/NPS/PTFE were exposed to NaCl solution to determine their corrosion resistance. The results revealed that the samples coated with PTFE and NPS (nanoparticles) considerably improved the anti-corrosion resistance due to the formation of super-hydrophobic films on their substrate.

In a similar work, the corrosion inhibition of carbon steel in hydrochloric acid using sodium dodecylsulfonate was investigated [128]. The inhibitors used were dodecylsulfonate, $NaC_{12}H_{25}SO_3$ (SDSO), and ethyletherpolyoxyethylene. The tests were conducted in acidic solutions. The addition of SDSO in various concentrations of hydrochloric acid, increased concentration of SDSO, strengthened the inhibiting efficiency in various concentrations of hydrochloric acid because of adsorption of sulfonate through oxygen atom of sulfonate, by forming a barrier layer on the cathodic sites. At an SDSO concentration of $5.33{\times}10^3$ mol.dm^2, an abrupt decrease in efficiency was observed due to change from a hydrophobic to hydro-philic state. The effect of addition of $FeCl_3$, KCl, and $CaCl_2$ was also observed on the inhibition efficiency. It was observed that $FeCl_3$ addition increased the inhibition efficiency. It was attributed to an increase in Zeta Potential on the surface of carbon steel which leads to an increase in the adsorption density. It was observed that K^+ and Cl, Ca^{2+} ions had a weaker influence. A commercial surfactant Brij58 added to SDSO increased the inhibitor efficiency because of the adsorption of two surfactants.

Produced water in oil and gas industry is highly corrosive because of the presence of chlorides, dissolved CO_2, salinity and H_2S contents. Use of high performance alloys such as 316L, 2205, 20Cb-3, C-22 is effective, but they may increase the cost of the system by as much as five fold. Silicone based materials provide a cost effective alternative as they have the capability not only to reduce corrosion but also to minimize water surface interaction due to their hydrophobic nature. Tests were conducted on coupons coated with amorphous silicon or carbo-silane by chemical vapor deposition. On conducting immersion tests according to ASTM G31, in 6M HCl, the highest weight loss was shown by 316 stainless steel and the least (3.29) by carbo-silane-coated steel (91.90 vs. 3.29) equivalent to an overall mpy improvement of 27.9X over 316 stainless steel. The corrosion potential (Ecorr), corrosion current (Icorr), pitting potential of 316L, a-316L, and 304 were determined electrochemically. The data obtained suggested a 50x reduction in corrosion rate for Si-coated coupon in 3000 ppm Cl^- solution. The Si-coated coupon also showed no corrosion in salt spray tests in 3.5 wt% of NaCl. The reduction in corrosion of Si-coated 316 stainless steel was attributed to its inherent property of hydrophobicity, which was determined by wet contact angle.

The carbo-silane coating determined a higher degree of hydrophobicity with lesser hysteresis indicating a reduction in wetted area. Comparative studies on adsorption of water on commercial stainless steels showed that they adsorbed most water whereas electro-polished stainless steel tubes minimized adsorption. It was shown that a Si-coated electro-polished steel reduced corrosion further by 50%. The results show that hydrophobic surfaces minimize water adsorption on steel surface. It was further shown that a-Si coated electro-polished stainless steel specimen dried in 83% less time compared to the drying time of 180 min taken by an uncoated 316 stainless steel [129].

One of the recent areas of interest is the application of hydrophobic surfaces in prevention of corrosion of steel in concrete (re-bar corrosion). A serious problem is encountered by corrosion damage under de-icing salts. The molecular attraction between water and concrete is weakened by application of hydrophobic agents such as silanes and siloxanes. One important siloxane used is polydimethylsiloxane (PDMS). The structure of monomer of PDMS is shown in Fig. 24.

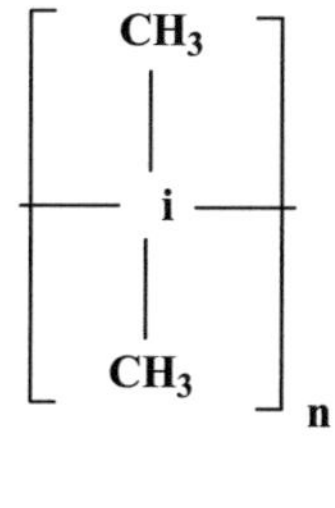

Figure 24. Structure of PDMS.

Silane particles are hydrophobic and they are functionalized by different silanes, for example fluoroalkyl-silanes, tri-chloroethylsilanes, perfluoroalkylsilanes, and several others.

The studies were carried out in concrete with W/C ratio of 0.50. Two types of concrete were used; portland cement and blast furnace slag cement, that is OPC and BFSC, respectively. Three commercially available hydrophobic products were used; a 99% silane (0 solvent), 100% silane (0 solvent), 20% silane/siloxane dispersed in water to determine the effect of hydrophobicity on the ingress of chloride after 52 weeks. It was observed that the chloride content was between 0.2% and 0.55 by mass of cement at a depth of 16- 20 mm in OPC and BFSC concrete, respectively, compared to 2.8% and 2.4% in untreated concrete, respectively, which was a significant reduction [130]. The question concerns the durability of the treated surface and it is feared that carbonation may occur after five years due to complete drying out of the surface. The water repellent surface minimizes carbonation of concrete.

7. Iceophobic surfaces

A new development of hydrophobic concretes has shown anti-icing properties (iceophobic). Iceophobic concrete reduces the ice adhesion on roadways and pavements. Super-hydrophobic admixtures, siloxanes, have been successfully developed. The admixtures are based on combination of siloxane- based hydrophobic liquid and small quantity of silica to form polymethyl-hydrogensiloxane. The super-hydrophobic surface reduces the ice bonding strength. The need for maintenance is hugely reduced with the use of iceo-phobic cementitious composites which, provide durability and strength [131].

Accumulation of ice is a serious problem in terms of life of equipment as it involves huge costs. It is a serious problem on aircraft, motor vehicles, power lines, decks, and bridge nuts. It is known that icing on aircraft can reduce their life by 30% and increase the drag by 10%. Heavy icing has led to air crashes in the past. De-icing of aircraft varies from country to country. In Netherlands, the cost of de-icing is 3000 Euros for Boeing 737. The cost of anti-icing treatment on private jets ranges from 250 Euros for a propeller jet to 10, 000 Euros. The price is based on the de-icing fluids required for removing ice. The cost of de-icing of motor vehicle ranges from 125 to 200 dollars depending on the size. Considering the number of vehicles affected by icing in the cold regions of the globe, the cost reaches highly alarming figures. Although the existing research on super-hydrophobic surfaces has not addressed the problem directly, it has offered hope for a future solution for de-icing. Research work has shown that the freezing time is substantially delayed on a super-hydrophobic surface [132]. Ice that accumulates on a super-hydrophobic surface in the Cassie Bexter state is weakly bonded and easy to remove. A new concept of Slippery Liquid Infused Porous Surface (SLIPS) [133] may be used to generate surfaces that are super-iceo-phobic. Attempts have been made to fabricate a network of Teflon nano-fiber and infuse in it a low surface energy perfluorinated fluid. Compounds such as 3 M fluorinate FC-80 and Du-pont Krytoxin (lubricating compound) have been used. SLIPS repelled water ~ 675 atm, which is hundred times the highest recorded pressure on super-hydrophobic surfaces. It showed very superior ice-repelling properties. The anti-icing properties of nanoparticle composites have been recently investigated [134]. Nanoparticles polymer composite has shown a high promise of iceophobicity. The anti-icing property depends not only on the nano-polymer composite but also on the particle size. Further development would need a greater understanding of the hydrodynamic structure of the water film and the mechanism of adhesion of ice in relation to surface morphology.

8. Conclusion

The phenomena of hydrophobicity bestowed by nature on several biological objects, such as the lotus flower and the wings of butterflies, has evoked a great academic and industrial interest. Super-hydrophobicity is now recognized as one of the top fifty technologies. Research in the last decade has clearly shown that it is addressing serious problems such as those related to corrosion and bio-fouling, as shown by successful applications of hydrophobic surfaces in engineering systems such as aircraft, automobiles, pipelines, and naval vessels. By reducing the drag force and bio-fouling, these surfaces are playing a vital role in fuel saving. The fabrication of hydrophobic surfaces has undergone a significant transformation from highly complex processes to simple one step facile processes. Several techniques such as physical vapor deposition, chemical vapor deposition, magnetic sputtering, layer by layer, spin coating, spraying, self-assembled monolayer, and sol-gel techniques, however, are simple and they have evoked a great interest in recent years. The interest of researchers is mainly focused on aluminum, copper, and steel alloys. Increasing attention is being given to fabricate hydrophobic surfaces on glasses to impart self-cleaning and anti-fogging properties due to its important applications in land, sea, and air transport, but also its applications in optical lenses. Super-hydrophobic surfaces have shown enhanced corrosion resistance on Al alloy and are widely used. The effect of nano-pores, that is pore diameter, inter-pore diameter, plays an important role on the efficiency of hydrophobic coatings on aluminum alloys.

The corrosion resistance of copper based alloys has been significantly enhanced by coating with myristic and stearic acid. Hydrophobic surface on copper is likely to play a leading role in energy saving because of the jumping droplets phenomena observed on hydrophobic surface deposited on copper and copper wires. Surface morphology of hydrophobic surface plays a crucial part in its stability and wet contact angle. Surface roughness and deposition of low energy compounds are the key parameters for hydrophobicity. Surface roughness has been achieved in copper and aluminum alloys by anodizing at specific voltages for specific times. Etching has been used to create a rough surface on steel. Several other techniques have been used such as shot blasting, sand blasting, ultra-shot penning, and cavitation shot penning to obtain a nano-structured surface with grain size in the nano-range. Hydrophobic agents such as perfluorosilane, trimethoxypropyl silane, polydimethyl siloxane, 3-glycidoxypropyl-trimethoxy silane (GTPMS), trimethoxysilane (TEOS) have been extensively used to fabricate hydrophobic surfaces.

More recently, hydrophobicity has ushered a new era in clean building construction industry to address pollution and energy conservation. The emerging area which began from the observation of water and dust repelling properties of the lotus flower has now been successfully mimicked by scientists on engineered surfaces and it has successfully extended to all phases of engineering applications with dramatic effects on corrosion and bio-fouling.

Author details

Zaki Ahmad*, Asad U. Khan, Robina Farooq, Naila Riaz Mastoi and Tahir Saif

*Address all correspondence to: zakiahmad100@gmail.com

Department of Chemical Engineering, COMSATS Institute of Information Technology, Lahore, Pakistan

References

[1] R. S. Razavi, G. R. Gordani, "Laser surface treatments of aluminum alloys" in "*Recent Trends in Processing and Degradation of Aluminum Alloys*", Ed. Zaki Ahmad, Intech open, Croatia, 2011, 115-154

[2] T.R. Davies (Ed), *Corrosion of Aluminum and Aluminum Alloy*, ASM International Materials Park, Ohio, USA, 1999, 44073-0002.

[3] K. Nisancioglu, H. Holter, Correlation of the open-circuit and electrochemical measurement for pitting corrosion of aluminum in chloride media, *Mater Corr*, 2004, Vol. 30(2), 115-113.

[4] Trung Hung Nguyen, R.T. Foley, On the mechanism of pitting of Aluminum, for electrochemical Soc *Electrochem Sci Technol*, 1979, Vol. 124 (11), 1855-1860.

[5] Saravanan Nagappa, Chang-Sik Ha, Emerging trends in superhydrophobic surface based magnetic materials: fabrications and their potential applications, *J Mater Chem A*, 2015, Vol. 3, 3224-3251.

[6] Ruomei Wu, Shuquan Liang, Jun Liu, Anqiang Pan, Y. Yu, Yan Tang, Fabrication of the micro/nano-structure superhydrophobic surface on aluminum alloy by sulfuric acid anodizing and polypropylene coating, *J Nanosci Nanotechnol*, 2013, Vol. 13, 2362–2368.

[7] Ying Huang, D.K. Sarkar, X-Grant Chen, Fabrication of superhydrophobic surfaces on aluminum alloy via electrodeposition of copper followed by electrochemical modification, *Nano-Micro Lett*, 2011, Vol. 3 (3), 160-165.

[8] Zhijin Gong, Jianli Wang, Limei Wu, Xiaoyu Wang, Guocheng Lu, Libing Liao, Fabrication of super hydrophobic surfaces on copper by solution-immersion, *Chin J Chem Eng*, 2013, Vol. 21 (8), 920–926.

[9] Xi Yao, Qinwen Chen, Liang Xu, Qikai Li, Yanlin Song, Xuefeng Gao, David Que´re, Lei Jiang, Bioinspired ribbed nanoneedles with robust superhydrophobicity, *Adv Funct Mater*, 2010, Vol. 20, 1–7.

http://dx.doi.org/10.5772/60815

[10] Jinming Xi, Lin Feng, Lei Jiang, A general approach for fabrication of superhydrophobic and super-amphiphobic surfaces, *Appl Phys Lett*, 2008, Vol. 92 (5), 053102-053103.

[11] Xia Zhang, Yonggang Guo, Yue Liu, Xue Yang, Jieqiong Pan, Pingyu Zhang, Facile fabrication of superhydrophobic surface with nanowire structures on nickel foil, *Appl Surf Sci*, 2013, Vol. 287, 299–303.

[12] Mei Li, Jianhai Xu, Qinghua Lu, Creating superhydrophobic surfaces with flowery structures on nickel substrates through a wet-chemical-process, *J Mater Chem*, 2007, Vol. 17, 4772-4776.

[13] Y.W. Hu, S. Liu, S.Y. Huang, W. Pan, Fabrication of superhydrophobic surfaces of titanium dioxide and nickel through electrochemical deposition on stainless steel substrate, *Key Eng Mater*, 2010, Vol. 434-435, 496-498.

[14] Zhongwei Wang, Qing Li, Zuxin She, Funan Chen, Longqin Li, Xiaoxu Zhang, Peng Zhang, Facile and fast fabrication of superhydrophobic surface on magnesium alloy, *Appl Surf Sci*, 2013, Vol. 271, 182–192.

[15] Jun Liang, Zhiguang Guo, Jian Fang, Jingcheng Hao, Fabrication of superhydrophobic surface on magnesium alloy, *Chem Lett*, 2007, Vol. 36 (3), 416-417.

[16] Qin Liu, Dexin Chen, and Zhixin Kang, One-step electrodeposition process to fabricate corrosion-resistant superhydrophobic surface on magnesium alloy, *ACS Appl Mater Interf*, 2015, Vol. 7 (3), 1859–1867.

[17] B.V. Mahesh, R.K. Singh Raman, Role of nanostructure in electrochemical corrosion and high temperature oxidation: a review, *Metall Mater Trans* A, 2014, Vol. 45A, 55799-55822.

[18] Rajeev K. Gupta, R.K. Singh Raman, Carl C. Koch, B.S. Murty, Effect of nanocrystalline structure on the corrosion of a Fe.20Cr alloy, *Int J Electrochem Sci*, 2013, Vol. 8 (5), 6791-6806.

[19] K.D. Ralston, N. Birbilis, Effect of grain size corrosion: a review, *Corrosion*, 2010, Vol. 66 (6), 0750051-07500513.

[20] X.Y. Wang, D.Y. Li, Mechanical and electrochemical behavior of nanocrystalline surface of 304 stainless steel, *Electrochem Aceta*, 2002, Vol. 47 (9), 3939-3947.

[21] T. Ishizaki, Y. Masuda, M. Sakamoto, Corrosion resistance and durability of superhydrophobic surface formed on magnesium alloy created with nanostructured cerium oxide film and fluoroalkylsilane molecules in corrosive NaCl aqueous solutions, ACS Publications, *Langmuir*, 2011, Vol. 27, 4780-4788.

[22] R.F. North, M.J. Pryor, The influence of corrosion product structure on the corrosion rate of Cu-Ni alloys, *Corr Sci*, 1970, Vol. 10 (5), 297-311.

[23] The nickel development institute, The Halloway, Alvechurch, Birmingham, B48, 7QB UK.

[24] K. Lu, Nanocrystalline metals crystallized from amorphous solids: nanocrystallization, structure and properties, *Mater Sci Eng*, R16, 1996, Figure 1, 162.

[25] Atul H. Chokshi, Triple junction limited grain growth in nanomaterials, *Script Mater* 2008, Vol. 59, Figure 1, 727.

[26] J. K. Yu, E. H. Han, L. Lu, X. J. Wei, Corrosion behaviors of nanocrystalline and conventional polycrystalline copper, *J Mater Sci*, 2005, Vol. 40, Figure 2, 1020.

[27] J.E. Switzer, G.J Shiflet, J.R. Scully, Localized corrosion of Al 90Fe 5Gd and Al 87Ni 8.7Y 4.3 alloys in the amorphous, nanocrystalline, and crystalline states: resistance to micrometer-scale pit formation, *Electrochem Aceta*, 2003, Vol. 48, 1223-1224.

[28] A. Barbucci, G. Farne, P. Matteazzi, R. Riccieri, G. Cerisola, Corrosion behavior of nanocrystalline Cu90Ni10 alloy in neutral solution containing chlorides, *Corr Sci*, 1999, Vol. 41, 463-475.

[29] F.J. Cornwell, G. Wildsmith, P.T. Gilbert, Pitting corrosion in copper tubes in cold water service, *Corr J*, 1973, Vol. 8 (5), 202-209.

[30] J.K. Yu, E.H. Han., L. Lu, X.J. Wei, Corrosion of nano-crystalline and conventional poly-crystalline copper, *J Mater Sci*, 2005, Vol. 40, 1019-1022.

[31] S. Virtanen, J. Brunner, J. Wioka, 28th ES meeting # 322, 1987.

[32] Zaki Ahmad, Abdul Aleem, Mohammad Abbas, Effect of scandium doping on the corrosion resistance and mechanical behavior of Al-3Mg alloy in neutral chloride solutions, *Mater Sci App*, 2011, Vol.2, 244-250

[33] Zaki Ahmad, The properties and applications of Scandium-reinforced aluminum, *J Min, Met Mater Soc*, 2003, Vol. 55 (2), 35-39.

[34] Gerhardus H. Koch, Michiel P.H. Brongers, Neil G. Thompson, *Cost of Corrosion and Prevention Strategies in United States*, C.C Technologies Laboratories, Inc., 2001, Dublin, Ohio, USA.

[35] Ahmad Z, Corrosion phenomena in coastal area of Arabian Gulf, *Brit Corr J*, 1996, Vol. 31 (2), 191-197.

[36] Maria Salta, Julian A. Wharton, Paul Stoodley, Simon P. Dennington, Liam R. Goodes et al., Designing biomimetic anti-fouling surfaces, *Philo Trans Royal Soc* A. 2010, Vol. 368, 4729-4754.

[37] R. Jafari, M. Farzaneh, A simple method to create super-hydrophobic aluminium surfaces, *Mater Sci For* 2012, Vol. 706-709, 2874-2879.

[38] Rolla G, Ellingsen, J.E, Herlofson B, Enhancement and inhibition of dental plaque formation – some old and new concepts, *Biofouling*, 1991, Vol. 3, 175-181.

[39] M. Quiryen, H.C. Van Der Mei, C.L. Bollen, G.I. Geertsema-Doornbusch, H.J. Busscher, D. Van Steenberghe, Clinical relevance of the influence of surface free energy and roughness on the supragingival and sub-gingival plaque formation in man, *Coll Surf B* 2, 1994, Vol. 3, 25-31.

[40] Jan Genzer, Kirill Efimenko, Recent developments in super-hydrophobic surfaces and their relevance to marine fouling; a review, *Biofouling: J Bio-adhes Biofilm Res*, 2006, Vol. 22 (5), 339-360.

[41] C.S. Gudipati, J.A. Finlay, J.A. Callow, Maureen E. Callow, K.L. Wooley, The anti-fouling and fouling release performance of hyper-branched fluoropolymers (HBFP) – polyethylene glycol (PEG) composites coatings evaluated by adsorption of bio-macromolecules and the green fouling alga (ulva), *Langmuir*, 2005, Vol. 21, 3044-3053.

[42] D.J. Gan, A. Mueller, K.L. Wooley, Amphiphilic and hydrophobic surface patterns generated from hyper-branched fluoropolymers/linear polymer networks: minimally-adhesive coatings via cross-linking of hyper-branched fluoropolymers, *J Polym Sci: A Polym Chem*, 2003, Vol. 41 (22), 3531-3540.

[43] Kenneth R. Trethewey, John Chamberlain, *Corrosion: For Students of Science and Engineering*, Longman Scientific and Technical, London. 1988.

[44] Mars J.G. Fontana, *Corrosion Engineering*, McGraw-Hill Book Company, 1985, ISBN-10: -0070214638.

[45] L.L. Shrier, *Corrosion* Vol. 1 & 2, Newnes-Butterworths, London, 1979.

[46] Zaki Ahmad, *Principles of Corrosion Engineering and Corrosion Control*, Butterworth-Heinemann, 2006, ISBN: 978-0-7506-5924-6.

[47] Anil Kumar Sinha, *Ferrous Physical Metallurgy*, Butterworth, Boston, 1989.

[48] Michael F. Ashby, *Material Selection in Mechanical Design*, 3rd edition, Butterworth-Heinemann, 2013.

[49] W. Barthlott, C. Neinhuis, The lotus-effect: nature's model for self cleaning surfaces. *Int Text Bull*, 2001, Vol. 1, 08-12.

[50] Robert J. Daniello, Nicholas E. Waterhouse, Jonathan P. Rothstein, Drag reduction in turbulent flows over super-hydrophobic surfaces, *Phys Flu*, 2009, Vol. 21 (8), 085103

[51] Xi Zhang, Feng Shi, Jia Niu, Yugui Jiang, Zhigiang Q. Wang, Super-hydrophobic surfaces: from structural to functional application, *J Mater Chem*, 2008, Vol. 18 (6), 621-633.

[52] Michael Nosonovsky, Bharat Bhushan, Super-hydrophobic surfaces and emerging applications: non-adhesion energy green engineering, *Curr Opin Coll Interf Sci*, Vol. 14 (4), 270-280.

[53] J.N. Israelachvili, *Intermolecular and Surface Forces*, 2nd edition, Academic Press, New York, 1991, 7665-7669.

[54] S. Shibuichi, T. Yamamoto, T. Onda, K. Tsujii, Super water and oil repellent surfaces resulting from fractal structure, *J Coll Interf Sci*, 1998, Vol. 208, 287-294.

[55] R.N. Wenzel, Resistance of solid surfaces to wetting by water, *Ind Eng Chem*, 1936, Vol. 28, 988-994.

[56] A.B.D. Cassie, S. Baxter, Wettability of porous surface, *Trans. Faraday Soc*, 1944, Vol. 40, 546-551.

[57] Liangliang Cao, *Superhydrophobic Surface: Design, Fabrication, and Applications*, PhD Thesis, Graduate Faculty of Swanson School of Engineering, University of Pittsburgh, 2010.

[58] K. Tsujii, T. Yamamoto, T. Onda, S. Shibuichi, Super oil repellent surface, *Angew Chem* (Int Ed Engl), 1997, Vol. 36 (9), 1011-1012.

[59] Lin Feng, Zhongyi Zhang, Zhenhong Mai, Yongmei Ma, Biqian Liu, Lei Jiang, Daoben Zhu, A super-hydrophobic and super-oleophilic coating mesh film for the separation of oil and water, *Angew Chem* (Int Ed), 2004, Vol. 43 (15), 2012-2014.

[60] Charged superhydrophobic condenser surface may make power plants more efficient, http://science-beta.slashdot.org/story/13/10/04/1836234/charged-superhydrophobic-condenser-surface-may-make-power-plants-more-efficient

[61] Koji Takahashi, Tatsuya Ikuta, Kunihito Nagayama, Yasuyuki Takaka, Tanemasa Asano, Simple fabrication of hydrophobic surface for high-temperature microsystems, 7th International Conference on Miniaturized Chemical and Biochemical Analysts System, 2003, Squaw Valley, California, U.S.A.

[62] J.T. Han, Y. Zheng, J. H. Cho, X. Xu, K. Cho, Stable superhydrophobic organic-inorganic hybrid films by electrostatic self-assembly, *J Phys Chem* B 2005, Vol. 109 (44), 20773–20778.

[63] Lianbin Zhang, Huan Chen, Jungi Sun, Jiacong Shen, Layer-by-layer deposition of poly (diallyldimethylammonium chloride and sodium silicate multi-layers on silica sphere coated substrate – facile method to prepare a super-hydrophobic surface, *Chem Mater*, 2007, Vol. 19, 948-953.

[64] X. Zhang, F. Shi, X. Yu, H. Liu, Y. Fu, Z. Wang, L. Jiang, X. Li, Polyelectrolyte multilayer as matrix for electrochemical deposition of gold clusters: towards super-hydrophobic surfaces, *J Am Chem Soc*, 2006, Vol. 126 (10), 3064-3065.

[65] F. Shi, Z. Wang, X. Zhang, Combining a layer-by-layer assembling technique with electrochemical deposition of gold aggregates to mimic the legs of water striders, *Adv Mater*, 2008, Vol. 17 (8), 1005-1009.

[66] N. Zhao, F. Shi, Z. Wang, X. Zhang, Combining layer-by-layer assembly with electrodeposition of silver aggregates for fabricating super-hydrophobic surfaces, *Langmuir*, 2005, Vol. 21 (10), 4713-4716.

[67] F. Shi, J. Niu, J. Liu, Z. Wang, X.Q. Feng, X. Zhang, Towards understanding why a superhydrophobic coating is needed by water striders, *Adv Mater*, 2007, Vol. 19 (17), 2257-2261.

[68] S. Wang, L. Feng, H. Liu, T. Sun, X. Zhang, L. Jiang, D. Zhu, Manipulation of surface wettability between super-hydrophobicity and super-hydrophilicity on copper films, *Chem Phys Chem*, 2005, Vol. 6 (8), 1475-1478.

[69] J.P. Youngblood, T.J. McCarthy, Ultra-hydrophobic polymer surfaces prepared by simultaneous ablation of polypropylene and sputtering of poly(tetrafluoroethylene) using radio frequency plasma, *Macro-molecules*, 1999, Vol. 32 (20), 6600-6806.

[70] I. Woodward, W.C.E. Schofield, V. Roucoules, J.P Badyal, Super-hydrophobic surfaces produced by plasma fluorination of polybutadiene films, *Langmuir*, 2003, Vol. 19 (8), 3432-3438.

[71] Kenneth K.S. Lau, J. Bico, Kenneth B.K. Teo, Manish Chhowalla, Gehan A.J. Amaratunga, W.I. Milne, G.H. Mckinley, Karen. K. Gleason, Super-hydrophobic carbon nano-tube forests, *Nano Lett*, 2003, Vol. 3 (12), 1701-1705.

[72] Y. Wu, H. Sugimura, Y. Inoue, O. Takai, Thin films with nanotextures for transparent and ultra water-reppelent coatings produced from trimethylmethoxysilane by microwave plasma CVD, *Chem Vap Depos*, 2002, Vol. 8 (2), 47-50.

[73] Y. Wu, M. Bekke, Y. Inoue, H. Sugimura, H. Kitaguchi, C. Liu, O. Takai, Mechanical durability of ultra-water-repellent thin film by microwave plasma-enhanced CVD, *Thin Solid Films*, 2004, Vol. 457, 122-127.

[74] J.M. Lim, G.R. Yi, J.H. Moon, C.J. Heo, S.M. Yang, Super-hydrophobic films of electrospun fibers with multiple-scale surface morphology, *Langmuir*, 2007, Vol. 23, 7981-7989.

[75] A. Greiner, and J.H. Wendorff, Electrospinning: A fascinating method for the preparation of ultrathin fibres, *Angew Chem* (Int. Ed.), 2007, Vol. 46(30), 5670-5703.

[76] J. Bravo, L. Zhai, Z. Wu, R.E. Cohen, M.F. Rubner, Transparent super-hydrophobic films based on silica nanoparticles, *Langmuir*, 2007, Vol. 23 (13), 7293-7298.

[77] Y. Xiu, F. Xiao, D.W. Hess, C.P. Wong, Super-hydrophobic optically transparent silica films formed with a eutectic liquid, *Thin Solid Films*, 2009, Vol. 517 (5), 1610-1615.

[78] G.R.J Artus, S. Jung, J. Zimmermann, H.P. Gautschi, K. Marquardt, S. Seeger, Silicon nano-filaments and their applications as super-hydrophobic coatings, *Adv Mater*, 2006, Vol. 18 (20), 2758-2762.

[79] Kamal K. Gupta, Manjeet Jassal, Ashwini K. Agarwal, Sol-gel derived titanium dioxide finishing of cotton fabric for self cleaning, *Ind J Fiber Text Res*, 2008, Vol. 33 (4), 443-450.

[80] Violeta Purcar, Dan Donescu, Ilie Catalin Spataru et al., Surface modification of sol-gel hybrid films using fluorinated silica nanoparticles, *Revue Roumanie de Chimie*, 2013, Vol. 58 (1), 37-42.

[81] R.J. Nuzzo, D.L. Allara, Adsorption of bifunctional disulfides on gold surfaces, *J Am Chem Soc*, 1983, Vol. 105 (3), 4481-4483.

[82] Deepak Prashar, Self-assembled monolayers: a review, *Int J ChemTech Res*, Vol. 4 (1), 258-265.

[83] J.C. Love, L.A. Estroff, J.K. Kriebel, R.J. Nuzzo, G.M. Whitesides, Self-assembled monolayers of thiolates on metals as a form of nanotechnology, *Chem Rev*, 2005, Vol. 105 (4), 1103-1170.

[84] F. Shi, X. Chen, L. Wang, J. Niu, J. Yu, Z. Wang, X. Zhang, Roselike microstructures formed by direct in situ hydrothermal synthesis: from superhydrophophilicity to superhydrophobicity, *Chem Mater*, 2005, Vol. 17 (24), 6177-6180.

[85] Jakub Sirc, Radka Hobzova, Nina Kostina, et al., Morphological characterization of nanofibers: methods and application in practice, *J Nanomater*, 2012, Vol. 2012.

[86] D. Oner, T.J. McCarthy, Ultrahydrophobic surface, effects of topography length scales on wettability, *Langmuir*, 2000, Vol. 16 (20), 7777-7782.

[87] J.Y. Shiu, C.W. Kuo, P. Chen, C.Y. Mou, Fabrication of tunable superhydrophobic surfaces by nano-sphere lithography, *Chem Mater*, 2004, Vol. 16 (4), 561-564.

[88] Lester Li, Victor Breedveld, Dennis W. Hess, Creation of super-hydrophobic stainless steel surfaces by acid treatment and hydrophobic film deposition, ACS, *Appl Mater Interf*, 2012, Vol. 4 (9), 4549-4556.

[89] Boisier, Grégory and Lamure, Alain and Pébère, Nadine and Portail, Nicolas and Villatte, Martine, Corrosion protection of AA2024 sealedanodic layers using the hydrophobic properties of carboxylic acids. *Surf Coat Technol*, 2009, Vol. 203 (22), 3420-3426.

[90] Liu et al., New application potential of the underground super-hydrophobic surface in the corrosion protection, *Adv Mater Res*, 2009, Vol. 79 (82), 1115-1118.

[91] Liu et al., Inhibition microbial adherence of super-hydrophobic surface on aluminium in sea water, *Adv Mater Res*, 2009, Vol. 79 (82), 1123-1126.

[92] Samer Darwich, *Corrosion Protection Concepts for Aluminium and Magnesium Alloy Coated with Silica Films Prepared by Water Based Sol-Gel Process*, PhD Thesis, 2012, Technischen Universitat Chemnitz, Germany.

[93] Tian He, Yuanchao Wang, Yijian Zhang, Quan L.V, Tugen Xu, Tao Liu, Super-hydrophobic surfaces treatment as corrosion protection for aluminium in sea-water, *Corr Sci*, 2009, Vol. 51, 1751-1763.

[94] J. Bico, U. Thiele, D. Quere, Wetting of textured surfaces, *Coll Surf* A: *Physiochem Eng Asp*, 2002, Vol. 206, 41-46.

[95] R.I. Joned, A.D. Warren, D.M. Rosenberg, V.J. Bellito, R. Park, M.R. Zachariah, Surface passivation of bare aluminium nanoparticles using perfluoroalkyl carboxylic acids, *Chem Mater*, 2008, Vol. 17 (11), 2087-2996.

[96] Rodrigo S. Neves, Daiane P. B. Silva, and Artur J. Motheo, Corrosion protection of AA7075 aluminium alloy by trimethoxy-silanes self-assembled monolayers, ISRN *Electrochemistry*, 2013, Vol. 2013, Article ID 142493.

[97] Paul E. Hintze, Luz Marina Celle, Electrochemical properties and corrosion properties of organo-silane self-assembled mono-layer on aluminium 2024-T3, *Electrochem Aceta*, 2009, Vol. 51, 1761-1766.

[98] V. Palanivel, Y. Huang, W.J. Van Ooij, Effect of addition of corrosion inhibitors to silane films on the performance of AA2024-T3 in 0.5M NaCl solution, *Prog Org Coat*, 2005, Vol. 53 (2), 153-168.

[99] A. Ulman, Formation and structure of self-assembled monolayers, *Chem Rev*, 1996, Vol. 96 (4), 1533-1554.

[100] Feiyue Li, Lan Zhang, Robert M. Metzger, On the growth of highly ordered pores in anodized aluminum oxide, *Chem Mater*, 1998, Vol. 10, Figure 1, 2471.

[101] J.P. O'Sullivan, G.C. Wood, Morphology and mechanism of formation of porous anodic films on aluminium, *Proc Royal Soc London*, Series A, 1970, Vol. 317 (1731), 511-543.

[102] G. Plumbo, S.J. Thorpe, K.T. Aust, On the contribution of triple junctions to the structure and properties of nanocrystalline materials, *Scrip Metall Mater* 1990, Vol. 24, 1347-1350.

[103] R. Rofagha, R.Langer, A.M. El-Sheikh, U. Erb, G. Palumbo. K.T. Aust, The corrosion behavior of nanocrystalline nickel, *Scrip Metall Mater*, 1991, Vol. 25 (12), 2867-2872.

[104] S.J. Thorpe, B. Ramaswami, K.T. Aust, Corrosion and auger studies of a nickel based metal-metalloid glass I. The effect of elemental interactions on passivity in the general corrosion of metglass 2826A, *J Electrochem Soc*, 1988, Vol. 135 (9), 2162-2170.

[105] J.K. Yu, E.H. Han, L. Lu, X.J. Wei, M. Leung, Corrosion behavior of nano-crystalline and conventional polycrystalline copper, *J Mater Sci*, 2005, Vol. 49 (4), 1019-1022.

[106] Y. Huang, D.K. Sarkar, X. Grant Chen, A one-step process to engineer superhydrophobic copper surfaces, *Mater Lett*, 2010, Vol. 64, Figure 1, 2723.

[107] Felios Athanasios, *Corrosion Behavior of Super-Hydrophobic Film on Copper in Sea Water*, PhD thesis, 2012, National Technical University of Athens, School of Naval Architecture and Marine Engineering.

[108] H. Baba, T.K. Odana, K. Mon, The corrosion inhibition of copper by potentiostatic anodization in triazinedithiol solutions, *Corr Sci*, Vol. 39 (3), 556-564.

[109] Saman Hosseinpour, Mats Gothelid, Christofer Leygraf, C. Magnus Johnson, Self-assembled mono-layers as inhibitors for the atmospheric corrosion of copper induced by formic acid; a comparison between hexanethiol and hexaneselenol, *J Electrochem Soc*, 2014, Vol. 161 (C50-C56).

[110] Zhiqing Yuan, Xian Wang, Jiping Bin, Chaoyi Peng, Suli Xing, Menglei Wang, Jiayu Xiao, Jingcheng Zeng, Yong Xie, Ximei Xiao, Xin Fu, Huifang Gong, Dejian Zhao, A novel fabrication of a superhydrophobic surface with highly similar hierarchical structure of the lotus leaf on a copper sheet, *Appl Surf Sci*, 2013, Vol. 285(B), 205-210

[111] Z. J. Wei, W. L. Liu, D. Tian, C. L. Xiao and X. Q. Wang, Preparation of Lotus-Like Superhydrophobic Fluoropo- lymer Films, *Appl Surf Sci*, 2010, Vol. 256, 3972-3976.

[112] Bimal P. Singh, Bikash Kumar Jena, Sarama Bhattacharjee, Laxmidhar Besra, Development of oxidation and corrosion resistance hydrophobic grapheme oxide-polymer composite coating on copper, *Surf Coat Technol*, 2013, Vol. 232, 475-481.

[113] G.X. Shen, Y.C. Chen, L. Lin, C.J. Lin, D. Scantlebury, Study on the hydrophobic nano-TiO_2 coating and its properties for corrosion protection of metals, *Electrochem Aceta*, 2005, Vol. 50, 5083-5089.

[114] E.O. Zayim, Effect of calcination and pH value on the structural and optical properties of titanium oxide thin film, *J Mater Sci*, 2005, Vol. 40 (6), 1345-1352.

[115] B. O'Regan, M. Gratzel, A low cost, high efficiency solar cell based on dye-synthesized colloidal TiO_2 films, *Nature*, 1991, Vol. 353.

[116] M. Joshi, A. Bhattacharya, N. Agarwal, S. Parman, Nanostructured coatings for super-hydrophobic textiles, *Bull Mater Sci*, 2012, Vol. 35 (6), 933-938.

[117] S. Takeda, S. Suzuki, H. Odaka, H. Hosono, Photocatalytic TiO_2 thin film deposited onto glass by DC magnetron sputtering, *Thin Solid Films*, 2001, Vol. 392 (2), 338-344.

[118] H.Y. Ha, S.W. Nam, T.H. Lim, In. H. Oh, S.A. Hong, Properties of the TiO_2 membranes prepared by CVD of titanium tetraisopropoxide, *J Memb Sci*, 1996, Vol. 111 (1), 81-92.

[119] L. Hu, T. Yoko, H. Kozuka, S. Sakka, Effect of solvent on properties of sol-gel derived TiO_2 coating film, *Thin Solid Films*, 1992, Vol. 219 (1-2), 18-23.

[120] C.J. Brinker, G.W. Scherer, *Sol-Gel Science: The Physics and Chemistry of Sol-Gel Processing*, Academic Press, Inc, New York, 1990, 163-169.

[121] Z. Zainal, C.Y. Lee, Properties and photoelectrocatalytic behavior of sol-gel derived TiO_2 thin films, *J Sol-Gel Sci Technol*, 2006, Vol. 37 (1), 19-25.

[122] N. Barati, M.A. Faghihi Sani, H. Ghasemi, Z. Sadeghian, S.M.M. Mirhoseini, Preparation of uniform TiO_2 nano-structure film on 316 stainless steel by sol-gel dip coating, *Appl Surf Sci*, 2009, Vol. 255, 8328-8333.

[123] P.V. Mahalakshmi, S.C. Vanithakumari, Judy Gopal, U. Kamachi Mudali, R. Baldeve, Enhancing corrosion and biofouling resistance through super-hydrophobic surface modification, *Curr Sci*, 2011, Vol. 101 (10), 1328.

[124] Li Juan Chen, Miao Chen, Hui Di Zhou, Jian Min Chen, Preparation of super-hydrophobic surface on stainless steel, *Appl Surf Sci*, 2008, Vol. 255, 3459-3462.

[125] Hu Yawei, Huang Siya, Liu Shan, Pan Wei, A corrosion resistance super-hydrophobic TiO_2 film, *Appl Surf Sci*, 2012, Vol. 258 (19), 7460-7464.

[126] J.A. Nasdi, D.E. Weibel, A.F. Michels, F. Horowitz, Super-hydrophobic and anti-corrosion properties of silane/silica nanoparticles/PTFE coatings on carbon steel. Instituto de Quimica and Pgcimat Universidad Fereraldo Rio Grande do sul, Porto Alegre, RS, Brazil, 2012.

[127] Shazly Salem, Corrosion inhibition of carbon steel in HCl using sodium dodecyl sulfonate: the synergistic effect of non-ionic co-surfactant and some inorganic ions, *J Eng Sci*, 2011, Vol. 39 (5), 1147-1156.

[128] M. Higgins, D. Smith, J. Mattzela, G. Barone, Protective chemical vapor deposition coatings for stainless steel surfaces used in produced water environments, Silico Tek. Corporation, 112, Benner Circle. Bellefonte, PA, 16823, http://www.silcotek.com/

[129] Rob. B. Polder, Huibert Boreje, Prevention of re-inforcement corrosion by hydrophobic treatment of concrete, HERON, 2001, Vol. 46 (4).

[130] Anti-icing and de-icing super-hydrophobic concrete to improve the safety on critical element of roadway pavements and bridges, Sept. 2013, National center for freight and infrastructure research and education University of Wisconsin, Milwaukee, USA.

[131] P. Tourkine et al., Delayed Freezing on water repellent materials, *Langmuir*, 2009, Vol. 25 (13), 7214-7216.

[132] T.S. Wong et al., Bio-inspired self-repairing slippery surface with pressure stable omniphobicity, Nature, 2011, Vol. 477, 443-447.

[133] Liangliang Cao, Andrew K. Jones, Xinod K. Sikka, Jianzhong Wu, D. Gao, Anti-icing super-hydrophobic coating, Langmuir, 2009, Vol. 21, 1244-1248.

Chapter 8

The Role of Hydrophobicity in the Development of Aluminum and Copper Alloys for Industrial Applications

Amir Farzaneh, Zaki Ahmad, Mustafa Can,
Salih Okur, Omer Mermer and Ali Kemal Havare

Additional information is available at the end of the chapter

http://dx.doi.org/10.5772/60724

Abstract

Improvement of corrosion resistance is a main challenge in surface and corrosion engineering. Water is a main part of corrosion mechanism, and the omission of this part will be helpful. By mimicking from nature, superhydrophobicity is a helpful method to decrease corrosion and water presence on the surface. Superhydrophobic surface and coating is a new type of coating that extension in recent decade and increase application of this area every day. These type of coatings are using on widespread of applications such as solar panels, displays, windows, paints and fabrics to obtain water-proof, anticorrosive, self-cleaning and stain-resistant surfaces. Many different studies have been reported to produce superhydrophobic surfaces from many diverse materials (polymers, metals and other inorganic materials, composites, textiles, paper). In this chapter, recent developments in the application of superhydrophobic coatings for corrosion protection of aluminum and copper alloys will be discussed. This chapter includes new production methods, theoretical works, and the limitations of superhydrophobic coatings.

Keywords: Superhydrophobic, coating, Aluminum, Copper, Corrosion

1. Introduction

Investigation and research about superhydrophobic (SH) surfaces have been done as early as the 1940s, but it is just recently that this field became a great point of interest [1]. At this time, the state of the technology is at a level to produce many possible applications. Also, there is a race to discover the best procedures, methods, and theories. During the past decade, researchers have published a lot of results in application and production methods. Polystyrene nanotube forests to nano-casting lotus leaves are examples of production methods, and the application ranges from lab-on-chip fluidics to self-cleaning paints [2]. There are widespread production methods to obtain different properties. There are two main challenging problems for researchers to solve: low effective lifetimes and high fabrication costs.

Water aggregation on the surface of metals in nature is a main problem that is responsible for many types of destruction of metals such as corrosion and erosion [3-5]. Corrosion of metals depends on the amount and state of water on the surface. The destruction of the passive layer of oxide through the impact of water, sand, and abrasive particles occurs with loss of mechanical strength and corrosion resistance [6, 7]. Concrete structures that are very prone to water penetration, as well as metallic structures, become sensitive to corrosion on coming in contact with water [8]. Aeroplanes, helicopters, all ground transports, and concrete structures are subjected to serious erosion problems [9, 10].

Over the past decade, a lot of progresses have been made on the development of artificial superhydrophobic surfaces based on lotus leaf [1, 11-13]. Lotus leaf is held as a symbol of purity in many countries, especially in some regions of Asia, due to its ability to remain clean when emerging from murky ponds [14]. The natural cleaning mechanism that originated from their microscopic structure and surface chemistry and that was first time reported by Botanists is called Lotus Effect. This is explained in terms of water contact angle [15, 16]wherein it is larger than 150º and the sliding angle is less than 10º. It is also hydrophobic as a result of a binary structure (micro- and nanoscale), as well as the wax layer present on the leaf surface [14, 17-20]. Recently, various researchers have successfully fabricated superhydrophobicity with some specific metallic substrates, such as Copper (Cu), and Aluminum (Al) [21-28].

The fabrication of engineered dust and water repellent surfaces is a novel method to eliminate destructive species and control corrosion [11, 29]. Nature has shown the way to keep the surfaces clean and the lotus flower is a good example with a hierarchical surface comprising of micro valleys and nano hills. The valleys are comprised of nanohairs and epicuticular wax particles while the hierarchical structure is responsible for dust and water repulsion. This type of surface with a high contact angle is called a superhydrophobic surface [30-34].

The wet contact angle on a flat solid surface is shown in figure 1. Surface chemical composition is the main factor which affects surface wettability. The contact angle is correlated by three interfacial free energies: free energies at the solid-air (γ_{SV}), solid-liquid (γ_{SL}), and liquid-air (γ_{LV}) interfaces, by the Young's equation [35]:

$$\cos\theta_{flat} = \frac{\gamma_{SV} - \gamma_{SL}}{\gamma_{LV}} \tag{1}$$

http://dx.doi.org/10.5772/60724

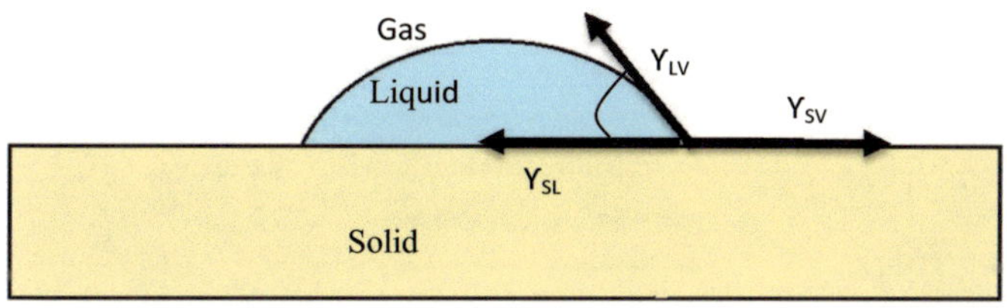

Figure 1. Young's equation equilibrium condition on a flat surface.

As shown in Equation (1), by decreasing the surface free energy of solid-air interface (γ_{SV}), hydrophobicity is increased. By the range of θ^{flat}, the surface is divided into two categories. First, when θ^{flat} is smaller than 90°, the solid surface is considered intrinsically hydrophilic. Second, when θ^{flat} is greater than 90°, the solid surface is considered intrinsically hydrophobic. According to the reports, with the lowest surface energy, the maximum water contact angle can be obtained such as in lotus flower to 120°. Further increase of the hydrophobicity can be achieved by increasing by surface roughness [36].

In 1936, Robert N. Wenzel reported the effect of surface roughness on wettability [37]. As shown in figure 2, Wenzel argued that if the droplet contours in the roughness, it should enhance the hydrophobicity with a linear relationship between the contact angle and roughness given by:

$$\cos\theta^{\omega} = r\cos\theta \tag{2}$$

In the equation 2, Static θ referred to equation (1) is the contact angle of a smooth flat, θ^{w} is the apparent contact angle enhanced by roughness and r is the roughness factor can be defined by

$$r = \frac{\textit{True Surface Area}}{\textit{Horizontally Projected Surface Area}} \tag{3}$$

The Wenzel equation shows that by increasing r (roughness factor), the hysteresis will increase but this equation is only applicable until r reaches 1.7, after which, the hysteresis begins to decrease. The Wenzel contact angle theory cannot be explained on a rough surface which signifies the transition to the Cassie-Baxter model. A. B. Cassie and S. Baxter further developed the theory to consider trapped air helping to suspend the droplet over the roughness contours. Figure (2) shows that the drop can lie on a surface and air composite state defined by the solid fraction (f), which is the fraction of solid in contact with the liquid droplet. With the Cassie-Baxter model, the apparent contact angle (θ^{c}) related by Young's contact angle (θ) and solid fraction (f) is given as [38]:

$$\cos\theta^{c} = f\cos\theta + f - 1 \tag{4}$$

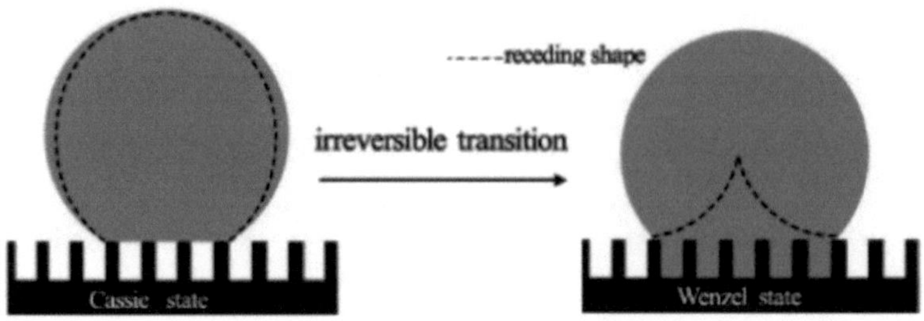

Figure 2. Wenzel and Cassie-Baxter theory of droplet formation on a rough surface suspension on a solid/air composite [38]

As seen in this equation, apparent contact angle (CA) is increased by decreasing the solid fraction (f) in high porous surface and increasing the CA of the smooth flat surface (θ) in low surface energy and it is going to be a superhydrophobic surface.

The Wenzel's and Cassie-Baxter's models represent the exact contact state between the liquid and the rough solid surface. These equations are applied for numerical demonstration on the superhydrophobic surface materials. The Cassie-Baxter model explains how extremely rough surfaces are energetically favourable to suspend the droplet on a composite surface of solid and air [39].

For more explanation, the dependence of surface roughness and contact angle will be further discussed. By placing a droplet on the solid surface, the three interfaces exist [40]: solid-liquid, solid-air, and liquid-air. The relationship between these three interface energies is shown in equation 1. Surfaces can be divided into two groups according to their affinity to water: hydrophilic (water attracting) and hydrophobic (water repellent). Water molecules that interact with the surface, which bear electric charges or polar groups typically are hydrophilic, while non-polar surfaces are generally hydrophobic. On hydrophilic surfaces, the water molecules suffer a loss of hydrogen bonding partners at the interface [36, 41].

For superhydrophobic surfaces, roughness by special properties plays a main role. For explaining the surface roughness, there are surface parameters such as Roughness Average (R_a), Root Mean Roughness (R_q), and Roughness Depths (R_z or R_{max}). The effects of these parameters on superhydrophobicity are not clear yet. The surface roughness in a superhydrophobic surface should be a combination of nano and micro structures. For a rough solid surface, which consists of asperities and valleys, with a typical size of roughness detail smaller than the size of the droplet, CA was calculated by the Wenzel model. Natural superhydrophobic surfaces also exhibit hierarchical roughness structure, therefore; a successful water-repellent surface should also have a multiscale roughness.

2. Development of alloys by superhydrophobic surfaces

The surface treatment of metals is most important in metal production. There are various surface treatment methods such as surface hardening, painting, and coating [4, 5, 38, 42].

http://dx.doi.org/10.5772/60724

Superhydrophobic coating is a favourable research area in various applications because of its natural water-resistant properties. Superhydrophobic surfaces have been achieved by the controlling of surface roughness and topography in hydrophobic materials [39, 43]. For superhydrophobic surface production, a variety of methods have been reported by mimicking the surface of lotus leaves such as electrodepositing [44-46], anodic oxidation [27, 47, 48], plasma and laser treatment [49-53], physical and chemical vapour deposition [54-59], electro-spining [60-62], sol gel techniques [63-65], and other methods. Various methods that have been used to prepare superhydrophobic surfaces according to the properties of metals and the method of superhydrophobic treatment for copper and aluminium will be discussed in this review.

2.1. Superhydrophobic coatings in copper and alloys

Copper and its alloy are important materials that are used in the industry because of their high electrical and thermal conductivities, good strength formability, and some others remarkable properties. They are widely used in many applications such as electronic industry and communications as a conductor in electrical power lines, pipelines for domestic and industrial water utilities including seawater, and heat conductors and heat exchangers [13, 66]. In environments that contain chloride ion such as marine environment, the copper alloys are susceptible to various corrosion types such as pitting, stress corrosion cracking and crevice corrosion. The protection of these alloys from this attack is a main concern for researchers [67-70]. One of best method for protection is the use of superhydrophobic coatings.

The physical and chemical properties of copper play an important role in superhydrophobic coating preparation. The information on the properties can help in choosing the right chemical agent and predicting the available mechanisms during the superhydrophobic surface creation.

2.1.1. Mechanism of hydrophobic coating on copper and copper based alloys

Copper has an atomic number of 29, with valance of 1 or 2, lattice structure FCC, and oxidation state of 1 and 2. The typical blue hexaaquacopper (II) ion ([Cu (H_2O) 6]$^{2+}$) is the simplest ion that copper forms in a solution. Hydroxide ions of NaOH remove hydrogen ions from the water ligands attached to the copper ion [71].

When the hydrogen ion has been removed from two of the water molecules, a complex with no charge complex has been obtained. This neutral complex is insoluble in water and a precipitate is formed.

$$\left[Cu(H_2O)_6\right]^{2+} + 2(OH)^- \longrightarrow \left[Cu(H_2O)_4(OH)_2\right] + 2H_2O$$

Ammonia acts as both a base and a ligand. To obtain the same neutral complex, a small amount of ammonia can be added to pull hydrogen ions off the hexaaqua ion exactly as in the hydroxide.

$$\left[Cu(H_2O)_6\right]^{2+} + 2NH_3 \longrightarrow \left[Cu(H_2O)_4(OH)_2\right] + 2NH_4^+$$

That precipitate can be dissolved if an excess of ammonia is added. The ammonia replaces water as a ligand to give tetraamminediaquacopper (II) ions. It must be noted that only 4 of the 6 water molecules are replaced.

$$\left[Cu(H_2O)_6\right]^{2+} + 4NH_3 \longrightarrow \left[Cu(NH_3)_4(H_2O)_2\right]^{2+} + 4H_2O$$

A precipitate of CuCO3 is formed as below.

$$Cu^{2+}_{(aq)} + CO^{2-}_{3(aq)} \longrightarrow CuCO_{3(s)}$$

If concentrated hydrochloric acid is added to a solution containing hexaaquacopper (II) ions, the six water molecules can be replaced by four chloride ions. The reaction taking place is reversible.

$$\left[Cu(H_2O)_6\right]^{2+} + 4Cl^- \longrightarrow \left[Cu(Cl)_4\right]^{2-} + 6H_2O$$

Due to the reversible reaction, mixtures of colours of both complex ions are obtained. The colour of the tetrachlorocuprate (II) ion is variously described as olive-green. If more water is added to the green solution, it returns to its blue colour.

Copper (II) ions can oxidise iodide ions to iodine. In this process, they are reduced to copper (I) iodide.

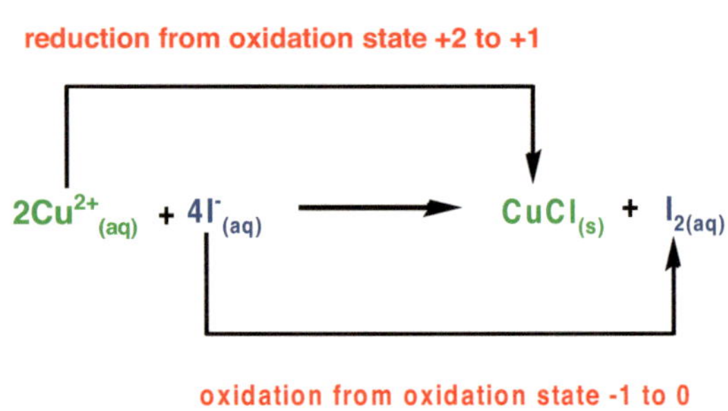

Copper (I) chemistry is limited by a reaction that occurs involving simple copper (I) ions in solution. Copper (I) ions in solution disproportionate to give copper (II) ions and a precipitate of copper.

The reaction is:

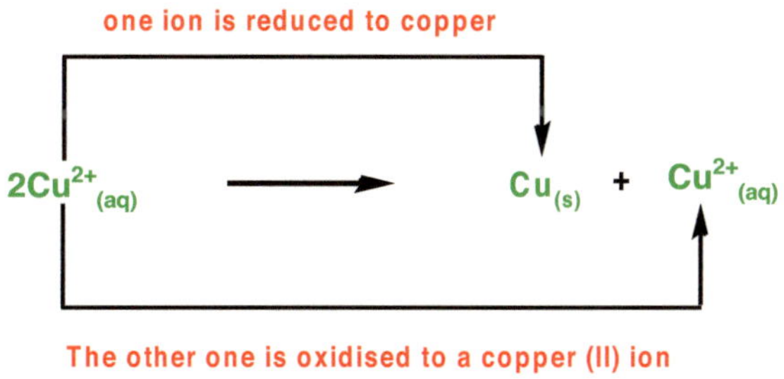

Any attempt to produce a simple copper (I) compound in solution results to as follows: for example, if copper (I) oxide reacts with hot dilute sulphuric acid, a solution of copper (I) sulphate and water may be produced. Because of the disproportionation reaction, a brown precipitate of copper and a blue solution of copper (II) sulphate are obtained.

$$Cu_2O + H_2SO_4 \longrightarrow Cu + CuSO_4 + H_2O$$

The copper (I) oxidation state is stabilized by the formation of copper (I) complexes. For example, both [Cu (NH3)2] + and [CuCl2] are copper (I) complexes that don't disproportionate. The chlorine-containing complex is formed if copper (I) oxide is dissolved in concentrated hydrochloric acid. It is possible to think that the reaction occurs in two steps. Initially, copper (I) chloride forms:

$$Cu_2O_{(s)} + 2HCl_{(aq)} \longrightarrow 2CuCl_{(s)} + H_2O_{(l)}$$

However, in the presence of excess chloride ions from the HCl, a stable and soluble copper (I) complex is formed.

$$CuCl_{(s)} + Cl^-_{(aq)} \longrightarrow \left[CuCl_{2(aq)}\right]^-$$

Subsequently, to obtain white precipitate of copper (I) chloride (mentioned above), some water is added to this solution to strip off the extra chloride ion [72, 73].

Preparations of superhydrophobic coating on copper surface initiates by surface modification and etching in acidic environment based on copper oxidation. Liu et al. [13] have fabricated the superhydrophobic coating on copper substrate by a simple immersion method. In their research, Cu substrates have been immersed in 13 wt. % HNO_3 for 15s to remove surface oxides, then immersed in solution (0.06 mL 37 wt. % HCl; 0.02 mL CH3COOH; 54 mL H_2O) for 12 hours to roughen the surfaces, and then dried in the N_2 flows. They were subsequently immersed in an ethanol solution of n-tetradecanoic acid (0.1 M) at 37 ∘C for 10 days after which, they could obtain a superhydrophobic film with a contact angle larger than 150∘. In this attempt they improved corrosion resistance of copper and claim that the copper corrosion is effectively inhibited by the formation of the superhydrophobic film.

The creation of superhydrophobic surfaces by n-alkanoic acid can be explained below:

1. oxidation of copper:

$$2Cu + O_2 + 4H^+ \longrightarrow 2Cu^{2+} + 2H_2O$$

2. forming copper carboxylate:

$$Cu^{2+} + 2CH_3(CH_2)_{12}COOH \longrightarrow Cu[CH_3(CH_2)_{12}COO]_2 + 2H^+$$

N-alkanoic acid molecules capture Copper ions after being released instantly. By the reaction of copper with the above acid, the reaction is transformed to copper carboxylate and appears as a self-assembled flower-like structure. The immersion time is an important factor for this process. The superhydrophobicity of the prepared surface can be explained by Cassie-Baxter model [13, 21].

The superhydrophobic surface on copper can be prepared by 4, 5-diphenyl-imidazole (DPI). Sheng Zhao et al. [74] have used this procedure to prepare copper superhydrophobic surface. They firstly cleaned the copper ultrasonically in ethanol, acetone, and dilute hydrochloric acid in turns for 15 min to remove the contamination, grease, and oxide etc. on the surface. The DPI solvent was with various solutions such as methanol, ethanol, acetone, acetonitrile with 0.02 M concentrations. The treated copper plate was subsequently suspended in DPI solutions, which were then put into a culture cabinet and deposited for 10–15 days at an ambient temperature of 25 ∘C. After these processes, the hierarchical DPI mesostructures assembled spontaneously at the solid–liquid interface.

Figure 3 shows the suggested mechanism in the previous process. The interface forces and $\pi - \pi$ interaction and van der Waals contacts between the building blocks is a main building mechanism for assembling mesostructures in the copper surface in DPI solutions.

It was also observed that copper easily reacts with fatty acids. Copper ions created the various bonds with ion solutions and grow flowerlike self-assembled film on the surface with low energy and rough morphology. There are various suggested mechanisms for superhydrophobic growth films on copper surface but it seems that more research is needed to provide an acceptable mechanism. Additionally, the more important properties in copper superhydrophobic coatings such as its durability and its condensing behaviour on copper tubes, are the promising future areas of research.

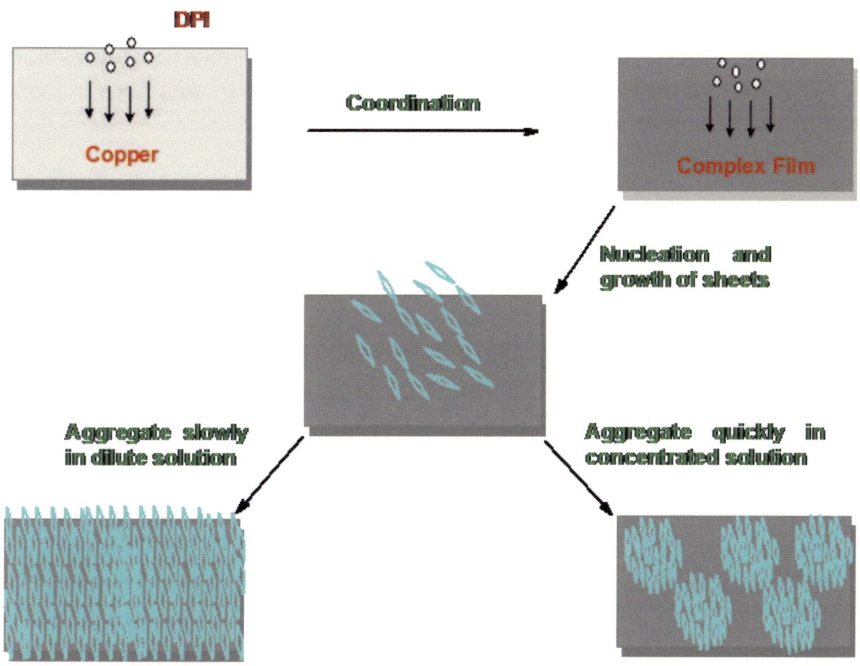

Figure 3. Suggested mechanism for hierarchical DPI mesostructures assembled at the solid–liquid interface.

For improvement of superhydrophobity and the durability of copper surface, the next method is to use additional polymeric compound with low surface energy properties. For the first time, the copper has been treated by fatty acids, and then modified and coated by a low surface energy agent. For example, by immersion, copper in oxalic acid aqueous solution at a concentration of 0.5 wt. % for about 5–7 days at ambient temperature, a superhydrophobic surface is obtained. Poly (dimethysiloxane) vinyl terminated (PDMSVT) is used as a surface modification agent, which is applied by spin coating methods and curing at in 120 °C for 2 h. water contact angle of 154°. The bis- [triethoxysilylpropyl]tetrasulfide (BTESPT) by ethanol solvent can be used as upper coating in treated copper [40].

The acceptable mechanism that can be used to create a superhydrophobic surface on copper can be explained by two strategies. First strategy is through the creation of a rough surface by etching copper in suitable acid and then coating of this rough template with low surface energy agent such as silane base materials. For choosing acid as an etchant, it is important that the copper can be reacted by the acid's ions to grow the self-assemble compound on the surface. As shown, the copper ion reacts with carboxyl radical, hence, the fatty type acids can be a good choice to play this role. However, other types of acids can also be used. Another strategy is by degreasing the copper after putting it into dilute nitric acid to get rid of the copper oxides on the surface. This should be followed by immersing in the aqueous solution of 2 M sodium hydroxide at room temperature for 24 hours, fully washing with deionized water, and drying in nitrogen stream. After that, the pre-treated copper is immersed in a formed aqueous solution of 0.8 M cupric nitrate trihydrate for 24 hours, allowing the generation of a blue slice-like Cu_2 (OH) $3NO_3$ crystals. A good rough surface can be created by the process described, followed by coating a blue slice-like Cu2 (OH)3NO3 film by immersing in an ethanol solution of 1H,1H,2H,2H-perfluorodecyltriethoxysilane for 24 hours. The specimens are washed with ethanol and dried in an oven at 120^0C for 1 hour [75].

After the formation of a superhydrophobic surface by simple immersion methods, a flowerlike morphology is observed. Copper oxide or sulphides or other suitable composition grow on the copper surface to produce a good micro-nano rough template. There are no special optimum conditions for the surface treatment of copper and in each report, different times and temperatures have been reported but the basic principle is that coating by the fatty acids is a good option for copper surface superhydrophobic treatment.

2.1.2. Techniques of deposition of superhydrophobic coating

There are various methods to create a superhydrophobic coating that still exists until now. Among these methods, the electrochemical method is favorable because there are many factors to control the film growth rate, morphology, and properties. The most common method, however, is the anodizing method where the copper plate acts as an anode in electrolyte. Caustic Soda solution can be used as an electrolyte at 3.0 mol/l concentration. The current density and temperature range is about 600 mA/dm^2 and 25 respectively. If the current density is decreased to 200-600 mA/dm^2 and the temperature enhanced to 60 centigrade, the CuO film will be converted to Cu $(OH)_2$films. Under these conditions, oxide film formation is controlled by applied current. Below is the general mechanism [76]:

$$Cu^{2+} + 2OH^- \longrightarrow Cu(OH)_2^- \quad [1]$$

$$Cu(OH)_2^- \longrightarrow Cu_2O + 2H_2O + 2OH^- \quad [2]$$

$$2Cu_2O + O_2 \longrightarrow 4CuO \quad [3]$$

$$Cu^{2+} + 4OH^- \longrightarrow Cu(OH)_4^- \quad [4]$$

$$Cu(OH)_4 \longrightarrow 4CuO + 2H_2O + 2OH^- \quad [5]$$

The first reaction captures the released Cu^+ ions and then the captured radical are decomposed to copper oxide. After passage of time, the film grows on the surface and the nano-micro surface is obtained. After this process, various low surface energy agents can be used for surface modification.

http://dx.doi.org/10.5772/60724

Ethanolic stearic acid is a second choice as an electrolyte. In the anodizing system, the copper also can be made as a cathode. The voltage could be about 30 V DC. By using this type of electrolytes, the copper can be reacted by the acid's ions to be able create the low surface energy compounds with micro-nano rough surfaces. In this process, where copper acts as an anode in electrolyte, the copper ion reacts with the electrolyte's free radicals and form slow surface energy films. By using the proper voltage and time, the morphology can be controlled. The voltage used in most of the electrolytes is near 30 V DC but by altering the concentration and temperature, it can be changed [77]. This procedure can be varied because there are lots of fatty acid that copper ions can react with their free ions in electrolyte, make the stable compound and deposit on the surface [78-80]. It seems that the mechanism, optimum conditions and electrolyte comparison is a question that should be answered in the future.

Electrodeposition is a well-known process that was invented by Alessandro Volta in the 1800s [81]. In this method, a current is passed through an electrochemical cell from an external source to reduce the dissolved metal cations, so that they form a coherent metal coating on an electrode [82]. This method has a lot of advantages, such as being low cast, facile, highly effective, and easily controllable by simply adjusting the electrodeposition parameters, such as DC voltage, current and electrolysis time [83]. In this method, various metals and compounds can be coated on a copper surface. For this reason, copper is used as a cathode in electrolyte to deposit the coating on the surface. Various types of electrolytes can be used. For example, nickel or cerium based electrolytes, copper based electrolytes, and the proper metals salts solution with acidic media can be used as electrolytes [84]. An important point to be noted is to use a low surface energy compound in electrolytes. For example, myristic types of acid can be a good choice. Chen et al. [85] in 2012 have prepared suitable superhydrophobic coatings on copper surface by the electrodeposition method. Nickel chloride (NiCl2 6H2O) and myristic acid with ethanol were used as an electrolytic solution. Anode and cathode both were copper plates. The DC current of 30 V has been used. Using this method, the superhydrophobic nickel coating deposited on the copper surface. To optimize electrodeposition conditions, the concentration of nickel chloride was kept to 0.08M and that of the myristic acid to 0.071M. The superhydrophobic surface showed a high water contact angle of 160^0. The mechanism in this process can be explained by the following equation:

$$Ni^{2+} + 2e^- \longrightarrow Ni$$

$$Ni^{2+} + 2CH_3(CH2)_{12}COOH \longrightarrow Ni\left[CH_3(CH2)_{12}COO\right]_2 + 2H^+$$

$$2H^+ + 2e^- \longrightarrow H_2$$

Nickel ions absorbs electron with two reactions and deposited on the surface. It can be seen that the nickel deposited had two types: nickel and Ni [CH3 (CH2)12COO] 2. By passing the current for 15 minutes, the water contact angle rises to about 160^0, but in further increasing the time, the superhydrophobicity decreased because of losing micro-nano rough properties. More time causes an increase in the film thickness. The film thickness can be chosen according to the application. For corrosion protection in aggressive environment, the thickness must be high. But, it is important that the coating thickness is not higher than a specific thickness and that it can be optimized by testing the deposition time and water contact angle. By growing the coating on the surface, the micro nano porosity by the new deposited compound and the rough surface will be destroyed.

Myristic acid with 0.038 M cerium chloride and 0.1 M myristic acid can also be a good electrolyte for the electrodeposition of superhydrophobic coatings on copper surface. The superhydrophobic properties of this deposition are because of the cerium compound (Ce [CH3 (CH2)12COO] 3 composition, which increase the water contact angle to 165^0 [85].

It is possible to use other methods to produce the superhydrophobic coating on the copper surface. If one process can make a rough surface with low surface energy properties, the superhydrophobic coating will be obtained. Hitherto, so many attempts have been done about superhydrophobic surfaces. Sol gel methods, electroless, electrodeposition and sputtering, etc. are available methods to surface modification in copper. But the main concern is the durability of the coating and the wear and erosion performance. Future research must focus on durability and application of the superhydrophobic coatings in aggressive media.

2.2. Superhydrophobic coatings in aluminium and alloys

Aluminium is widely used in various industrial fields as a basic material because of its excellent properties such as low density, excellent thermal and electrical conductivity, high specific strength, and good castability [86, 87]. However, the application and development of aluminium alloy are limited due to high metal reactivity and poor corrosion resistance in aqueous solution, especially in corrosive environment such as in aqueous solution containing Cl^- ions [88]. So, it is necessary to enhance the corrosion resistance property of aluminium in corrosive environment, which will greatly extend their applications as engineering materials [89]. For this reason, making an aluminium surface superhydrophobic has many application possibilities. By anticorrosive coating, the corrosion performance of aluminium alloys can be improved. Feng et al. [90] introduced facile and environment-friendly methods to fabricate hydrophobic surfaces to mitigate corrosion.

Aluminium has special properties due to its oxide layer on the surface that protects aluminium. Anodizing of aluminium can achieve a very durable surface that may be unaffected by weather for many years and many chemicals, if done properly. In this process, a thick film of aluminium oxide is built up on the surface of the aluminium, through the use of a direct current electrical supply. During the process, the aluminium located in the electrochemical cell is used as anode in an acidic media. The electrolyte can be a solution of sulphuric acid, phosphoric acid, oxalic acid, etc. and a direct current is applied to start the anodizing process [91].

The important point is it to optimize parameters which control the oxide layer thickness. This oxide layer contains hexagonal pores. The aluminium porous oxide layer consists of two layers that are a thick porous outer layer and a thin inner layer of oxide called barrier layer. The outer oxide layer of barrier-type alumina strongly depends on the type of electrolyte, the concen-

http://dx.doi.org/10.5772/60724

tration of adsorbed anions, and the faradaic efficiency of film growth. The oxide layer can be achieved with pores of various sizes and shapes which are important for roughness as roughening depends on pores [92, 93]. Several attempts have been made to create superhydrophobic coatings on aluminium surface by anodizing method.

Shibuichi et al. [94] have reported a new research about super water- and oil-repellent surfaces in 1998. The rough aluminum surface was fabricated in o.5 M of acid sulfuric solution, 10 mA/ cm^2 the current density and 3 hours anodizing time. Anodizing is a good method for surface roughening of Al. The chemical additives in the anodizing electrolyte affect the surface morphology and pore size. Holes and pores can be ordered by configuration of anodic porous alumina with two types of processes: naturally occurring long-range ordering, and ordering with ideally arranged hole configuration formed using anodization of pretextured [95].

In the pretexturing methods, the ordered convexes are generated as a mold by nanoindentation. The ordered sites are an initial nucleation location for holes during the anodizing process.

During anodizing for each application, the pore size and distance can be controlled. For superhydrophobic applications, obtaining suitable rough surface with micro-nano specifications is an important part. By increasing the voltage in each media, the distance of interpores is maintained between 0-500nm. The pH and conductivity of electrolyte lead to smaller pore size. There is no clear result on the effects of temperature to the pore size but it is obvious that temperature is very important. In addition, obtaining room or lower temperature depends on the electrolyte and choosing suitable temperature types provide uniformity on oxide film with low defects. The aluminum alloy types have various impurity caused to be defects in the film. It seems that the research in this area is so rare and needs to concentrate on the effect of alloy elements on pore formation and surface roughening. So many parameters can affect the final surface morphology during anodizing, such as grain size, residual stress, stirring the electrolyte, and so many additives to electrolyte [47, 48]. Here, we will focus on the effect of additives on the sulphuric acid electrolyte to obtain optimum conditions.

In anodizing with sulphuric acid, the concentration of sulphuric acid can be around 165-225 g/l but a concentration of 185g/l has shown better results. The current density can be varied between 0.8-1.8 A/dm2. There are many additives which can be used for changing pore size and morphology. Sodium chloride, by delivering of Cl^- ions in electrolyte, changes the pore size. The possible mechanism is that the Cl^- ions create a corrosive media attack on to the surface and generate pits on the oxide film. There are many reports on the effect of chloride ions on the pore geometry [47, 48, 96].

Wu et al. [47] have reported the effects of various additives in sulphuric acid electrolyte on superhydrophobicity of aluminum. The aluminum was placed in a sulphuric acid bath at 30 ^{0}C temperature. The additives were oxalic acid, potassium dichromate, chloride sodium, and glycerol.

The pore growth may be due to a field-assisted hydrogen ion attack on the oxide layer. By decreasing the hydrogen ion concentration, this attack will be slowed down or stopped and a barrier-type film layer will be formed. But it should be noted that the hydrogen ion attack must be fundamental to pore nucleation and development. So, the oxalic acid addition can increase the film thickness by retarding the dissolution rate of aluminum oxidation film and widen the oxidation temperature range without prominent effect on the morphology variation of the oxidation film and structural properties, which is in favor of controlling the anodic oxidation process. The optimal concentration of oxalic acid is at a range of 5–10 g/L.

Chloride ion is a highly active ion that reacts with Al ion to create the pits. The mechanism of pitting can be explained as below:

$$Al^{3+} + H_2O \longrightarrow Al(OH)^{2+}$$

$$Al(OH)^{2+} + Cl^{-} \longrightarrow Al(OH)Cl^{+}$$

After production of chloride complex, the final reaction is acid production and this process is a main reason of pitting production. This reaction is:

$$Al(OH)Cl^{+} + H_2O \longrightarrow Al(OH)_2Cl^{+}H^{+}$$

Acidic condition is corrosive than the other sides. This reaction creates a corrosive location by which holes and pits are generated in the oxide film, resulting in bigger porous structures occurring on the surface of the aluminum alloy. Therefore, the roughness further increases and leads to a good superhydrophobic surface. Presence of glycerol in electrolyte can minimize the corrosive effect of electrolyte and also may increase the ductility of the oxidation film. The chromate components such as potassium dichromate can improve uniformity of oxide films. The chromate plays the role of an inhibitor in aluminum corrosion and pitting, and by control, the film pitting may improve uniformity [27, 47, 48, 55, 91-93]. The mechanism and real effect of glycerol and chromate is not clear and needs more research.

In the first 5 minutes during anodizing, the pores are heterogeneous. With the passage of time, the film formation becomes homogenous and the pore size and shape becomes suitable for superhydrophobic surfaces with adequate roughness. The optimum time can be changed from 10 to 60 minutes according to the electrolyte composition.

In another report, Liu et al. [48] also prepared superhydrophobic coatings by anodizing and depositing polymeric coating on the surface of the aluminum alloy. They fabricated a superhydrophobic surface by polypropylene (PP) coating after anodizing, and obtained a superhydrophobic surface on aluminum with a contact angle (CA) of 162º and a sliding angle of 2º. They chose sulfuric acid concentration at a constant value of about 170 g/L and the oxalic acid and sodium chloride concentration as variable. Results showed that the optimum concentration of oxalic acid was 10 g/l and thess results were confirmed in Ruomei Wu's report. The optimum concentration for sodium chloride reached about 1.2 g/l. This electrolyte consist of a solution with acid sulphuric, oxalic acid, and sodium chloride

http://dx.doi.org/10.5772/60724

[49].The concentration of chloride ion could not be increased like the electrolyte which contained potassium dichromate because this additive inhibits the corrosive effect of chloride ions. It could increase only up to 50 g/L.

As a conclusion, the current, time, temperature can be varied by additive agents and it should be noted that the additive amount is also affected by each other's presence at the same time. Anodizing process is a good method to create suitable rough surface by controlling the pore size and morphology. After anodizing the surface modification by low surface energy agents such as propylene, superhydrophobic coating can be obtained.

Surface modification is a final process to fabricate superhydrophobic surfaces after anodizing. In this process, various low energy agents can be used but it should be noted that these agents should have good bonding characteristics with aluminum oxide and also do not destroy the rough templates [14, 19, 22-27, 63, 65, 95, 97-99].

In a study, anticorrosive superhydrophobic surfaces have been fabricated by melting myristic acid (CH3 (CH2)12COOH) adsorbed onto the anodized aluminum [97]. The static contact angle for seawater on the surface was about 154º. The predictive reaction can be shown as:

$$Al^{3+} + 3(CH_2)_{12}COOH \longrightarrow \left[3(CH_2)_{12}COO\right]_3Al + 3H^+$$

The myristic acid is a fatty acid and its component by metal ions on the surface generated low energy agents. As it is shown in the previous formula, the released Al ions, during the process of chemical modification, can be immediately captured by coordination with n-tetradecanoic acid molecules, forming aluminum carboxylate. This is the final coating that makes the anodized aluminum surface superhydrophobic.

Polypropylene is another inexpensive polymer with low surface energy properties to prepare superhydrophobic surfaces on anodized aluminum. The sample can be coated by dipping it in a polymer solution. Acetone or dimethyl-benzene can be used as a solvent and the concentration can be around 6-7 g/l [48, 100].

All low surface energy polymers and agents, by ability of absorbing on anodized aluminum surface, can be used to surface modification like methacrylate polymer, silane coupling agent, etc.

Surface pretreatment or clearly roughening of aluminum can be done by other chemical etching. It is important to choose one chemical agent to have the ability to have aluminum corroded in this environment to obtain a rough surface by presenting pores. Strong mineral acid solutions like hydrochloric acid in the presence of some additives are a good etchant for surface roughening. Aluminum, in normal conditions, is an active metal so it can be oxidized by acids and aggressive conditions. For this reason, various methods can be used to surface roughening by corrosion of surface. Cl^- ions form complex with Al ion, hydrogen, and hydroxide ions. The concentration, temperature, etching time, and additives are important parameters for aluminum roughening by chemical etching. Hydrofluoric acid is one of additives to the acid solution. Acid nitric and sodium hydroxide solutions are other options to use as an etchant to roughen the aluminum surface [101, 102]. Putting aluminum in boiling water can also change the surface morphology [89, 90]. Results show that the lotus like structure has been obtained after surface treatment in boiling water. By increasing the boiling time, denser and more interconnected network structures form.

The following reactions are possible in this condition:

$$2Al + 6H_2O \longrightarrow 2Al(OH)_3 + 3H_2$$

$$2Al + 4H_2O \longrightarrow 2AlO(OH) + 3H_2$$

$$2Al + 3H_2O \longrightarrow Al_2O_3 + 3H_2$$

When aluminum is placed in boiling water, the intrinsic oxid layer starts to dissolve and react with water. After aluminum starts to react with water, hydrogen gas is released after the oxidation reaction of aluminum. It can change the film growth mechanism like nanograss.

Here, we will review recent attempts to fabricate superhydrophobic coatings on aluminium. Feng et al. [90] introduced facile and environment-friendly methods to produce excellent corrosion resistance. They treated the aluminum alloy surface in boiling water and immersed it in a stearic acid (STA)–ethanol–H_2O solution. They fabricated a porous and rough structure surface by superhydrophobic performance with a water contact angle of 155º and a rolling angle of 5º. The superhydrophobic aluminum alloy also gained excellent corrosion resistance. [89] in another research, Feng et al., fabricated the anticorrosion superhydrophobic coating on aluminum alloy with boiling water treatment and stearic acid modification. They showed that a superhydrophobic surface with a contact angle of 154.1º can be obtained when the aluminum alloy was treated in boiling water for 30 seconds and modified with 5 mmol/L of stearic acid for 24 hours at room temperature. Also, the corrosion resistance of the aluminum alloy improved with the increase of the water contact angle, and the superhydrophobic aluminum alloy has the best corrosion resistance as compared to the other aluminum alloys. Also, Han et al. [103] made aluminum surface as superhydrophobic by the same process (boiling in water and modification by low surface energy agent) and they evaluated an icephobic behaviour of this coating. It was found that the superhydrophobic durability is remarkably affected by the preparation conditions, even if the surfaces initially showed almost identical superhydrophobicity. The best result was observed for the Al substrate etched for 2 hours and octadecyltrichlorosilane (OTS) coating with 0.09 wt. % concentration in toluene. Zubayda et al. [104] improved this method by the two simple and environmentally friendly techniques of mechanical sand blasting and boiling water treatment, and lowering of surface energy by 2H-perfluorodecyltriethoxysilane agent. The highest contact angles of 158º for water, 156º for ethylene glycol, and 154º for peanut oil were obtained after forming hierarchical structures involving shapes of microgrooves obtained by one directional sanding and nano grass by

http://dx.doi.org/10.5772/60724

immersing in boiling water for 1 minute. Guo et al. [14] made aluminum and its alloy surfaces superhydrophobic by immersing them in sodium hydroxide for several hours followed by the spin coating with a layer of perfluorononane or poly (dimethysiloxane) vinyl terminated(PDMSVT). This treatment increases the water contact angle from about 67º to above 160º. In similar reports, they obtained a stable biomimetic superhydrophobic surface on aluminum alloy by wet chemical etching following modification with cross-linked silicone elastomer, perfluorononane (C9F20), and perfluoropolyether (PFPE), respectively. They claimed that "The trapped air with the binary structure plays a key role in fabricating superhydrophobic surface on aluminum alloy." In other words, the unusual structure on the surface, which has a binary structure consisted of micro protrusions and nanoparticles, plays a very vital role in the constructing of a stable biomimetic superhydrophobic surface on the aluminum alloy [14, 19, 20]. Zhang et al. [105], with a simple method, rendered a stable superhydrophobic surface on the aluminum alloy by a low-cost, one-step method, which is simply immersing the substrates in a solution containing hydrochloric acid and fatty acid molecules. High water contact angle was about 163º and low adhesion was achieved on the treated surfaces. Cho et al. [106] provided a superhydrophobic surface with uniformly superhydrophobic surface where nano-scale structures were fabricated through an alkali surface modification method and self-assembled monolayer coating. The fabrication process for dual-scale hierarchical structures combining both microstructure via sandblasting techniques and the nano-structured aluminum hydroxide layer was done to enhance the mechanical durability of the superhydrophobicity. They remarked that the resulting hierarchical structures are suitable for diverse applications of aluminum in various industrial areas, including self-cleaning, anti-frosting, and microfluidic devices for rigorous environments. In another attempt, a superhydrophobic surface with a water contact angle of 166º was fabricated on an aluminum substrate by electrochemical machining using neutral NaNO3 electrolytes, followed by fluorination [26]. In another similar work, a superhydrophobic surface on an aluminum substrate by one-step electrochemical machining using sodium chloride (NaCl) aqueous solution containing fluoroalkylsilane as the electrolyte have been fabricated [86]. Static water contact angle obtained was about 166º and a tilting angle of about 10º. Saleema et al. [23]with a simple one-step process made aluminum alloy surfaces superhydrophobic by immersing the aluminum alloy substrates in a solution containing NaOH and fluoroalkyl-silane (FAS-17) molecules. In this report, the surfaces provided a water contact angle as high as ~162º and a contact angle hysteresis as low as ~4º. The study indicates that it is possible to fabricate superhydrophobic aluminum surfaces easily and effectively without involving the traditional two-step processes.

Liu et al. [99]coated aluminum surface by the following procedure. First, the aluminum foil was immersed in a sealed autoclave containing an aqueous solution (dissolved AlCl3 and triethanolamine (TEA, 0.75 M) in deionized water). And then it is heated in the oven at 100 ºC for 5 hours washing the as-obtained aluminum foil several times with deionized water and drying at ambient temperature. Finally, the film was modified by dipping the substrate in an ethanol solution of 5 mM stearic acid (STA) for 10 hours followed by rinsing the sample in absolute ethanol. The obtained surface demonstrates superhydrophobicity with a water CA of 169º and a sliding angle of about 4º after STA modification. The non-sticking behaviour, long-term storage stability, and relatively good mechanical strength were also obtained.

Some other attempts were conducted by anodizing process and showed very interesting result [27, 47, 48, 100, 107]. Zhang et al. [105] simply coated superhydrophobic alumina films with a 2D close-packed array of nano-spheres on industrial aluminum foils by an anodization method and following modification. The water contact angle of the resultant surfaces was about 154.6º ± 1.4º and the adhesion of water droplets was very weak. Wu et al. [48]obtained a superhydrophobic surface of the aluminum alloy micro/nano film with porous structures anodized and polypropylene coating under the difference test parameters. They obtained the best superhydrophobic film on aluminum alloy with the contact angle (CA) of 162º and the sliding angle of 2º. Liu et al. [47]also prepared a superhydrophobic coating by anodizing and polymeric coating on the surface of aluminum alloy. They fabricated a superhydrophobic surface by polypropylene (PP) coating after anodizing and the best superhydrophobic surface on aluminum alloy achieved was with the contact angle (CA) of 162º and the sliding angle of 2º. He et al. [97] fabricated underwater anticorrosive superhydrophobic surface by melting myristic acid (CH3 (CH2)12COOH) adsorbed onto the anodized aluminum. The static contact angle for seawater on the surface was about 154º. They showed that the superhydrophobic surface significantly improved the corrosion resistance of aluminum in sterile seawater. Liu et al. [24] prepared a superhydrophobic surface by anodizing and surface modification to evaluate reducing microbiologically influenced corrosion of aluminum. They showed that the superhydrophobic surface significantly reduced the corrosion rate of aluminum. Some others also reported the same by fabricating superhydrophobic coatings with chemical modification on the anodized aluminum and this has been shown that the aluminum alloy corrosion resistance is effectively improved by the formation of a stable superhydrophobic coating and this can be widely applied in the corrosion protection of various engineering materials [48]. In some others researches, the superhydrophobic coating was prepared by using additional deposition methods like CVD and dip coating with a combination of anodic oxidation [20, 108-111]. As a conclusion, it can be said that superhydrophobic coatings with excellent properties and long term durability can be a good selection for various industrial applications.

3. Corrosion prevention through superhydrophobic surfaces for various alloys

All metals have a tendency to corrode in most of atmospheric conditions. The metal corrosion accrues because of the interaction between water and a metal substrate. Deposition of soluble salt particles at a certain relative humidity can accelerate the corrosion rate most of the time [111].

Salt particles tend to adsorb moisture and form saline droplets, which can in turn accelerate atmospheric corrosion due to the presence of corrosive ions [112]. In principle, the presence of a water droplet (or a film) over a metal surface, electrochemically changes its surface condition from homogeneity to heterogeneity. Improvement of corrosion resistance is a main challenge of surface and corrosion engineering [113-115]. Water is a main part of the corrosion mechanism and the absence of water will be helpful. By mimicking from nature, superhydrophobicity is a helpful method to decreasing corrosion and water presence on the surface. Superhydrophobic surfaces and coatings are the focus of attention in the recent decade and their applications are increasing every day [116, 117].

The advantage of superhydrophobic coatings in corrosion protection can be explained by two observations. Firstly, the hydrophobicity prevents water absorbing on the surface while the air gets trapped in the grooves of a superhydrophobic film, by which water droplets comes into contact with the film through composite interface of both air and solid. Secondly the superhydrophobic coating acts as a protective coating against corrosion. The self-cleaning properties of superhydrophobic coating prevent deposition of corrosive salts water droplet, which easily rolls off and takes away salt particles adhered to the surface [118]. In this section, the principle of corrosion prevention by superhydrophobic coatings would be discussed.

The relative humidity (RH) of the atmosphere is an important factor for generating water droplets on the surface and absorbing by solid salt particles. The salt deposition on the surface followed by water absorption creates a suitable condition for corrosion. Deliquescence happens when water molecules absorbed on a salt particle surface form a film of saline solution until the salt particles are dissolved completely. This process is an important factor that induces atmospheric corrosion. It is important that the salt particle present on superhydrophobic surfaces will be removed after it transforms into saline droplets as a preventive measure.

In one study, the corrosion behavior of superhydrophobic surfaces on copper has been investigated under humid condition. The RH of the atmosphere was 90% and the NaCl particles were located on the surface. The samples were inclined at an angle about 200 º and the behavior of salt particles during water absorption was monitored. The observed results were interesting because the superhydrophobic surface acted as a lubricant by reducing friction between two solid particles and caused water absorption by salt and condensing the water. The water did not spread over the surface compared to hydrophilic surface. The water droplets located in the spherical shape by low surface came into contact with the superhydrophobic surface. It showed a significant improvement in the corrosion resistance of copper alloys by superhydrophobic coatings in humid environment which is more aggressive than dry environment.

The trapped air played an important role to create superhydrophobic surfaces because the trapped air decreased the solid surface connection by water droplets. In the same work, the effects of the trapped air on corrosion performance of copper superhydrophobic coatings have been evaluated. The corrosion performance was electrochemically measured. Three samples, hydrophilic copper plates, degassed superhydrophobic, and superhydrophobic coatings in the same condition were tested. Results showed that the absence of air decreased the corrosion resistance. It can also be found that the air trapped in the film is the main contributor for the barrier performance and stability of the superhydrophobic film. The trapped air inhibited electrochemical reactions between the substrate and corrosive medium. Also, air can play as an isolation film between the corrosive medium and Cu substrate.

Aluminum alloys are tendentious to corrosion and a superhydrophobic coating is the best choice for the improvement of corrosion resistance. From the literature, it can be seen that the superhydrophobic coating can improve the corrosion performance of aluminum in simulated marine environment. He et al. [97]studied the corrosion protection for aluminum in seawater, where the results showed that the superhydrophobic coating has highly protective properties against aluminum corrosion in sea water. They claim that "this method should be easily applied to large-scale production of superhydrophobic engineering materials with ocean industrial applications." Also in another work, Liu et al. [97] showed that the superhydrophobic coating can reduce biofilm formation on the surface because of the low binding strength between bacteria and the surface, and probably due to a cohesive failure within the conditioning layer.

4. Conclusion

In conclusion, it is established that the superhydrophobic surface is the best choice as a coating and protection layer against environmental effects such as surface corrosion, biofouling, dust and water interaction by mitigating the effect of humidity and other gasses, and accumulation of particulate matters on the surface. The superhydrophobic coatings offer the unique advantage of ease of fabrication, and their capability to withstand harsh environment which deteriorates the properties of metals and non-metals.

These surfaces are also limited in withstanding erosion, high flow rates, and elevated temperatures under aggressive conditions. However, the ongoing research is likely to resolve the above mentioned problems. These coatings also possess multiple qualities and properties such as anti-corrosion performance, self-cleaning and other properties which depend on the fabrication method.

The major limitation of these coatings is their durability because these coatings lose their properties with the passage of time. From an economical view, this coating is expensive and this concern may possibly be solved in the future by adopting cheaper fabrication methods and using cheaper superhydrophobic materials for coating.

Author details

Amir Farzaneh[1,2*], Zaki Ahmad[3,4], Mustafa Can[5], Salih Okur[2], Omer Mermer[7] and Ali Kemal Havare[8]

*Address all correspondence to: amir.frz@gmail.com

1 Department of Materials Engineering, Faculty of Mechanical Engineering University of Tabriz, Iran

2 Department of Material Science and Engineering, Faculty of Engineering and Architecture, Izmir Katip Celebi University, Turkey

3 King Fahad University of Petroleum and Minerals, Dhahran, Saudi Arabia

http://dx.doi.org/10.5772/60724

4 COMSATS, Lahore, Pakistan

5 Department of Engineering Science, Faculty of Engineering and Architecture, Izmir Katip Celebi University, Turkey

6 Department of Electrical Engineering and Electronics, Ege University, İzmir, Turkey

7 Department of Electrical Engineering and Electronics, Toros University, Mersin, Turkey

References

[1] Li, X.-M., D. Reinhoudt, and M. Crego-Calama, *What do we need for a superhydrophobic surface? A review on the recent progress in the preparation of superhydrophobic surfaces.* Chemical Society Reviews, 2007. 36(8): p. 1350-1368.

[2] Roach, P., N.J. Shirtcliffe, and M.I. Newton, *Progess in superhydrophobic surface development.* Soft Matter, 2008. 4(2): p. 224-240.

[3] Malik, M.A., et al., *Anti-corrosion ability of surfactants: a review.* Int. J. Electrochem. Sci, 2011. 6(6): p. 1927-1948.

[4] Farzaneh, A., et al., *Electrochemical and structural properties of electroless Ni-P-SiC nanocomposite coatings.* Applied Surface Science, 2013. 276: p. 697-704.

[5] Farzaneh, A., M. Ehteshamzadeh, and M. Mohammadi, *Corrosion performance of the electroless Ni–P coatings prepared in different conditions and optimized by the Taguchi method.* Journal of Applied Electrochemistry, 2011. 41(1): p. 19-27.

[6] Wood, R.J., *Erosion-corrosion interactions of copper and aluminium alloys.* European Federation Of Corrosion Publications, 2007. 50: p. 19.

[7] Ahmad, Z. and B. Abdul Aleem, *Erosion-Corrosion Behavior of Plasma-Sprayed Nanostructured Titanium Dioxide Coating in Sodium Chloride-Polystyrene Slurry.* Corrosion, 2009. 65(9): p. 611-623.

[8] Bertolini, L., et al., *Corrosion of steel in concrete: prevention, diagnosis, repair.* 2013: John Wiley & Sons.

[9] McKinney, M., R. Kirby, and W. Newsom, *Aerodynamic Factors To Be Considered In The Design Of Tilt-Wing V/Stol Airplanes.* Annals of the New York Academy of Sciences, 1963. 107(1): p. 221-248.

[10] Hammitt, F.G., et al., *Cavitation and droplet impingement damage of aircraft rain erosion materials.* 1970.

[11] Mohamed, A.M.A., A.M. Abdullah, and N.A. Younan, *Corrosion behavior of superhydrophobic surfaces: A review.* Arabian Journal of Chemistry, (0).

[12] Ma, M. and R.M. Hill, *Superhydrophobic surfaces.* Current Opinion in Colloid & Interface Science, 2006. 11(4): p. 193-202.

[13] Liu, T., et al., *Corrosion behavior of super-hydrophobic surface on copper in seawater.* Electrochimica Acta, 2007. 52(28): p. 8003-8007.

[14] Guo, Z., et al., *Effects of system parameters on making aluminum alloy lotus.* Journal of Colloid and Interface Science, 2006. 303(1): p. 298-305.

[15] Barthlott, W. and C. Neinhuis, *Purity of the sacred lotus, or escape from contamination in biological surfaces.* Planta, 1997. 202(1): p. 1-8.

[16] Yu, Y., Z.-H. Zhao, and Q.-S. Zheng, *Mechanical and superhydrophobic stabilities of two-scale surfacial structure of lotus leaves.* Langmuir, 2007. 23(15): p. 8212-8216.

[17] Neinhuis, C. and W. Barthlott, *Characterization and distribution of water-repellent, self-cleaning plant surfaces.* Annals of Botany, 1997. 79(6): p. 667-677.

[18] Baker, E. and G.M. Hunt, *Erosion of waxes from leaf surfaces by simulated rain.* New Phytologist, 1986. 102(1): p. 161-173.

[19] Guo, Z. and W. Liu, *Biomimic from the superhydrophobic plant leaves in nature: Binary structure and unitary structure.* Plant Science, 2007. 172(6): p. 1103-1112.

[20] Guo, Z., W. Liu, and B.-L. Su, *Superhydrophobic surfaces: From natural to biomimetic to functional.* Journal of Colloid and Interface Science, 2011. 353(2): p. 335-355.

[21] Liu, T., et al., *Super-hydrophobic surfaces improve corrosion resistance of copper in seawater.* Electrochimica Acta, 2007. 52(11): p. 3709-3713.

[22] Fu, X. and X. He, *Fabrication of super-hydrophobic surfaces on aluminum alloy substrates.* Applied Surface Science, 2008. 255(5, Part 1): p. 1776-1781.

[23] Saleema, N., et al., *Superhydrophobic aluminum alloy surfaces by a novel one-step process.* ACS Appl Mater Interfaces, 2010. 2(9): p. 2500-2.

[24] Liu, T., et al., *Investigations on reducing microbiologically-influenced corrosion of aluminum by using super-hydrophobic surfaces.* Electrochimica Acta, 2010. 55(18): p. 5281-5285.

[25] Park, B.G., et al., *Superhydrophobic fabrication of anodic aluminum oxide with durable and pitch-controlled nanostructure.* Colloids and Surfaces A: Physicochemical and Engineering Aspects, 2010. 370(1–3): p. 15-19.

[26] Xu, W., et al., *Fabrication of superhydrophobic surfaces on aluminum substrates using NaNO3 electrolytes.* Journal of Materials Science, 2011. 46(18): p. 5925-5930.

[27] Menini, R., Z. Ghalmi, and M. Farzaneh, *Highly resistant icephobic coatings on aluminum alloys.* Cold Regions Science and Technology, 2011. 65(1): p. 65-69.

[28] Kulinich, S.A. and M. Farzaneh, *On ice-releasing properties of rough hydrophobic coatings*. Cold Regions Science and Technology, 2011. 65(1): p. 60-64.

[29] Yao, L. and J. He, *Recent progress in antireflection and self-cleaning technology–From surface engineering to functional surfaces*. Progress in Materials Science, 2014. 61: p. 94-143.

[30] Solga, A., et al., *The dream of staying clean: Lotus and biomimetic surfaces*. Bioinspiration & biomimetics, 2007. 2(4): p. S126.

[31] Bhushan, B., *Biomimetics: lessons from nature–an overview*. Philosophical Transactions of the Royal Society A: Mathematical, Physical and Engineering Sciences, 2009. 367(1893): p. 1445-1486.

[32] Sun, M., et al., *Artificial lotus leaf by nanocasting*. Langmuir, 2005. 21(19): p. 8978-8981.

[33] Kwon, Y., et al., *Design of surface hierarchy for extreme hydrophobicity*. Langmuir, 2009. 25(11): p. 6129-6136.

[34] Saifaldeen, Z.S., et al., *Superamphiphobic aluminum alloy surfaces with micro and nanoscale hierarchical roughness produced by a simple and environmentally friendly technique*. Journal of Materials Science, 2014. 49(4): p. 1839-1853.

[35] Mortazavi, V. and M. Nosonovsky, *Wear-Induced Microtopography Evolution and Wetting Properties of Self-Cleaning, Lubricating and Healing Surfaces*. Journal of Adhesion Science and Technology, 2011. 25(12): p. 1337-1359.

[36] Kim, S.H., *Fabrication of Superhydrophobic Surfaces*. Journal of Adhesion Science and Technology, 2008. 22(3-4): p. 235-250.

[37] Wenzel, R.N., *Resistance Of Solid Surfaces To Wetting By Water*. Industrial & Engineering Chemistry, 1936. 28(8): p. 988-994.

[38] Amir Farzaneh, et al., *Aluminium Alloys in Solar Power – Benefits and Limitations,* Aluminium Alloys – New Trends in Fabrication and Applications, 2012. InTech.

[39] Mengnan Qu, J. He, and J. Zhang, *Superhydrophobicity, Learn from the Lotus Leaf, Biomimetics Learning from Nature* InTech publisher, 2010.

[40] Guo, Z., et al., *Fabrication of superhydrophobic copper by wet chemical reaction*. Thin Solid Films, 2007. 515(18): p. 7190-7194.

[41] Sendner, C., et al., *Interfacial Water at Hydrophobic and Hydrophilic Surfaces: Slip, Viscosity, and Diffusion*. Langmuir, 2009. 25(18): p. 10768-10781.

[42] Farzaneh, A., et al., *Investigation and optimization of SDS and key parameters effect on the nickel electroless coatings properties by Taguchi method*. Journal of coatings technology and research, 2010. 7(5): p. 547-555.

[43] Bisetto, A., et al. *Dropwise condensation on superhydrophobic nanostructured surfaces: literature review and experimental analysis*. in *Journal of Physics: Conference Series*. 2014. IOP Publishing.

[44] Zhao, N., et al., *Combining layer-by-layer assembly with electrodeposition of silver aggregates for fabricating superhydrophobic surfaces.* Langmuir, 2005. 21(10): p. 4713-4716.

[45] Darmanin, T. and F. Guittard, *Superhydrophobic fiber mats by electrodeposition of fluorinated poly (3, 4-ethyleneoxythiathiophene).* Journal of the American Chemical Society, 2011. 133(39): p. 15627-15634.

[46] Hang, T., et al., *Super-hydrophobic nickel films with micro-nano hierarchical structure prepared by electrodeposition.* Applied Surface Science, 2010. 256(8): p. 2400-2404.

[47] Liu, W., et al., *Fabrication of the superhydrophobic surface on aluminum alloy by anodizing and polymeric coating.* Applied Surface Science, 2013. 264(0): p. 872-878.

[48] Wu, R., et al., *Fabrication of the micro/nano-structure superhydrophobic surface on aluminum alloy by sulfuric acid anodizing and polypropylene coating.* J Nanosci Nanotechnol, 2013. 13(3): p. 2362-8.

[49] Balu, B., V. Breedveld, and D.W. Hess, *Fabrication of "roll-off" and "sticky" superhydrophobic cellulose surfaces via plasma processing.* Langmuir, 2008. 24(9): p. 4785-4790.

[50] Fadeeva, E., et al., *Bacterial retention on superhydrophobic titanium surfaces fabricated by femtosecond laser ablation.* Langmuir, 2011. 27(6): p. 3012-3019.

[51] Ramos, S., et al., *Wettability control on vertically-aligned multi-walled carbon nanotube surfaces with oxygen pulsed DC plasma and CO 2 laser treatments.* Diamond and Related Materials, 2010. 19(7): p. 752-755.

[52] Wu, B., et al., *Superhydrophobic surfaces fabricated by microstructuring of stainless steel using a femtosecond laser.* Applied Surface Science, 2009. 256(1): p. 61-66.

[53] Wagterveld, R.M., et al., *Ultralow hysteresis superhydrophobic surfaces by excimer laser modification of SU-8.* Langmuir, 2006. 22(26): p. 10904-10908.

[54] Zhang, X., et al., *Superhydrophobic surfaces: from structural control to functional application.* Journal of Materials Chemistry, 2008. 18(6): p. 621-633.

[55] Fujii, T., Y. Aoki, and H. Habazaki, *Superhydrophobic hierarchical surfaces fabricated by anodizing of oblique angle deposited Al–Nb alloy columnar films.* Applied Surface Science, 2011. 257(19): p. 8282-8288.

[56] Chung, J.W., et al., *Self-assembled perpendicular growth of organic nanoneedles via simple vapor-phase deposition: one-step fabrication of a superhydrophobic surface.* Chemical Communications, 2008(26): p. 2998-3000.

[57] Li, Y., F. Liu, and J. Sun, *A facile layer-by-layer deposition process for the fabrication of highly transparent superhydrophobic coatings.* Chemical communications, 2009(19): p. 2730-2732.

[58] Crick, C.R. and I.P. Parkin, *CVD of copper and copper oxide thin films via the in situ reduction of copper (ii) nitrate—a route to conformal superhydrophobic coatings*. Journal of Materials Chemistry, 2011. 21(38): p. 14712-14716.

[59] Li, Y., L. Li, and J. Sun, *Bioinspired Self-Healing Superhydrophobic Coatings*. Angewandte Chemie, 2010. 122(35): p. 6265-6269.

[60] Ma, M., et al., *Superhydrophobic fabrics produced by electrospinning and chemical vapor deposition*. Macromolecules, 2005. 38(23): p. 9742-9748.

[61] Han, D. and A.J. Steckl, *Superhydrophobic and oleophobic fibers by coaxial electrospinning*. Langmuir, 2009. 25(16): p. 9454-9462.

[62] Miyauchi, Y., B. Ding, and S. Shiratori, *Fabrication of a silver-ragwort-leaf-like super-hydrophobic micro/nanoporous fibrous mat surface by electrospinning*. Nanotechnology, 2006. 17(20): p. 5151.

[63] Tadanaga, K., et al., *Superhydrophobic-superhydrophilic micropatterning on flowerlike alumina coating film by the sol-gel method*. Chemistry of materials, 2000. 12(3): p. 590-592.

[64] Manca, M., et al., *Durable superhydrophobic and antireflective surfaces by trimethylsilanized silica nanoparticles-based sol– gel processing*. Langmuir, 2009. 25(11): p. 6357-6362.

[65] Tadanaga, K., J. Morinaga, and T. Minami, *Formation of superhydrophobic-superhydrophilic pattern on flowerlike alumina thin film by the sol-gel method*. Journal of Sol-Gel Science and Technology, 2000. 19(1-3): p. 211-214.

[66] Syrett, B.C., *Erosion-corrosion of copper-nickel alloys in sea water and other aqueous environments-a literature review*. Corrosion, 1976. 32(6): p. 242-252.

[67] Gilbert, P., *Corrosion Resisting Properties of 90/10 Copper–Nickel–Iron Alloy with Particular Reference to Offshore Oil and Gas Applications*. British Corrosion Journal, 1979. 14(1): p. 20-25.

[68] Schweitzer, P.A., *Corrosion Engineering Handbook, -3 Volume Set*. 1996: CRC Press.

[69] Li, J., J. Mayer, and E. Colgan, *Oxidation and protection in copper and copper alloy thin films*. Journal of applied physics, 1991. 70(5): p. 2820-2827.

[70] Kear, G., B. Barker, and F. Walsh, *Electrochemical corrosion of unalloyed copper in chloride media—a critical review*. Corrosion science, 2004. 46(1): p. 109-135.

[71] Smith, W.F., *Structure and properties of engineering alloys*. 1993.

[72] Manuals, M., *Online medical library*. Copper. [URL: http://www. en. wikipedia. org/wiki/Copper# Toxic ity] accessed, 2012.

[73] http://www.chemguide.co.uk/.

[74] Zhao, Y.S., et al., *A hierarchical self-assembly of 4, 5-diphenylimidazole on copp*er. Colloids and Surfaces A: Physicochemical and Engineering Aspects, 2006. 277(1): p. 111-118.

[75] Kong, L., et al., *Preparation and characterization of slice-like Cu 2 (OH) 3 NO 3 superhydrophobic structure on copper foil.* Applied Surface Science, 2008. 254(22): p. 7255-7258.

[76] Tan, C., et al., *Fabrication of color-controllable superhydrophobic copper compound coating with decoration performance.* Applied Surface Science, 2015. 328: p. 623-631.

[77] Wang, P., et al., *Green approach to fabrication of a super-hydrophobic film on copper and the consequent corrosion resistance.* Corrosion Science, 2014. 80: p. 366-373.

[78] Xi, W., et al., *The preparation of lotus-like super-hydrophobic copper surfaces by electroplating.* Applied surface science, 2009. 255(9): p. 4836-4839.

[79] Wang, P., et al., *Fabricated super-hydrophobic film with potentiostatic electrolysis method on copper for corrosion protection.* Electrochimica Acta, 2010. 56(1): p. 517-522.

[80] Huang, Y., et al., *Corrosion resistance properties of superhydrophobic copper surfaces fabricated by one-step electrochemical modification process.* Applied Surface Science, 2013. 282: p. 689-694.

[81] Kalvoda, R. and R. Parsons, *Electrochemistry in research and development.* 1985: Springer.

[82] Brenner, A., *Electrodeposition of alloys: principles and practice.* 2013: Elsevier.

[83] Crick, C.R. and I.P. Parkin, *Preparation and Characterisation of Super-Hydrophobic Surfaces.* Chemistry-A European Journal, 2010. 16(12): p. 3568-3588.

[84] Liu, Y., et al., *Fabrication of biomimetic superhydrophobic surface with controlled adhesion by electrodeposition.* Chemical Engineering Journal, 2014. 248: p. 440-447.

[85] Chen, Z., et al., *A rapid one-step process for fabrication of superhydrophobic surface by electrodeposition method.* Electrochimica Acta, 2012. 59: p. 168-171.

[86] Song, J., W. Xu, and Y. Lu, *One-step electrochemical machining of superhydrophobic surfaces on aluminum substrates.* Journal of Materials Science, 2012. 47(1): p. 162-168.

[87] Liu, Y., et al., *Biomimetic superhydrophobic surface of high adhesion fabricated with micronano binary structure on aluminum alloy.* ACS applied materials & interfaces, 2013. 5(18): p. 8907-8914.

[88] Yin, B., et al., *Novel strategy in increasing stability and corrosion resistance for super-hydrophobic coating on aluminum alloy surfaces.* Applied Surface Science, 2011. 258(1): p. 580-585.

[89] Feng, L., et al., *Superhydrophobic aluminum alloy surface: Fabrication, structure, and corrosion resistance.* Colloids and Surfaces A: Physicochemical and Engineering Aspects, 2014. 441(0): p. 319-325.

[90] Feng, L., et al., *Fabrication of superhydrophobic aluminium alloy surface with excellent corrosion resistance by a facile and environment-friendly method.* Applied Surface Science, 2013. 283(0): p. 367-374.

[91] Grubbs, C.A., *Anodizing of aluminum*. Metal Finishing, 1999. 97(1): p. 476-493.

[92] Grubbs, C.A., *Anodizing of aluminum*. Metal Finishing, 2002. 100: p. 463-478.

[93] Soekrisno, S. and B. Anggoro, *Effect of Anodizing in Surface Finishing on Speed Boat Impeller Made of Aluminum*. Advanced Materials Research, 2014. 896: p. 253-256.

[94] Shibuichi, S., et al., *Super water-and oil-repellent surfaces resulting from fractal structure*. Journal of Colloid and Interface Science, 1998. 208(1): p. 287-294.

[95] Masuda, H., *Highly ordered nanohole arrays in anodic porous alumina*, in *Ordered Porous Nanostructures and Applications*. 2005, Springer. p. 37-55.

[96] Wu, R., et al., *Fabrication of nano-structured super-hydrophobic film on aluminum by controllable immersing method*. Applied Surface Science, 2012. 258(16): p. 5933-5937.

[97] He, T., et al., *Super-hydrophobic surface treatment as corrosion protection for aluminum in seawater*. Corrosion Science, 2009. 51(8): p. 1757-1761.

[98] Menini, R. and M. Farzaneh, *Elaboration of Al2O3/PTFE icephobic coatings for protecting aluminum surfaces*. Metal Finishing, 2009. 107(7–8): p. 40-46.

[99] Liu, L., et al., *Fabrication of superhydrophobic surface by hierarchical growth of lotus-leaf-like boehmite on aluminum foil*. Journal of Colloid and Interface Science, 2011. 358(1): p. 277-283.

[100] Jiang, H., et al., *Manufacturing a durable superhydrophobic polypropylene coating on aluminum alloy substrate by adding nano-titania nanoparticles*. Journal of nanoscience and nanotechnology, 2014. 14(7): p. 5421-5425.

[101] Kumar, C.S., et al., *Studies on white anodizing on aluminum alloy for space applications*. Applied Surface Science, 1999. 151(3): p. 280-286.

[102] Xie, D. and W. Li, *A novel simple approach to preparation of superhydrophobic surfaces of aluminum alloys*. Applied Surface Science, 2011. 258(3): p. 1004-1007.

[103] Han, S., J. Jeong, and D. Lee, *Ice-phobic behavior of superhydrophobic Al surface undervarious etching conditions*. Journal of Electroceramics, 2014: p. 1-7.

[104] Saifaldeen, Z., et al., *Superamphiphobic aluminum alloy surfaces with micro and nanoscale hierarchical roughness produced by a simple and environmentally friendly technique*. Journal of Materials Science, 2014. 49(4): p. 1839-1853.

[105] Zhang, Y., et al., *Facile fabrication of superhydrophobic nanostructures on aluminum foils with controlled-condensation and delayed-icing effects*. Applied Surface Science, 2012. 258(20): p. 8253-8257.

[106] Cho, H., et al., *A simple fabrication method for mechanically robust superhydrophobic surface by hierarchical aluminum hydroxide structures*. Current Applied Physics, 2013. 13(4): p. 762-767.

[107] Liang, J., et al., *Fabrication and Corrosion Resistance of Superhydrophobic Hydroxide Zinc Carbonate Film on Aluminum Substrates.* Journal of Nanomaterials, 2013. 2013: p. 6.

[108] Jafari, R., R. Menini, and M. Farzaneh, *Superhydrophobic and icephobic surfaces prepared by RF-sputtered polytetrafluoroethylene coatings.* Applied Surface Science, 2010. 257(5): p. 1540-1543.

[109] Nguyen, D.D., et al., *Superhydrophobic and superoleophilic properties of graphene-based sponges fabricated using a facile dip coating method.* Energy & Environmental Science, 2012. 5(7): p. 7908-7912.

[110] Zhang, G., et al., *Fabrication of superhydrophobic surfaces from binary colloidal assembly.* Langmuir, 2005. 21(20): p. 9143-9148.

[111] Farhadi, S., M. Farzaneh, and S. Kulinich, *Anti-icing performance of superhydrophobic surfaces*. Applied Surface Science, 2011. 257(14): p. 6264-6269.

[112] Wang, P., D. Zhang, and Z. Lu, *Advantage of super-hydrophobic surface as a barrier against atmospheric corrosion induced by salt deliquescence.* Corrosion Science, 2015. 90: p. 23-32.

[113] Roberge, P., *Atmospheric corrosion*. Uhlig's Corrosion Handbook, 2011. 51: p. 299.

[114] Ahmad, Z., A. Ul-Hamid, and B. Abdul-Aleem, *The corrosion behavior of scandium alloyed Al 5052 in neutral sodium chloride solution.* Corrosion Science, 2001. 43(7): p. 1227-1243.

[115] Ahmad, Z. and B.A. Aleem, *Degradation of aluminum metal matrix composites in salt water and its control.* Materials & design, 2002. 23(2): p. 173-180.

[116] Ahmad, Z. and F. Patel, *Development of Novel Corrosion Techniques for a Green Environment*. International Journal of Corrosion, 2012. 2012.

[117] Ahmad, Z., I. Ahmad, and F. Patel. *Fog Collection by Mimicking Nature*. in *Journal of Biomimetics, Biomaterials and Biomedical Engineering*. 2010. Trans Tech Publ.

[118] Ahmad, Z., S. Rehman, and M. Ahsan, *Methods of Fabrication of Superhydrophobic Engineering Surfaces: A Review.* International Review of Mechanical Engineering, 2009. 3(3).